Climate Change in Asia and the Pacific

Thank you for choosing a SAGE product! If you have any comment, observation or feedback, I would like to personally hear from you. Please write to me at <u>contactceo@sagepub.in</u>

—Vivek Mehra, Managing Director and CEO,
SAGE Publications India Pvt Ltd, New Delhi

Bulk Sales

SAGE India offers special discounts for purchase of books in bulk. We also make available special imprints and excerpts from our books on demand.

For orders and enquiries, write to us at

Marketing Department
SAGE Publications India Pvt Ltd
B1/I-1, Mohan Cooperative Industrial Area
Mathura Road, Post Bag 7
New Delhi 110044, India
E-mail us at <u>marketing@sagepub.in</u>

Get to know more about SAGE, be invited to SAGE events, get on our mailing list. Write today to <u>marketing@sagepub.in</u>

This book is also available as an e-book.

Climate Change in Asia and the Pacific

How Can Countries Adapt?

Edited by

Venkatachalam Anbumozhi
Meinhard Breiling
Selvarajah Pathmarajah
Vangimalla R. Reddy

www.sagepublications.com
Los Angeles • London • New Delhi • Singapore • Washington DC

First published in 2012 by

 SAGE Publications India Pvt Ltd
B1/I-1 Mohan Cooperative Industrial Area
Mathura Road, New Delhi 110 044, India
www.sagepub.in

SAGE Publications Inc
2455 Teller Road
Thousand Oaks, California 91320, USA

SAGE Publications Ltd
1 Oliver's Yard, 55 City Road
London EC1Y 1SP, United Kingdom

SAGE Publications Asia-Pacific Pte Ltd
33 Pekin Street
#02-01 Far East Square
Singapore 048763

Published by Vivek Mehra for SAGE Publications India Pvt Ltd, typeset in 10/13 pt Minion Pro by Star Compugraphics Private Limited, Delhi and printed at G. H. Prints Pvt Ltd, New Delhi.

Library of Congress Cataloging-in-Publication Data Available

ISBN: 978-81-321-0894-8 (HB)

The SAGE Team: Sharel Simon, Puja Narula Nagpal, Amrita Saha and Anju Saxena

Contents

List of Tables

List of Figures

List of Abbreviations

ABC	atmospheric brown cloud
ADB	Asian Development Bank
ADBI	Asian Development Bank Institute
AEGIS	agricultural and environmental geographic information systems
AEZ	agro-ecological zone
AF	Adaptation Fund
AFD	French Development Agency
AHNIP	Appropriate Hydrological Network Improvement Project
AIDADS	an implicitly direct additive demand system
AIM	action impact matrix
AIT	Asian Institute for Technology
APFED	Asia-Pacific Forum for Environment and Development
APWF	Asia-Pacific Water Forum
AR4	Fourth Assessment
AWG-LCA	Ad hoc Working Group on Long-term Cooperative Action
BAP	Bali Action Plan
BAPW	Buenos Aires Programme of Work
BOE	barrel oil equivalent
BQA	Better Air Quality
CASA	Church's Auxillary Social Action
CAT bond	catastrophe bond
CC	climate change
CBO	community-based organisations
CBT	capacity building and training
CCA	climate change adaptation
CCCS	Centre for Climate Change Studies
CCS	Climate change strategy
CDD	Cooling Degree Days
CDM	Clean Development Mechanism
CES	constant elasticity of substitution
CFC	chlorofluorocarbon
CGE	computable general equilibrium
COP	Conference of the Parties
CRED	Center for Research on the Epidemiology of Disasters

DAC	Development Assistance Committee
DHM	distributed hydrological model
DSS	decision support systems
DST	Department of Science and Technology, Government of India
EC	European Commission
EEA	European Environmental Agency
EIT	economies in transition
EMRI	Emergency Management and Research Institute
EPOC	Environment Policy Committee
EU	European Union
EV	equivalent variation
F&SP	flood and storm prevention
FAO	Food and Agriculture Organization
FAR	First Assessment Report
FFSP	Fund for Flood and Storm Prevention
FMM	flood management and mitigation
FTP	file transfer protocol
GCM	general circulation model
GDP	gross domestic product
GEF	Global Environment Facility
GFAS	global flood alert system
GHG	greenhouse gas
GIS	geographic information system
GLOF	glacial lake outburst flood
GPRS	general packet radio service
GTAP	Global Trade Analysis Project
HFC	hydrofluorocarbon
HYCOS	Hydrological Cycle Observing System
IC	indifference curve
ICCAP	Climate Change on Agricultural Production System in Arid Areas
ICHARM	International Centre for Water Hazard and Risk Management
ICLEI	Local Governments for Sustainability
IDI	Infrastructure Development Institute
IFAS	Integrated Flood Analysis System
IFNet	International Flood Network
IFPRI	International Food Policy Research Institute
IGES	Institute for Global Environmental Strategies
IIASA	International Institute for Applied Systems Analysis
IIED	International Institute for Environment and Development
IMD	Indian Meteorological Department
IMF	International Monetary Fund

IMPAM	Irrigation Management Performance Assessment Model
IPB	Institut Pertanian Bogor
IPCC	Intergovernmental Panel on Climate Change
IRRI	International Rice Research Institute
IRS	Indus River System
IRWR	internal renewable water resources
IWRM	Integrated Water Resources Management
JICA	Japan International Cooperation Agency
JMA	Japan Meteorological Agency
JSCE	Japan Society of Civil Engineers
KfW	German Development Bank
KP	Kyoto Protocol
LAM	Local Area Model
Lao PDR	Lao People's Democratic Republic
LDC	least developed countries
LIPI	The Indonesian Institute of Sciences
LMMA	locally managed marine areas
LP	linear program model
LSIP	Lower Seyhan Irrigation Project
LUT	Land Utilization Types
MASSCOTE	Mapping System and Services for Canal Operation Techniques
MDG	Millennium Development Goal
MLIT	Ministry of Land, Infrastructure, Transport and Tourism
MoEF	Ministry of Environment and Forests
MoWR	Ministry of Water Resources
MRC	Mekong River Commission
MSE	Madras School of Economics
MSSRF	M S Swaminathan Research Foundation
NAPA	national adaptation programmes of action
NAPCC	National Action Plan on Climate Change
NARBO	Network of Asian River Basin Organizations
NATCOM	India's Initial National Communication to the UNFCCC
NCRC	NGO Coordination and Rehabilitation Center
NDPRCC	National Development Planning Response to Climate Change
NGO	nongovernmental organization
NOAA	National Oceanic and Atmospheric Administration
NPM	non-pesticidal management
NWP	Nairobi Work Programme
ODA	official development assistance
OECD	Organisation for Economic Co-operation and Development

PAGASA	Philippine Atmospheric, Geophysical and Astronomical Services Administration
PDC	Population and Development Consolidation
PDHM	PWRI-distributed hydrological model
PPP	public–private partnership
PRA	participatory rural appraisal method
PRC	People's Republic of China
PSC	Project Steering Committee
PTFCC	Presidential Task Force on Climate Change
PWRI	Public Works Research Institute
RAG	Research Advisory Group
RAP	Rapid Appraisal Process
RBO	River Basin Organizations
RCM	Regional Climate Model
REDD	reducing emissions from deforestation and forest degradation
RIHN	Research institute for Humanity and Nature
SAARC	South Asian Association for Regional Cooperation
SAR	Second Assessment Report
SCCF	Special Climate Change Fund
SD	sustainable development
SEDP	Socio-Economic Development Plans
SEI	Stockholm Environment Institute
SENSA	Swedish Environmental Secretariat for Asia
SIDS	small island developing states
SNC	Second National Communication
SOLUS	sustainable options for land use
SOM	soil organic matter
SPA	strategic priority on adaptation
SRES	Special Report on Emission Scenarios
SRI	system of rice intensification
SSARR	stream flow synthesis and reservoir regulation model
SST	sea surface temperature
TAR	Third Assessment Report
TFP	total factor productivity
TMD	The Meteorological Department
UN	United Nations
UNDP	United Nations Development Programme
UNEP	United Nations Environment Programme
UNEPRRC.AP	United Nations Environment Programme Regional Resource Centre for Asia and the Pacific

UNESCAP	United Nations Economic and Social Commission for Asia and the Pacific
UNFCCC	United Nations Framework Convention on Climate Change
UN/ISDR	United Nations Strategy for Disaster Reduction Secretariat
UNU	United Nations University
URBS	system based on the unified river basin simulator model
US	United States
USAID	United States Agency for International Development
USDA–ARS	United States Department of Agriculture–Agricultural Research Service
US$	US dollar
VIA	vulnerability, impacts, and adaptation
WBGT	wet bulb globe temperature
WFD	Water Framework Directive
WHA	World Health Assembly
WHO	World Health Organization
WKC	WHO Kobe Centre
WMO	World Meteorological Organization
WRF	weather research forecast
WUA	water users associations
°C	degree Celsius
CH_4	methane
cm	centimeter
CO_2	carbon dioxide
km	kilometer
km^2	square kilometer
kWh	Kilowatt hour
m^3	cubic meter
$MtCO_2$-eq	metric ton of carbon dioxide equivalent
maf	million acre-feet
Mha	million hectares
mm	millimeter
MW	Megawatt
N_2O	nitrous oxide
ppb	parts per billion
ppm	parts per million
ppt	parts per trillion
W	Watt

Foreword

Developing countries of the Asia and Pacific region are particularly vulnerable to the impacts of climate change. The poor in these countries are at especially high risk, given their heavy dependence on agriculture, strong reliance on ecosystem services, rapid growth and intense concentrations of population, and relatively poor health services. Developing countries are usually also characterized by insufficient capacity to adapt to climate change impacts, inadequate infrastructure, meager household income and savings, and limited support from public services. There is a real danger that climate change impacts may derail the significant progress countries in Asia and the Pacific have made toward achieving the Millennium Development Goals.

To cope with impacts already locked into the climate system, the Asian Development Bank Institute (ADBI), in collaboration with the Regional and Sustainable Development Department and other operational departments of the Asian Development Bank, has put in place integrated adaptation solutions to address the causes and consequences of climate change in the Asia and Pacific region. This book is based on papers presented at two ADBI workshops on climate change adaptation held in Tokyo, Japan, in 2009, and Colombo, Sri Lanka, in 2010. The main objective of these workshops was to bring together leading academics and policymakers from the Asia and Pacific region to share good practices. The two workshops generated common insights and understanding, assessed policy implications, and identified further capacity building needs, all of which are captured by this book.

This book is being published as part of ADBI's efforts to produce knowledge products that can be used to promote inclusive and sustainable growth, one of our three priority themes. I am confident that this book will contribute to policy development and academic understanding in an area where new insights are badly needed. I hope this book will also help countries in Asia and the Pacific to set up and implement robust institutional frameworks for mainstreaming climate change concerns into development planning so that they can adapt to uncertainties, and sustainably manage their critical resources for the long-term development of their people.

Masahiro Kawai
Dean and CEO
Asian Development Bank Institute

Acknowledgments

The book is based on papers presented at two workshops on climate change adaptation held in Tokyo, Japan, and Colombo, Sri Lanka. We would like to thank the workshop participants for making the discussions intellectually stimulating and for sharing their experiences. We are grateful for the contributions of all resource persons who not only shared their wisdom, experiences, and perspectives, but also were willing to spend many days in writing them in an easily understandable, useful, and usable way. This book would not have been possible without generous support of the time, energy, and intellectual analysis of contributors at our partner institutions. Special thanks to our colleagues in ADB headquarters Katsuji Matusnami, Robert Dobias, Newin Sinsri, Suphachol Suphalsai, and Jung Tae Yong, who offered us challenging and sometimes provoking thoughts that helped us to shape our ideas and workshop contents. This book owes its existence to the energy and efforts of the ADBI team. Masahiro Kawai, ADBI Dean, provided vision and intellectual leadership. A special word of thanks in this respect is also owed to Worapot Manupipatpong and Mario Lamberte, former ADBI Directors, for their never-ending support to design the workshops and provide us with challenging feedback on this book. Tadashige Kawasaki and Alastair Dingwall played a major role in bringing this book to completion, as did Ainslie Smith, who was largely responsible for the editing and production. We are grateful to Mari Kimura, Joana Portugal, Apsara Chandanie, and Yuko Ichikawa for their research assistance and administrative support.

Introduction

Worapot Manupipatpong and Venkatachalam Anbumozhi

Climate change is one of the most significant challenges to global economic development. Every country contributes to growing greenhouse gas emissions, and every country will bear the socioeconomic consequences of global warming. Climate is a component of the natural environment within which and against whose bounds human civilization has developed and prospered. Left unchecked, continued global warming could cause worldwide social and environmental disruptions. The consequences of certain levels of climate change, such as a rise in average temperatures, are well understood and widely accepted. However, climate change is likely to include greater variations in climate phenomena, including droughts and floods, as well as more frequent and severe weather events, such as cyclones and storms, and greater seasonal variability from mild and severe winters to dry and very wet summers.

The Asia and Pacific region is more vulnerable to climate change risks than other regions, given its dependence on the natural resources and agriculture sectors, densely populated coastal areas, weak institutions, and the poverty of a considerable proportion of its population. So, adaptation—making adjustments in natural or human systems in response to actual or expected climatic stimuli or their effects which moderate negative or exploit beneficial opportunities—becomes a key strategy for sustaining economic growth. Failure to adapt could stall development, particularly in countries that depend on natural resources. Adaptive capacity entails the ability to change behaviors, shift priorities, produce necessary goods and services, and to plan and respond in ways that will reduce harmful climate change impacts or transform them into no-regret economic opportunities.

Climate Change in Asia and the Pacific: How Can Countries Adapt? examines the framework conditions for integrating climate change adaptation measures into agriculture, water, and natural resources management activities. Based on the review of country experiences, the book describes key dimensions, suggests interventions for further exploration, and serves as a basis for planning and mainstreaming climate change adaptation into sectoral planning. This volume draws mainly on two workshops organized by the Asian Development Bank Institute (ADBI) in Tokyo from April 14 to 18, 2009, and in Colombo from June 8 to 11, 2010. These workshops focused on mainstreaming climate change adaptation considerations in development planning and gathered experts from academia, think tanks, public service organizations, the private sector, and international organizations. The basis of our analysis also comes from post-workshop surveys, updated literature from the field, as well as interviews and consultations with experts.

Enhancing adaptive capacity is important to safeguard existing and future development progress in the light of current climate variability and to reduce its vulnerability to extreme weather events. Nevertheless, adaptation to climate change has not yet become a high-priority policy issue in most of the Asia and Pacific region, as barriers exist at both strategic and implementation levels. The main obstacles are the availability of credible climate information, the ability to transform the scientific information into usable form, lack of communication, absence of knowledge of successful measures, lack of available no-regret strategies, and deficient financial resources and funding mechanisms. There are also possible trade-offs between development priorities and adaptation measures. A country's adaptive capacity needs to be enhanced through building regional, national, and institutional commitment, as well as technical and scientific capacity.

Enhancing adaptive capacity involves several stages, beginning with assessing the causes and potential impacts of vulnerabilities and identifying the multiple benefits of best practices beyond adaptation. Hence, overriding priorities include generating accurate specific climate information and communicating it to policymakers as a basis for decisions, building awareness to understand potential climate impacts and devising responsive strategies, and creating communication channels to support knowledge transfer at sectoral level. In later stages, pilot programs are critical to test measures, to identify improvement opportunities, and to adjust implementation strategies.

Mainstreaming adaptive capacity of the most vulnerable sectors into development planning entails the full engagement of relevant stakeholders, including policymaking agencies, research institutions, the private sector, and civil society. The specific role in catalyzing adaptation strategies, in a cost-effective way, depends on the context of the issue being addressed. In a generic sense, local and regional governments are likely to be the government entities most directly involved in adaptation activities, both in the coordination across several actors and in the implementation at local level. Economic planning agencies also play a major coordinating role across different sectoral agencies, whereas sectoral agencies can facilitate implementation and identify trade-offs in specific sectors. Research and education networks operating across and within sectors are essential for cross-sectoral and international learning and knowledge generation. These networks play an essential role in inventorying specific climate data, and developing models that predict climate change and assess its impacts. Involvement of the private sector—particularly organizations involved in designing adaptation infrastructure and financing adaptation practices—is central to achieving impacts on a large scale. Civil society, local-based communities, and nongovernmental organizations also play an important role in the innovation of new adaptation measures, strategies, and pilot initiatives to support climate change adaptation that can then be replicated at scale. Involving them will also ensure due consideration of local specific conditions and social protection initiatives, as well as facilitate the successful implementation of adaptation actions. International organizations that combine knowledge and financial capacity can support proactive adaptation measures with the explicit focus on vulnerable populations. Finally, due to the great number of climate change issues and their cross-sectoral character, partnerships that involve diverse stakeholders are important in supporting adaptation at local and national levels.

The six parts presented in this book address these issues in detail.

Part I identifies the main risks associated with extreme weather events, quantifies their impacts, and reviews the available risk assessment tools and planning instruments. While some regions will actually benefit from climate change, the Asia and Pacific region is very likely to face a reduction in average annual gross domestic product (GDP) growth rates and negative impacts on the agriculture and natural resources sectors. Stressing the importance of enhancing the scientific capacity of the region, the authors suggest that the first step toward adaptive capacity should be to predict climate change impacts, by assessing the likely exposure of natural resources and examining the follow-on impacts of changes on human, economic, and social systems.

Part II discusses various driving forces to strengthen resilience to climate change. Although a wide range of adaptation measures are being undertaken by the Asian and Pacific countries, the effectiveness of those measures depends on specific factors and actors, including the magnitude of the impact, location, sectoral capacity, and socioeconomic situation. Based on an extensive review of reactive and proactive adaptation measures in the natural resources sectors, the authors address questions of how to design adaptation measures, which stakeholders to be involved, and how to prioritize investments. Emphasis is given to the importance of including long-term climate change considerations, such as extreme weather events, existing local coping mechanisms, and integrating these considerations into sectoral development plans.

Part III addresses the changes in adaptation when it becomes an issue in the decision-making process, leading to conceptual changes in the administrative systems at the subnational level. Analyzing successful experiences in both developed and developing countries, the authors recommend that adaptation measures should be tailored to local needs, be gender specific, and involve other key stakeholders such as private sectors in risk sharing. They argue that scientific information on climate change and climate change impact assessment tools need to be refined to reflect impacts on a local scale and reduce uncertainty about its risks. They also claim that attention should be given to financing instruments, including public and private insurance programs and catastrophic bonds, which could increase adaptive capacity of the most affected households.

In Part IV, the authors reinforce the relevance of promoting public awareness of climate change risks and developing local capacity to assess climate change impacts in various sectors at different levels. In their view, information dissemination regarding climate risks, impacts, vulnerabilities, and adaptation choices should occur among both high-level policymakers across the sectors and individuals in the affected communities. Additionally, they suggest that financial mechanisms should guarantee ex ante, self-reliant, and appropriate support to households who are at risk. Overall, this part discusses the role of predictions and adaptation responses in the decision-making process in reducing vulnerability and suggests financial mechanisms that can be operationalized at different levels.

Part V discusses key operational questions confronting policymakers when mainstreaming adaptive strategies into development policies and highlights the choice of policy instruments and the timing of interventions. In many countries, policymakers face enormous difficulties in integrating adaptation concerns into development planning. These barriers are mainly concerned with complex interaction among various development sectors; lack of accurate and timely information on climate-induced hazards; limited communication between relevant stakeholders, including policymakers, academics, private sector, and civil society; insufficient best practices information; and inadequate financial mechanisms. To overcome these barriers, the authors argue that efforts need to be channeled to develop new programs for structural and nonstructural measures, improve cross-sectoral coordination, and augment financial resources.

The concluding part reflects the book's findings and provides a comprehensive analysis of how the Asia and Pacific region can build its adaptive capacity in strategic ways in both the short and medium term. The authors envision development plans that include long-term national programs on climate change adaptation strategies that include crosscutting sectoral strategies and local-level actions and where all relevant stakeholders will be involved to create a shared understanding on the adaptation agenda. Effective mechanisms will be created to foster cooperation between sectoral, environmental, and financial authorities, as well as academia and private-sector stakeholders. Equally important is the support to capacity building at local levels to help communities understand the climate risks and links to sector activities. In the authors' perspective, significant attention should be given to reduce the uncertainty associated with impacts on local systems.

Overall, the book analyzes responses to climate change, introduces concepts and successful cases, and provides a framework for building adaptive capacity at local, national, and regional levels. No single volume can supply a comprehensive view of such a complex and continuously evolving subject. We have not attempted such a book. Instead, this book adopts diverse, intersecting angles of approach and experience sharing, with each part charting actions at local, sectoral, national, and regional levels. Individually and collectively, their contributions insightfully explore necessary and sufficient conditions for mainstreaming climate change adaption into development planning and suggest possible paths through which the resilience capacity of the region could be improved.

PART I

Climate Change Challenges, Scenarios, Risks, and Planning Tools

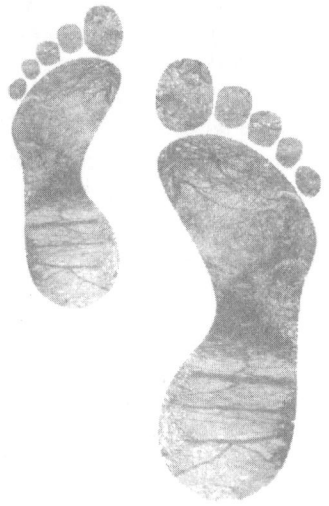

PART I

Climate Change, Challenges,
Scenarios, Risks, and Planning Tool

Key Messages

Production and activities in the agriculture, forestry, and fisheries sectors are inherently affected by variability in climate. While climate change may generate economic opportunities in some parts of the world, adapting to and accepting it are urgent issues for developing economies of Asia. The Asia and Pacific region is more vulnerable to climate change due to its geography that exposes the region to forces of nature, its weak institutional infrastructure, and a considerable portion of the population being poor.

Climate change predictions are often expressed in terms of anticipated patterns of temperature and rainfall, frequency of extreme events and sea-level rise, the impact on agricultural production, and gross domestic product (GDP). Assessing the exposure is the first step in developing strategies for climate change adaptation. These draw heavily on the outputs of global and (downscaled) regional models for predicted climate change impacts across a country.

Assessing sensitivity is a two-step process. The first step is to assess the likely exposure to natural resources by using climate model outputs. The second step is to examine the follow-on impacts of change on human economic and social systems, and ecosystems. Changes in agricultural production directly impact human welfare in many countries. The impact of change on ecological systems is equally important since they influence the availability of food, fiber, and medicine. Once a sensitivity assessment is made, it is possible then to highlight the other areas of harmful impacts.

Options for adaptation to climate change in natural-resource management include developing new crop varieties, maximizing water-use efficiency, formulating new standards for infrastructure design, exploiting co-benefit approaches, building institutional capacity, and changing the policymaking environment under which all other adaptation activities typically occur. However, there are formidable information, attitudinal, technological, and economic barriers to implementing adaptation measures.

Chapter 1

Review of the Economics of Climate Change on Southeast Asia[1]

Tae Yong Jung

Within Asia, Southeast Asia is the region that is most vulnerable to climate change. Climate change is happening now, and the worst is yet to come. If not addressed adequately, it could seriously hinder sustainable development and poverty eradication efforts in the region. There is no time for delay.

This chapter identifies factors that explain why the region is particularly vulnerable. Southeast Asia's 563 million people (World Bank 2010) are concentrated along coastlines measuring 173,251 km, leaving them exposed to rising sea levels.

The region relies heavily on agriculture for livelihoods—the sector accounted for 27% of total employment in 2007[2] (World Bank 2010) and contributed about 18% of gross domestic product (GDP) in 2009[3] (World Bank 2010)—making it vulnerable to droughts, floods, and tropical cyclones associated with warming. Its high economic dependence on natural resources and forestry—as one of the world's biggest providers of forest products—also puts it at risk. An increase in extreme weather events and forest fires arising from climate change jeopardizes vital export industries.

Poverty incidence remains high—as of 2005, about 93 million (18.8%) Southeast Asians still lived below the US$1.25-a-day poverty line (World Bank 2008)—and the poor are the most vulnerable to climate change.

This review also assessed evidence of climate change and its impact in Southeast Asia from 1990 to 2100. It tells a clear story—mean temperatures increased 0.1°C–0.3°C per decade between 1951 and 2000, rainfall trended downward during 1960–2000, and sea levels rose by 1–3 mm per year.

Heat waves, droughts, floods, and tropical cyclones have become more intense and more frequent, causing extensive damage to property, assets, and human life. The number of recorded floods/storms has risen dramatically, particularly in

---

[1] This review is based on ADB (2009). It focuses on Indonesia, the Philippines, Thailand, and Viet Nam, with Singapore participating in policy consultations.

[2] Excludes Brunei Darussalam, Cambodia, Lao People's Democratic Republic, Myanmar, and Viet Nam.

[3] Excludes Brunei Darussalam, Lao People's Democratic Republic, Myanmar, and Singapore.

the Philippines where the number rose from under 20 between 1960 and 1969 to nearly 120 between 2000 and 2008 (Center for Research on the Epidemiology of Disasters [CRED] 2008).

This review also examined studies that have attempted to predict climate change impact in the region, with all studies suggesting that it will intensify with dire consequences. Modeling work undertaken in this review covering Indonesia, the Philippines, Thailand, and Viet Nam confirms many of these findings. Indeed, it suggests that if no action is taken, on an average, the region is likely to suffer more from climate change than the rest of the world.

Annual mean temperatures are projected to rise up to 4.8°C on average by 2100 from the 1990 level. Mean sea levels are projected to rise by 70 cm during the same period, following the global trend. Indonesia, Thailand, and Viet Nam are expected to experience increasingly drier weather conditions in the next two to three decades, although this trend is likely to reverse by the middle of this century.

Global warming is likely to cause a decline in rice yield potential up to 50% on an average by 2100 compared to 1990 in the four countries; and a large part of the dominant forest and woodland could be replaced by tropical savanna and shrub with low or no carbon sequestration[4] potential.

For the four countries covered in the modeling work, the potential economic cost of inaction is huge—if the world continues the "business as usual" emission trends, considering market and non-market impacts and catastrophic risks of rising temperatures—the cost to these four countries each year could be equivalent to a loss of 6.7% of their combined GDP by 2100, more than twice the world average.

Southeast Asia is among the regions with the greatest need for adaptation, which is critical to reducing the impact of changes already locked into the climate system.

The review demonstrates that a wide range of adaptation measures is already being applied, but much more needs to be done. Adaptation requires building adaptive capacity and taking necessary technical and nontechnical measures in climate-sensitive sectors.

Strengthening adaptive capacity in Southeast Asia requires mainstreaming climate change adaptation in development planning, that is, making it an integral part of sustainable development, poverty reduction, and disaster risk management strategies. Some immediate priorities are:

- Step up efforts to raise public awareness of climate change and its impact.
- Undertake more research to better understand climate change, its impact, and solutions, especially at local levels.

---

[4] The process of removing carbon from the atmosphere and depositing it in a reservoir.

- Enhance policy and planning coordination across ministries and different levels of government for climate change adaptation.
- Adopt a more holistic approach to building the adaptive capacity of vulnerable groups and localities and their resilience to shocks.
- Develop and adopt more proactive, systematic, and integrated approaches to adaptation in key sectors that are cost effective and that offer durable and long-term solutions.

Many sectors have adaptation needs, but water, agriculture, forestry, coastal and marine resources, and health care require particular attention. While many countries have made significant efforts, this chapter identified the following priorities for further action:

1. Water resources: Scale up water conservation and management, and widen the use of integrated water management, including flood control and prevention schemes, flood early warning systems, irrigation improvement, and demand-side management.
2. Agriculture: Strengthen local adaptive capacity through better climate information, research, and development on heat-resistant crop varieties, early warning systems, and efficient irrigation systems, and explore innovative risk-sharing instruments such as index-based insurance schemes.
3. Forestry: Enhance early warning systems and awareness-raising programs to prepare for more frequent forest fires and implement aggressive public–private partnerships for reforestation and afforestation.
4. Coastal and marine resources: Implement integrated coastal-zone management plans, including mangrove conservation and planting.
5. Health: Expand or establish early warning systems for disease outbreaks, and develop health surveillance systems, awareness-raising campaigns, and infectious disease-control programs.
6. Infrastructure: Introduce "climate proofing" in transport-related investments and infrastructure, starting with public buildings.

1.1 Southeast Asia Has Great Mitigation Potential

In 2000, the region contributed 12% to the world's greenhouse gas (GHG) emissions, amounting to 5,187 metric tons of carbon dioxide equivalent ($MtCO_2$-eq), up 27% from 1990. The land-use change and forestry sector was the biggest source of emissions, contributing 75% of the region's total, followed by the energy sector contributing 15%, and the agriculture sector 8% (Intergovernmental Panel on Climate Change [IPCC] 2007a). There is considerable scope for mitigation measures that would contribute to a global solution to climate change and bring significant co-benefits to Southeast Asia.

As the largest contributor to the region's emissions, the forestry sector is the most critical. Major mitigation measures include reducing emissions from deforestation and forest degradation (REDD), increasing afforestation and reforestation, and improving forest management.

The region's energy sector—as the fastest growing contributor to emissions—also holds untapped potential for mitigation. Although Southeast Asian countries together contributed about 3% of global energy-related carbon dioxide (CO_2) emissions in 2000 (IPCC 2007b), if no action is taken, this share is expected to rise significantly, given the relatively higher economic and population growth compared to the rest of the world.

Win–win options that would reduce GHG emissions at a relatively low or even negative net cost on the supply side could include efficiency improvements in power generation, fuel switching from coal to natural gas, and renewable energy use (including biomass, solar, wind, hydro, and geothermal resources). Options on the demand side could include energy efficiency improvements and conservation in buildings (efficient lighting and electrical appliances, energy conservation, and better insulation), the industry sector (efficient equipment, heat/power recovery, and recycling), and the transport sector (cleaner fuels, energy-efficient vehicles, hybrid and/or electric vehicles, and public transport).

In case of the four countries (Indonesia, the Philippines, Thailand, and Viet Nam) covered in the modeling work, such win–win options could mitigate up to 40% of their combined energy-related CO_2 emissions per year by 2020. Another 40% could potentially be mitigated by using positive-cost options such as fuel switching from coal to gas and renewable energy in power generation, at a total cost below 1% of GDP.

In the agriculture sector, the region is estimated to have high technical potential to sequester carbon. Major mitigation options in agriculture include better land and farm management that will help reduce non-CO_2 emissions, reverse emissions from land-use change, and increase carbon sequestration in the agro-ecosystem.

Climate change mitigation is a global public good and requires a global solution built on common but differentiated responsibility. Addressing climate change requires all countries—developed and developing—to work together toward a global solution.

However, there is significant variation among countries in their capacity and affordability when undertaking adaptation and mitigation, and climate change and its impact to date are largely the result of past emissions from developed countries. This raises the important issue of equitable division of responsibilities.

An essential component of an effective global solution would, therefore, involve adequate transfers of financial resources and technological knowledge from developed to developing countries. Estimates of the additional investment needed for mitigation and adaptation in developing countries suggest that hundreds of billions of dollars per year are needed for several decades to come, far greater than the resources currently committed globally.

Global climate change cannot be tackled without the participation of developing countries. In the coming decades, their GHG emissions will grow faster than those of the developed countries and the developing countries hold significant potential for cost-effective emissions reductions.

As a highly vulnerable region with considerable need for adaptation and great potential for mitigation, Southeast Asia should play an important role in a global solution.

Most countries in Southeast Asia have developed their own national plans or strategies, established ministries or agencies as focal points, and implemented programs supporting adaptation and mitigation. This review identifies a number of policy priorities for future action.

1.1.1 Adaptation

The priority is to enhance climate change resilience by building adaptive capacity and taking technical and non-technical adaptation measures in climate-sensitive sectors. While a country's adaptive capacity depends on its level of development, greater effort in raising public awareness, more research to fill in gaps in knowledge, better coordination across sectors and levels of government, and more financial resources will enhance its adaptive capacity.

In the key climate sensitive sectors, including water resources, agriculture, coastal and marine resources, and forestry, the priority is to scale up action by adopting a more proactive approach and integrating adaptation in development planning.

1.1.2 Mitigation

While adaptation is important, the region should also make greater mitigation efforts. Low carbon growth brings significant co-benefits, and the costs of inaction far outweigh the costs of action. Implementing mitigation measures requires the development of comprehensive policy frameworks, development and availability of low-carbon technology, incentives for private sector action, elimination of market distortions, and significant flows of finance, among others. Some specific policy recommendations have been made in the forestry, energy, and agricultural sectors.

1.1.2.1 Forestry Sector

There is a need to strengthen the region's technical and institutional capacities to undertake forest carbon inventories and implement policies and measures to benefit from future global funding mechanisms for reducing emissions from deforestation and degradation. Countries should also increase efforts to avoid deforestation, to encourage reforestation and afforestation, and to enhance national and local governance systems for sustainable forest management, including monitoring and controlling illegal logging. Since forests are home to many indigenous communities, countries must design policies to recognize and respect their rights and priorities, and ensure their participation in the design and implementation of the REDD policies.

1.1.2.2 Energy Sector

To promote the adoption of win–win mitigation options in Southeast Asia, the priority is to identify and relax the binding constraints to the adoption of these options. These constraints could include information, knowledge, and technology gaps; market and price distortions; policy, regulatory, and behavioral barriers; lack of necessary finance for upfront investment; and other hidden transaction costs. Governments should work to gradually eliminate subsidies on the use of fossil fuels and provide targeted transfers only for the poor and vulnerable.

1.1.2.3 Agriculture Sector

The priority is to reduce emissions through better land and farm management, supported by a combination of market-based programs (such as taxes on the use of nitrogen fertilizers and reform of agricultural support policies), regulatory measures (such as guidelines on the use of nitrogen fertilizers and cross compliance of agricultural support to environmental objectives), voluntary agreements (such as, better farm management practices and labeling of green products), and international programs that support technology transfer in agriculture.

1.1.3 Funding and Technology Transfer

International funding and technology transfers are essential for the success of adaptation and mitigation efforts in Asia. The region should enhance institutional capacity to make better use of existing and potential international funding resources. Existing funding sources, albeit inadequate in view of the vast task at hand, provide initial support and can be used as a catalyst to raise cofinancing. Southeast Asia has not yet made full use of these funding sources, and its representation in the global carbon market is still limited. Governments need to facilitate access to these current and potential funds through better information dissemination and technical assistance. There is a need to increase the region's presence in making use of clean development mechanisms (CDM), REDD-related, and other financing mechanisms.

1.1.4 Regional Cooperation

Since most countries in the region experience similar climate hazards, regional strategies are likely to be more cost effective than national or subnational actions in dealing with transboundary issues. These issues include integrated river basin and water resources management, forest fires, extreme weather events, threatened coastal and marine ecosystems, climate change induced migration, and outbreaks of vector-borne diseases, such as dengue fever and malaria. Regional cooperation is also effective in pursuing some mitigation measures, for example—promoting power trade; using different peak times among neighboring countries to minimize the need for building new power generation capacity in each country; developing renewable energy sources; promoting clean energy and technology transfer; and regional benchmarking of clean energy practices and performance. In the long term, a regional voluntary emissions trading system could also be considered.

1.1.5 Policy Coordination

Climate change is an issue that cuts across all parts of the government, and therefore there is a need for involving not only environment ministries and related offices, but also economic and finance ministries, and for strong intergovernmental agency policy coordination. There is also a need for putting in place or enhancing central government–local authority coordination mechanisms (for planning and funding) to encourage local and autonomous adaptation actions, and to strengthen local capacity in planning and implementing initiatives addressing climate change. For effective coordination, government agencies responsible for formulating and implementing development plans and strategies should take the lead. Addressing climate change requires leadership at the highest level of government.

1.1.6 Research

More research is required to better understand climate-change challenges and cost-effective solutions at the local level and to fill knowledge gaps. Despite the emergence of more regional and country specific studies on climate change in Southeast Asia, there is still a large knowledge gap.

References

Asian Development Bank (ADB). 2009. *The Economics of Climate Change in Southeast Asia: A Regional Review*. Manila: ADB.

Center for Research on the Epidemiology of Disasters (CRED). 2008. EM-DAT: The OFDA/ CRED International Disaster Database. Brussels: CRED, Université Catholique de Louvain, www.emdat.be (accessed February 10, 2012).

Intergovernmental Panel on Climate Change (IPCC). 2007a. *Climate Change 2007: Impacts, Adaptation and Vulnerability—Contribution of Working Group II to the Fourth Assessment Report of the Intergovernmental Panel on Climate Change.* Cambridge: Cambridge University Press.

———. 2007b. *Climate Change 2007: Contribution of Working Groups I, II, and III to the Fourth Assessment Report of the Intergovernmental Panel on Climate Change.* Cambridge and New York: Cambridge University Press.

World Bank. 2008. PovcalNet Database. iresearch.worldbank.org/PovcalNet/povDuplic. html (accessed February 10, 2012).

———. 2010. World Development Indicators Online. www.worldbank.org (accessed February 10, 2012).

Chapter 2

Agricultural Impact of Climate Change: A General Equilibrium Analysis with Special Reference to Southeast Asia

Fan Zhai and Juzhong Zhuang

2.1 Introduction

Scientific research in the past two decades has concluded that the increased atmospheric concentration of greenhouse gases (GHGs) will have significant impacts on the global climate in the coming decades. Assuming no emission-control policies are implemented, the Intergovernmental Panel on Climate Change (IPCC) predicted that average global surface temperatures will increase by 2.8°C on average this century, with best-guess increases ranging from 1.8°C to 4.0°C (IPCC 2007a). Global warming would alter natural climate and environmental systems in many ways, leading to an increased frequency of extreme weather events, rising sea levels, reversal of ocean currents, and changes in precipitation patterns. These changes could impact social and economic activities, with serious implications for the well-being of humans long into the future.

Agriculture is one of the most vulnerable sectors to be affected by the anticipated climate change. Despite the technological advances in the second half of the 20th century, including the Green Revolution, weather and climate are still key factors in determining agricultural productivity in most areas of the world. The predicted changes in temperatures and rainfall patterns, as well as their associated impacts on water availability, pests, diseases, and extreme weather events are all likely to affect substantially the potential of agricultural production. Literature on the economics of climate change suggests that although global crop production may be boosted slightly by global warming in the short term (before 2030), it will ultimately turn negative in the long term (Bruinsma 2003; IPCC 2007b). Moreover, the impact of climate change on agricultural production is unlikely to be evenly distributed across regions. Low latitude and developing countries are expected to suffer more from the agricultural effects of global warming, reflecting their disadvantaged geographic location, greater agricultural share in their economies, and limited ability to adapt to climate change. In contrast, crop production in high latitude regions will generally benefit from climate change. In a recent global comprehensive estimate for over 100 countries, Cline (2007) predicted that global agricultural productivity would fall by 15.9% in the 2080s if global warming continues unabated, with developing countries experiencing a disproportionately larger decline of 19.7%.

Agriculture plays an important role in Southeast Asia, contributing to more than 10% of gross domestic product (GDP) in most regional economies, and providing jobs for over one-third of the working population in the region. As is the case in other developing regions of the world, nearly three-fourths of the poor in Southeast Asia reside in rural areas, and a large majority of them are dependent on agriculture. Consequently, agricultural development has important implications for the reduction of poverty in Southeast Asia. Moreover, the increased exposure of Southeast Asia's agriculture sector to international trade means that any climate change-related shocks in international markets for agricultural products will be easily transmitted to the region through trade channels.

This study used a dynamic computable general equilibrium (CGE) model of the global economy to investigate the potential impacts of climate change on agriculture and the world economy, with a special focus on Southeast Asia. The CGE model is an economy-wide model that elucidates interactions among industries, consumers, and governments across the global economy. The detailed region and sector disaggregation of the model makes it possible to capture the spillover effects of sector- or country-specific shocks. Climate changes impact an economy directly through the effects on that economy's agricultural outputs and indirectly through changes in the agricultural production of other countries. We established this distinction by comparing the scenario of agricultural productivity shrinkage in Southeast Asia to the scenario of agricultural productivity shrinkage in the rest of the world.

The role of productivity growth in adapting to the climate change was also examined. Section 2.2 discusses the relationship between climate change and agricultural production by reviewing the existing literature in which various modeling approaches have been employed to estimate the impacts of climate change on agricultural productivity. We then describe the specifications of the CGE model used in this study in Section 2.3. Section 2.4 assesses the impacts of climate change–induced global agricultural productivity decline on agricultural production, trade, and macro-economy. The final section offers conclusions.

2.2 Climate Change and Agriculture

Climate affects agricultural productivity in a variety of ways. Temperature, radiation, rainfall, soil moisture, and carbon dioxide (CO_2) concentration are all important variables to determine productivity, and their relationships are not simply linear. Research confirms that there are thresholds for these climate variables above which crop yields decline (Challinor et al. 2005; Porter and Semenov 2005). For example, the modeling studies discussed in IPCC reports indicate that moderate to medium increases in mean temperature (1°C–3°C), along with associated CO_2 increases and rainfall changes, are expected to benefit crop yields in temperate regions. However, in low latitude regions, moderate temperature increases (1°C–2°C) are likely to have negative yield impacts for major cereals. Warming of more than 3°C would have negative impacts in all regions (IPCC 2007b).

The interaction of temperature increases and changing rainfall patterns determines the impact of climate change on soil moisture. With rising temperatures, both evaporation and precipitation are expected to increase. The resulting net effect on water availability would depend on which force is more dominant. The IPCC reports project that by the middle of the 21st century, water availability will increase as a result of climate change at high latitudes and in some wet tropical areas, and decrease over some dry regions at mid latitudes and in the dry tropics (IPCC 2007b). Some regions that are already drought prone may suffer more severe dry periods.

Increases in atmospheric CO_2 concentration can have a positive impact on crop yields by stimulating plant photosynthesis and reducing water loss via plant respiration. This carbon fertilization effect is strong for so-called C3 crops,[1] such as rice, wheat, soybeans, fine grains, legumes, and most trees, which have a lower rate of photosynthetic efficiency. For C4 crops like maize, millet, sorghum, sugarcane, and many grasses, these effects are much smaller. Other factors such as a plant's growth stage, or the application of water and nitrogen, can also impact the effect of elevated CO_2 on plant yield. Research based on experiments with the free air-concentration-enrichment method suggests a much smaller CO_2 fertilization effect on yield for C3 crops and little or no stimulation for C4 crops, in comparison with past estimates from studies conducted under enclosed test conditions (Ainsworth and Long 2005; Long et al. 2006). Based on analysis of recent data, the IPCC reports suggest that yields may increase by 10%–25% for C3 crops and by 0%–10% for C4 crops when CO_2 levels reach 550 parts per million (ppm) (IPCC 2007b). However, as a number of limiting factors were not included in the modeling and experiment analysis, considerable uncertainties still surround the estimates of carbon fertilization effect.

Besides temperature and carbon concentration, some other ecological changes brought on by global warming will have an impact on agriculture. For example, the patterns of pests and diseases may change with climate change, leading to reductions in agricultural production. Moreover, agricultural productivity will be depressed by increased climate variability and increased intensity and frequency of extreme events such a drought and floods. These factors further contribute to the difficulties in estimating the impacts of climate change on agricultural productivity.

Quantitative estimates of the agricultural impact of climate change have predominantly relied on three approaches: crop-simulation models, agro-ecological zone (AEZ) models, and cross-sectional (Ricardian) models. Crop-simulation models draw on controlled experiments where crops are grown in fields or under

---

[1] Crops are generally divided into two groups, C3 and C4, depending on their efficiency of use of CO_2 during photosynthesis.

laboratory settings simulating different climates and levels of CO_2 in order to estimate yield responses of a specific crop variety to certain climates, and other variables of interest.[2] These models do not include farmer adaptation to changing climate conditions in the estimates. Consequently, their results tend to overstate the damages of climate change to agricultural production (Mendelsohn and Dinar 1999). The second approach, AEZ analysis, combines crop-simulation models with land-management decision analysis, and captures the changes in agro-climatic resources (Darwin et al. 1995; Fischer et al. 2005). An AEZ analysis categorizes existing lands by AEZs, which differ in the length of growing period and climatic zone. The length of growing period is defined based on temperature, precipitation, soil characteristics, and topography. The changes of the distribution of the crop zones along with climate change are tracked in AEZ models. Crop modeling and environmental matching procedures are used to identify crop-specific environmental limitations under various levels of inputs and management conditions, and provide estimates of the maximum agronomically attainable crop yields for a given land resource unit. However, as the predicted potential attainable yields from AEZ models are often much larger than current actual yields, the models may overestimate the effects of autonomous adaptation. Cline (2007) observed that AEZ studies tend to attribute excessive benefits to the warming of cold high latitude regions, thereby overstating global gains from climate changes.

The Ricardian cross-sectional approach explores the relationship between agricultural capacity (measured by land value) and climate variables (usually temperature and precipitation) on the basis of statistical estimates from farm surveys or country-level data. This approach automatically incorporates efficient climate change adaptations by farmers. The major criticisms of the Ricardian approach are its ignorance of price changes and that it fails to fully control the impact of other variables that affect farm incomes (Mendelsohn and Dinar 1999; Cline 1996).

Cline (2007) used both Ricardian statistical models and crop models to develop a set of consensus agricultural impact estimates through the 2080s for over 100 countries. He first developed geographically detailed projections for changes in temperature and precipitation through the 2080s based on a baseline emission projection from the IPCC's emission scenarios. Next, these climatic change projections were applied to the agricultural impact models to assess the effects of climate change on agricultural productivity. The final consensus estimates were the weighted average of the Ricardian estimates and the crops model estimates. Table 2.1 presents the major results of Cline's estimates.

---

[2] For more information on crop-simulation models, see Adams et al. (1990), Rosenberg (1993), and Rosenzweig and Parry (1994).

Table 2.1	Projected Climate Changes and Their Impacts on Agricultural Productivity, 2080	
Climate variables	**Land area**	**Farm area**
Base levels		
Temperature (°C)	13.15	16.2
Precipitation (mm per day)	2.2	2.44
By 2080s		
Temperature (°C)	18.1	20.63
Precipitation (mm per day)	2.33	2.51
Impacts on agricultural productivity (%)	Without carbon fertilization effect	With carbon fertilization effect
World (output weighted)	−15.9	−3.2
Industrial countries	−6.3	7.7
Developing countries	−21.0	−9.1
Africa	−27.5	−16.6
Asia	−19.3	−7.2
Middle East and North Africa	−21.2	−9.4
Latin America	−24.3	−12.9

Source: Cline (2007).
Note: mm = millimeter.

The climate models used in Cline's study predicted that under the IPCC's scenario A2,[3] atmospheric concentrations of CO_2 would increase to 735 ppm by 2085 from a current level of 380 ppm, and that global mean temperature would rise by 3.3°C. Land areas would warm more than oceans, with the average surface temperature increasing by 5.0°C weighting by land area and 4.4°C weighting by farming area. By the 2080s, global agricultural productivity would decline by about 3% with carbon fertilization effect and by about 16% if the carbon fertilization effect does not materialize. These losses would be disproportionately concentrated in developing countries, which would suffer losses of 9% with carbon fertilization effect and 21% without the carbon fertilization effect, in contrast to an 8% gain (with carbon fertilization effect) and 6% loss (without carbon fertilization effect) in industrial countries. The detailed estimates by country and region reported in Table 2.2 indicate that South Asia and Africa would be the two regions most harmed by climate change. In Southeast Asia, the damages of climate change to agriculture would also be severe, ranking from 15.1% for Viet Nam to 26.2% for Thailand if the carbon fertilization effect did not materialize.

---

[3] Scenario A2 is the second highest emission scenario among the six scenarios considered by the Third and Fourth Assessments Reports of the IPCC. Cline (2007) argued that scenario A2 should be viewed as an intermediate emission path as IPCC scenarios are biased toward underestimation of future emissions.

Table 2.2	Regional Impacts of Climate Change on Agricultural Productivity, 2080	
	Without carbon fertilization effect	With carbon fertilization effect
Canada	-2.2	12.5
US	-5.9	8.0
Latin America	-23.6	-12.2
EU	-5.5	8.6
Australia	-26.6	-15.6
New Zealand	2.2	17.5
PRC	-7.2	6.8
Japan	-5.7	8.4
Korea	-9.3	4.3
Indonesia	-17.9	-5.6
Malaysia, Singapore	-22.5	-10.9
Philippines	-23.4	-11.9
Thailand	-26.2	-15.1
Viet Nam	-15.1	-2.0
India	-38.1	-28.8
Other South Asia	-25.3	-14.1
Central Asia	-0.8	13.9
Rest of Asia	-25.6	-15.6
Sub-Sahara Africa	-28.3	-17.6
Rest of the world	-14.5	-1.7

Source: Authors' calculations based on Cline (2007).
Notes: EU = European Union; PRC = People's Republic of China; US = United States.

2.3 The Model

The model used in this study was a dynamic CGE model of the global economy. It was built on the LINKAGE model developed at the World Bank (van der Mensbrugghe 2005; Anderson, Martin, and van der Mensbrugghe 2006), and has its intellectual roots in the group of multi-country applied general equilibrium models used over the past two decades to analyze the global trade and environmental issues (Shoven and Whalley 1992; Hertel 1997).

This section describes the major features of the model. Production in each economic sector was modeled using nested constant elasticity of substitution (CES) functions and constant returns to scale was assumed. There were three types of production structures, depending on activities. Crop sectors reflected the substitution possibility between extensive and intensive farming. Livestock sectors reflected the substitution possibility between pasture and intensive feeding. All other sectors reflected the standard capital–labor substitution. The study assumed differentiation of products by regions of origin; that is, the Armington assumption

(Armington 1969). Top-level aggregate Armington demand was allocated between goods produced domestically and an aggregate import following a CES function. In the second level, the aggregate import was further disaggregated across the various trade partners using an additional CES nest. On the export side, it was assumed that firms treat domestic markets and foreign markets indifferently. Thus the law of one price would hold; that is, the export price was identical to that of domestic supply.

Incomes generated from production were assumed to accrue to a single representative household in each region. Households maximized utility using an implicitly direct additive demand system (AIDADS) (Rimmer and Powell 1996). AIDADS is a demand system that allows the marginal budget shares to vary as a function of total expenditure. Work by Yu et al. (2004) demonstrated the superiority of AIDADS over other demand systems in projecting food demand, especially for long-term projections involving a wide range of countries.

All commodity and factor markets were assumed to clear through prices. There are five primary factors of production: agricultural land, skilled labor, unskilled labor, capital, and natural resources. Agricultural land and skilled and unskilled labor were assumed to be fully mobile across sectors within a region. Some adjustment rigidities in capital markets were introduced through the vintage structure of capital, under which the "new" capital was fully mobile across a sector, while "old" capital in a sector could be disinvested only when this sector was in decline. In the forestry, fishing, and mining sectors, a sector-specific factor was introduced into the production function to reflect the resource constraints. These sector-specific factors were modeled using upward-sloping supply curves. For other primary factors, stocks were fixed for any given year. The numéraire of the model was defined as the manufactured export index of the high-income countries, which was held fixed.

The model was recursive dynamic, beginning with the base year of 2004 and being solved annually through 2080. Dynamics of the model were driven by exogenous population and labor growth and technological progress, as well as capital accumulation, which were driven by savings. Population and labor force projections were based on the medium-variant forecast of the United Nations (UN). As the UN population forecast covers only 2005–2050, the growth rates of population and labor forces were assumed to decline exponentially at a rate of 2% per year. The household savings rate was set as a function of economic growth and demographic changes, which were drawn from a global cross-country analysis by Bosworth and Chodorow-Reich (2006). Technological progress was assumed to be labor augmented, so the model could reach a steady state in the long run.

The model was calibrated to the GTAP version 7, using 21 countries and regions and 19 sectors. There was a heavy emphasis on agriculture and food, which accounted for 10 of the 19 sectors. Six Southeast Asian countries are explicitly modeled as individual regions in the model.

2.4 Simulations and Results

A baseline scenario from 2004 to 2080 was constructed under the assumption that there would be no climate change impacts on economic activities. The baseline scenario provided a reference growth trajectory for examining the effects of climate change-induced agricultural damages. In the baseline, GDP growth up to 2013 was exogenous, derived from the medium-based projection of the International Monetary Fund (IMF). For each region, an economy-wide, labor-augmented productivity grew endogenously over the simulation period of 2005–2013 to match the pre-specified GDP growth path. After 2013, the productivity growth rate was held fixed at the level of 2013 up to 2040, and then declined by 1% per year afterwards. The supply of agricultural land was assumed to be fixed in high-income countries and to grow by 0.12% annually in Asia and 0.2% annually in Latin America, Africa, and other regions.

The baseline scenario projected a high rate of world economic growth over the next seven decades, with global GDP growing by an average of 3.1% per year over the period of 2010–2050, and slowing down to 2.5% per year between 2050 and 2080. The average annual growth of Southeast Asia over 2010–2080 was 1.1 percentage points higher than that of the world average, and its share in global GDP increased from less than 2% in 2004 to 4.1% in 2080. Growth was accompanied by rapid structural change in developing countries. The share of agricultural value added, in volume terms, would decline from nearly 10% in 2004 to 3.8% in 2080 in Southeast Asia. Even though some Asian countries like India and Viet Nam had trade surpluses in agricultural products in the base year, they would become net importers in the next decade because of the combined effects of economic growth, industrialization, and land constraints. However, Thailand, the Philippines, and Central Asia were expected to maintain surpluses in agricultural trade over the projection period.

In the counterfactual scenario with agricultural damages, it was assumed that productivity in four crop agricultural sectors (paddy rice, wheat, other grains, and other crops) would be lower than that in the baseline scenario because of the projected changes in climate. Crop productivity shocks, which were Cline's estimates without carbon fertilization effect as reported in the first column of Table 2.2, were imposed gradually over 2009–2080. The crop productivity shocks were assumed to be uniform across sectors. The impacts of climate change were assessed by a comparison of the counterfactual scenario with the baseline scenario.

2.4.1 Global Impacts

Table 2.3 presents the simulated impacts on global welfare, GDP, and agricultural production, which are reported as percentage deviation from the "no damage" baseline. The table indicates that global real GDP would decline by 1.4% by 2080

Table 2.3 | Impact on Global Welfare and Production, 2080 (% change)

| | GDP | Welfare (EV as % of GDP) | Terms of Trade | Crop Agriculture | | | | Sectoral Output | | |
				Paddy rice	Wheat	Other grains	Other crops	Livestock	Processed food
World	-1.4	-1.3	-7.4	-9.1	-6.8	-7.8	-7.3	-5.9	-4.6
Australia	-0.3	-0.6	-42.9	-12.8	-66.7	-42.5	-40.6	7.1	-0.2
New Zealand	0.2	1.5	140.6	31.4	38.2	12.0	156.2	-11.0	-3.8
Japan	0	-0.2	1.9	-4.7	6.8	43.7	3.5	0.5	2.2
PRC	-1.3	-1.1	-0.1	-0.5	4.2	-0.5	-0.2	-1.9	-3.6
Korea	-0.2	-0.6	-5.1	-4.8	0.4	-10.6	-5.6	-1.4	-0.4
Southeast Asia 6*	-1.4	-1.7	-17.3	-16.5	-36.3	-12.6	-17.9	-1.4	-4.5
India	-6.2	-5.2	-24.0	-11.5	-24.7	-36.7	-24.1	-19.1	-29.1
Rest of South Asia	-1.9	-2.7	-19.5	-16.0	-29.0	-24.6	-19.2	-3.1	-10.8
Central Asia	-1.9	-1.5	49.7	12.8	66.9	5.1	48.9	-10.9	-0.5
Rest of Asia	-0.4	-0.7	-18.4	-20.8	-46.9	-40.5	-14.3	1.0	-5.2
Canada	-0.2	0.2	22.1	0.9	17.7	5.1	34.6	-15.3	-1.6
US	-0.1	0	5.1	21.3	10.5	0.9	6.9	-7.0	-0.3
EU	-0.2	0	21.4	12.9	32.0	17.0	20.7	-10.1	3.6
Latin America	-1.7	-2.1	-24.3	-12.2	-40.5	-23.4	-24.3	-2.7	-5.2
Sub-Saharan Africa	-2.2	-3.2	-29.6	-23.6	-61.6	-22.2	-31.3	-0.8	-4.3
Rest of the world	-1.0	-1.2	-10.1	-5.0	-16.1	-13.1	-7.9	-4.7	-2.1

Source: CGE model simulation results.
Notes: EU = European Union; EV = equivalent variation; PRC = People's Republic of China; US = United States.
*Including Indonesia, Malaysia, Philippines, Singapore, Thailand, and Viet Nam.

as a result of the predicted impacts of climate change on agricultural productivity. India would suffer the largest GDP loss of 6.2%, followed by Sub-Saharan Africa, other South Asian countries, and Central Asia. Although the estimated productivity losses from Cline's study were modest for the overall Central Asia region, high agricultural shares in some of the region's national economies account for the relatively large loss of GDP in Central Asia. Southeast Asia would see a drop in real GDP of 1.4%, similar to that of the world's average. New Zealand is the only country in the model that would experience a real GDP increase in response to the climate change-induced global agricultural adjustment.

Aggregate welfare effects, which were measured by the sum of equivalent variation of the households and real investment, generally followed the changes in real GDP. However, international price adjustment played a role in determining the distribution of global welfare losses. After incorporating agricultural damage, international prices of crop products were expected to increase by 16%–22% relative to the price of manufacturing exports of high-income countries, reflecting the inelastic demand structure of agricultural products (Figure 2.1). The resulting changes in terms of trade would benefit net agricultural exporting countries, but damage net agricultural importing countries. As shown in the second column of Table 2.3, New Zealand's welfare gained as much as 1.5% of GDP, much higher than its GDP

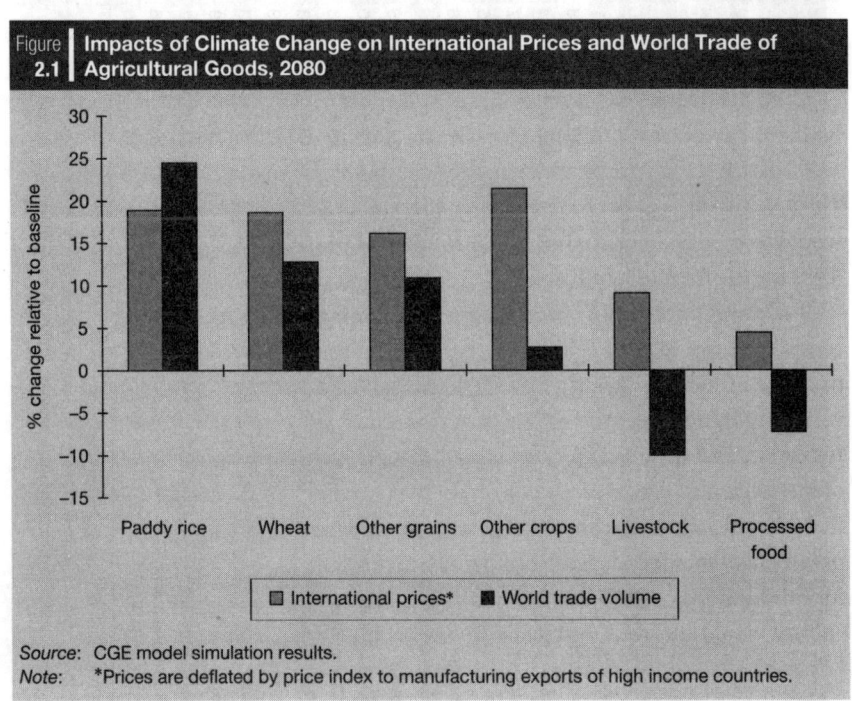

Figure 2.1 Impacts of Climate Change on International Prices and World Trade of Agricultural Goods, 2080

Source: CGE model simulation results.
Note: *Prices are deflated by price index to manufacturing exports of high income countries.

expansion, due to its improved terms of trade. In Canada and the European Union (EU), improvements in terms of trade more than offset the direct losses from agricultural productivity reduction, leading to slight welfare gains. Central Asia would benefit from changes in terms of trade. However, for other regions the deterioration of their terms of trade would amplify the effects of agricultural damage. Generally, the resulting welfare losses would be larger than GDP decline.

The detailed world agricultural production simulation results suggest that global crop production would shrink by 7.4% by 2080, which is less than half of Cline's estimate. This is partly due to the declining weight of developing countries, which would be more adversely impacted by climate change than developed countries, in global agricultural production over 2004–2080. In Cline's original estimate, agricultural output values in 2003 were used as weights to obtain the estimate for global impact. The reallocation of resources across sectors also partially offset the direct impact of agricultural productivity slowdown, contributing to the smaller magnitude of crop-output contraction. In regions where the impacts on agricultural productivity are small or positive, crop production would expand. New Zealand's crop output would increase the most, by 141%, because of its higher agricultural productivity under climate change and relatively small crop share in its economy. Central Asia, the EU, US, and Japan, would see crop production rise by 5%–50% in response to the crop price hikes. In general, the crop production expansion would come at the expense of the livestock sector, with land and other production resources being diverted toward crops sectors.

Crop production in South Asia, Latin America, and Sub-Saharan Africa would be the most adversely affected by climate change. The decline of crop output in Southeast Asia would be more moderate, but still significant at 17.3% by 2080. The negative impact of climate change on crop production in East Asian countries would be modest, ranging from 0.1% for the People's Republic of China (PRC) and 5.1% for the Republic of Korea.

As downstream sectors of crop agriculture, the production of livestock and processed food would also decline with rising input costs. World output of livestock and processed food would shrink by 5.9% and 4.6%, respectively. Again, cross-region variation exists. The production of these two sectors would drop significantly in India, but rise in Japan. Australia and the EU would also see output expansion of livestock and processed food, respectively, reflecting their stronger comparative advantage in these products as a result of climate change. The shifting comparative advantage induced by climate change would have important implications for international patterns in agricultural commodities. Global trade in crop agriculture would increase, but trade in livestock and processed food would shrink (Figure 2.1).

2.4.2 Impacts on Southeast Asian Countries

Table 2.4 reports the macroeconomic effects of the projected slowdown in agricultural productivity on six Southeast Asian countries. It is not surprising that the impact on real GDP was very modest for Singapore, given the small agricultural sector in its economy. However, the GDP contractions in Thailand, Viet Nam, and the Philippines were much more significant, ranging from 1.7% to 2.4%. The welfare losses were generally larger than GDP reductions, except for Viet Nam, which would experience a slight improvement in terms of trade. Both consumption and investment would decline compared to the baseline scenario. The incorporation of agricultural productivity damage would hamper agricultural exports of Southeast Asian countries, leading to a reduction of their aggregate exports. Consequently, aggregate imports would also decline to maintain the current account balance.

Table 2.4	Macroeconomic Impacts of Climate Change on Southeast Asian Countries, 2080 (% change)					
	Indonesia	Malaysia	Philippines	Singapore	Thailand	Viet Nam
Real GDP	−1.4	−0.9	−1.7	−0.3	−2.4	−1.7
Welfare (EV as % of GDP)	−1.7	−1.6	−1.9	−0.7	−2.7	−1.2
Terms of Trade	−0.5	−0.7	−0.9	−0.2	−0.3	0.1
Consumption	−1.9	−1.8	−2.5	−0.8	−3.0	−1.9
Investment	−0.9	−2.2	−2.4	−0.8	−2.5	−0.9
Exports	−0.9	−0.7	−0.7	0	−2.5	−1.7
Imports	−1.4	−1.5	−1.6	−0.3	−2.7	−1.5
Factor prices						
Capital	−2.0	0.3	0.2	−0.2	−0.9	−1.5
Unskilled labor	−1.5	−1.6	−2.0	−1.0	−4.0	−1.6
Skilled labor	−2.8	−1.8	−2.6	−1.2	−3.3	−2.3
Land	9.6	4.9	0.9	−8.7	−4.3	3.9

Source: CGE model simulation results.
Notes: EV = equivalent variation; GDP = gross domestic product.

To get a sense of the contribution of agricultural production slowdown in other regions to welfare losses in Southeast Asia, we ran two additional scenarios in which the climate change-induced agricultural productivity shocks were applied to Southeast Asia and other regions separately. The welfare effects of these two scenarios are presented in Figure 2.2. It is clear that that domestic productivity reduction would be the major source of welfare losses of Indonesia, the Philippines, Thailand, and Viet Nam. Actually, Indonesia, Thailand, and Viet Nam would benefit slightly from the agricultural production contraction in rest of the world.

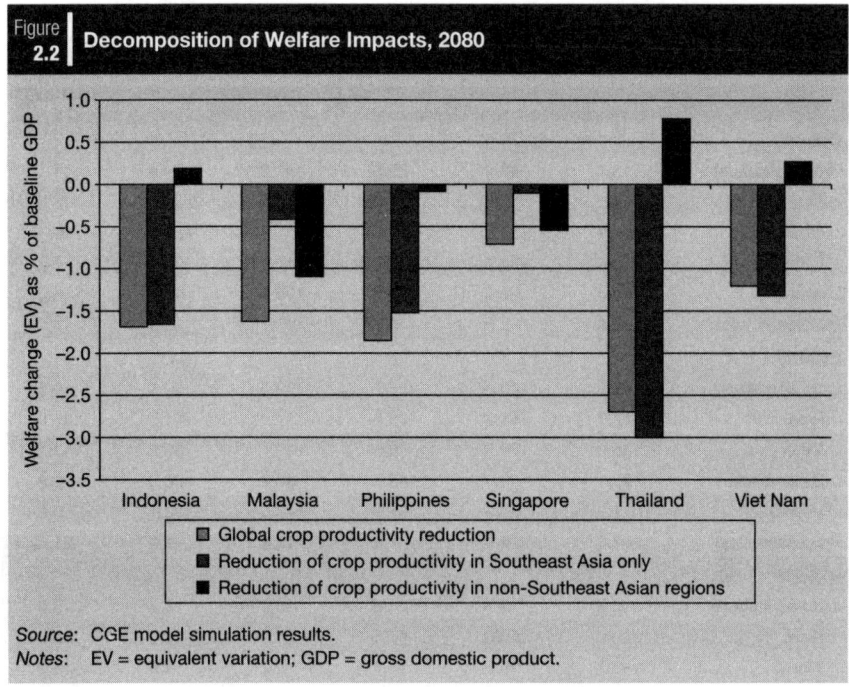

Figure 2.2 Decomposition of Welfare Impacts, 2080

Source: CGE model simulation results.
Notes: EV = equivalent variation; GDP = gross domestic product.

However, in Malaysia and Singapore the shocks from rest of the world would dominate total welfare effects because of their small agricultural sectors and their high dependence on imports for agricultural supply.

The pattern of changes in production factor gains and losses is specific to each country. In general, following negative agricultural productivity shocks, the average return to agricultural factors of production would rise relative to non-agricultural production factors, because of the inelastic demand of agricultural products. This is evident from the smaller wage decline received by unskilled labor than skilled labor, and the rising rate of return to agricultural land in most Southeast Asian countries. Singapore and Thailand are two exceptions with declining rates of return to land, mainly due to their high use of intermediate crop inputs in their crop production.

The impact on agricultural and food production and trade is shown for each Southeast Asian country in Table 2.5. All countries would see output losses in all crop sectors, except for rice production in Malaysia. Livestock output would increase in Thailand and Singapore, partly because declining land returns in the crops sectors would lead to the conversion of some arable lands to pastures. The production of the processed-food sector would expand in Malaysia and Singapore, reflecting their relatively higher efficiency in the use of crop inputs in production.

Table 2.5	Impacts on Agricultural Production and Trade in Southeast Asian Countries, 2080 (% change)					
	Indonesia	Malaysia	Philippines	Singapore	Thailand	Viet Nam
Output						
Crop agriculture	−13.4	−13.4	−22.5	−47.6	−29.4	−11.1
Rice	−15.0	1.6	−11.9		−36.3	−13.6
Other grain	−9.9	−52.6	−13.0		−26.5	−0.1
Other crops	−13.4	−31.1	−25.6	−47.6	−27.4	−7.4
Livestock	−4.4	−2.6	−0.3	105.1	12.6	−5.0
Processed food	−6.4	5.5	−4.2	12.7	−0.9	−14.2
Exports						
Crop agriculture	−25.3	−49.2	−56.7	−49.2	−59.4	10.3
Rice	−17.1	−51.2	−73.2		−41.5	46.8
Other grain	−39.9	−74.6	−48.8		−58.2	−11.2
Other crops	−25.1	−49.1	−56.7	−49.2	−60.3	9.8
Livestock	1.9	21.9	57.5	117.6	82.1	20.6
Processed food	−7.3	4.8	−7.4	13.8	−1.0	−21.6
Imports						
Crop agriculture	8.7	4.7	24.3	−0.4	11.9	−9.3
Rice	15.0	50.6	34.1	1.5	13.9	32.8
Wheat	−2.7	15.6	17.7	2.2	4.0	−15.3
Other grain	30.8	3.3	42.8	7.4	69.0	−27.6
Other crops	13.6	3.2	34.1	−0.6	12.1	−6.8
Livestock	−9.9	−16.4	−25.2	−4.2	−24.3	−12.2
Processed food	−13.6	−14.0	−12.4	−1.9	−16.1	−16.7

Source: CGE model simulation results.

As a result of the rising prices of production relative to other regions in the world, crop exports would shrink significantly for all Southeast Asian countries except Viet Nam. Viet Nam would experience export expansion in rice and other crop products due to its stronger comparative advantage in crop production and smaller reduction in agricultural productivity relative to other Southeast Asia countries. Similarly, the imports of crop agricultural products would rise for Southeast Asian economies. As a consequence, the import dependence of Southeast Asia's crops sector in 2080 would rise from 23.3% of baseline to 25.8% under the climate change scenario. Southeast Asia's grain self-sufficiency ratio in 2080 would decrease by 2.4 percentage points to 84.1% (Figures 2.3 and 2.4).

2.4.3 Sensitivity to the Baseline Agricultural Productivity

In the baseline scenario, agricultural productivity was assumed to grow at the same rate as the manufacturing and services sectors. However, in recent decades,

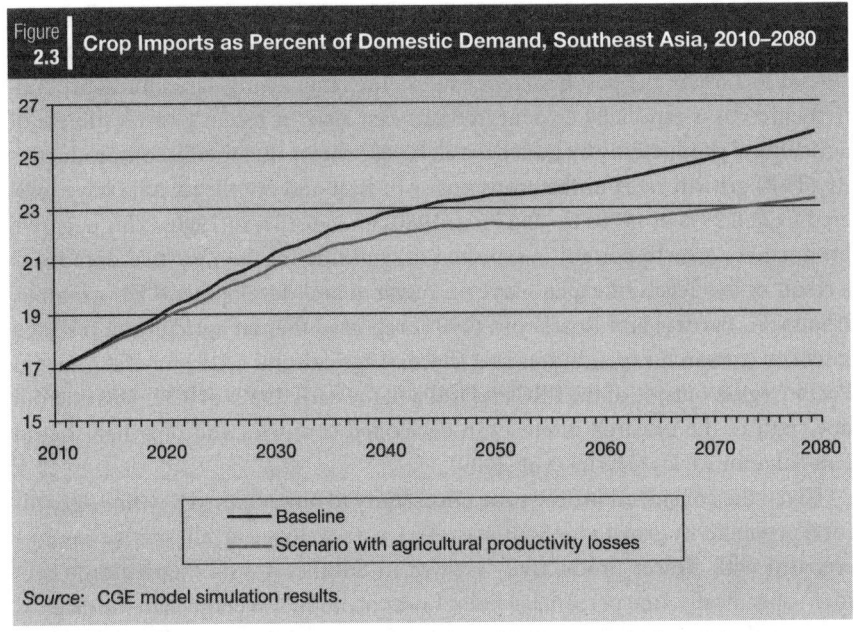

Figure 2.3 Crop Imports as Percent of Domestic Demand, Southeast Asia, 2010–2080

— Baseline
— Scenario with agricultural productivity losses

Source: CGE model simulation results.

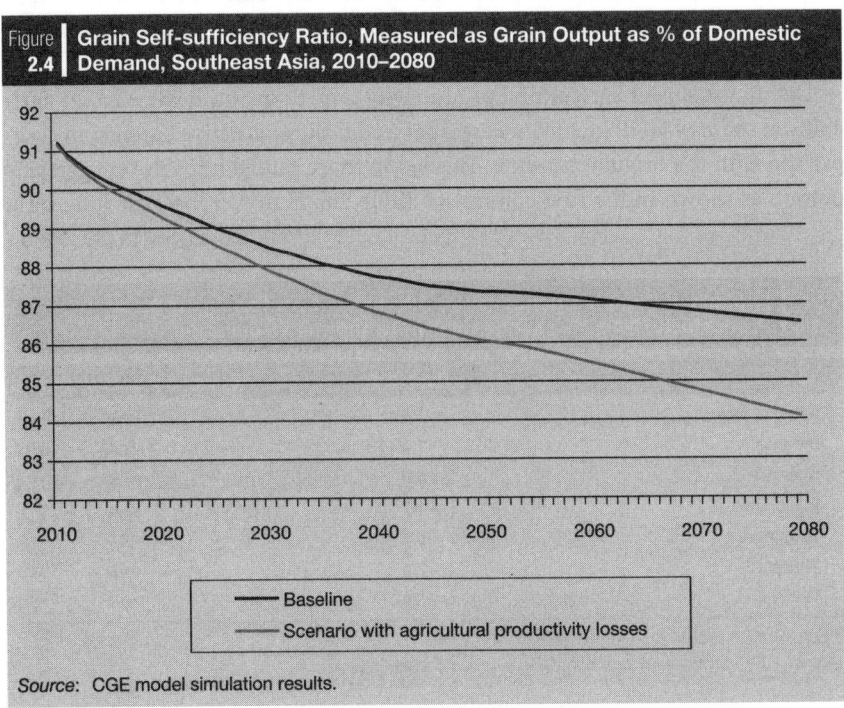

Figure 2.4 Grain Self-sufficiency Ratio, Measured as Grain Output as % of Domestic Demand, Southeast Asia, 2010–2080

— Baseline
— Scenario with agricultural productivity losses

Source: CGE model simulation results.

there has been significant slowdown in agricultural technological progress. In the 1960s and 1970s, world grain yields rose at an annual rate of 2.7%. This rate has slowed to 1.6% in the past quarter century. The languishing agricultural productivity growth is especially evident in Southeast Asia. A recent global estimate of agricultural productivity by Ludena et al. (2007) shows that total factor productivity (TFP) growth rates in the crops sector in East and Southeast Asia have lowered from 0.99% in 1970s to –0.67% in 1980s and –0.48% in 1990s. This negative productivity growth pattern is expected to continue for the next two decades as a result of low levels of expenditure on research and development. For example, Anderson, Pardey, and Roseboom (1994) reported that an agricultural research intensity (research expenditures as a share of agricultural GDP) for the Asia and Pacific region outside of the PRC and India in the early 1980s of 0.32. This is about one-sixth of the research intensity in developed countries and only half that in Sub-Saharan Africa (Hertel et al. 2009).

Given the considerable downside uncertainty to our assumed baseline agricultural productivity growth for Southeast Asia, we developed an alternative baseline scenario with slower productivity growth in Southeast Asia's agricultural sectors—specifically, one percentage point lower on annual average than the original baseline, thereupon repeating the scenario of incorporating agricultural damages. The key simulation results are presented in Table 2.6. Since the results for non-Southeast Asian regions are little changed from our original results, only revised results on GDP and welfare of Southeast Asian countries are reported.

Due to the slower agricultural productivity growth in Southeast Asia, its agricultural share of GDP in 2080 was smaller under the alternative baseline in comparison with the original baseline. This led to more muted impacts on aggregate output, as shown in the first column of Table 2.6. However, because long-term

Table 2.6	Impacts of Climate Change under Alternative Baseline Agricultural Productivity, 2080 (% change relative to alternative baseline)	
	Real GDP	Welfare (EV as % of GDP)
Southeast Asia	–1.3	–2.0
Indonesia	–1.5	–2.4
Malaysia	–1.0	–1.8
Philippines	–1.7	–2.4
Singapore	–0.3	–0.7
Thailand	–2.0	–2.8
Viet Nam	–0.9	–1.4

Source: CGE model simulation results.
Notes: EV = equivalent variation; GDP = gross domestic product.

agricultural import dependence was larger as a result of slower agricultural productivity improvement in the alternative baseline, most Southeast Asian economies were more vulnerable to the rise in world prices of agricultural products. Southeast Asian economies' losses in terms of trade, and thus welfare, were generally larger. Therefore, the results from the alternative simulations suggested that agricultural technological progress would be important for Southeast Asia to cope with the potential risks from global climate change.

2.5 Conclusion

Climate change is an increasingly significant global challenge and its negative impacts have already been felt in some regions of the world. This chapter uses a global CGE model to assess the long-term economic effects of climate change. The results suggest that the aggregate impacts of agricultural damages caused by climate change on the global economy are moderate. However, the impacts are not evenly distributed across the world. Developing countries would bear disproportionately large losses arising from climate change. Some significant adjustments in global agricultural production and trade, and consequently the distribution of income, may be accompanied by the changes of climate.

Southeast Asia is an important agricultural producer and consumer and plays a major role in the world market. With the anticipated decline in the agriculture share of GDP, the aggregate output losses from climate change-related agricultural productivity reduction would be modest for most Southeast Asian countries. However, import dependence on crop products would rise for Southeast Asia in the coming decades. This increasing exposure to world agricultural markets would make Southeast Asian economies suffer more welfare losses through the deterioration of terms of trade. This effect is especially significant for Malaysia and Singapore.

It is important to mention that there are great uncertainties in both the scientific projections and technical, social, and economic prospects. Therefore the results presented in this chapter are only illustrative. Their purpose is to provide insights on the direction and order of magnitude of the potential medium- and long-term impacts, and reveal some key potential driving forces in determining these impacts. They do not represent forecasts for the future.

One major uncertainty is the technological progress in agriculture. Agricultural productivity growth has been, and will remain to be, the most important line of defense for global food security. However, in the past two decades, productivity gains from Green Revolution have shown signs of being exhausted. If the rising demand of agricultural products, driven by population and income growth, runs a close race with technological progress in the future, the impacts of agricultural damage arising from climate change could be substantial (Zilberman et al. 2004; Cline 2007). This is especially pronounced in Southeast Asia, where productivity growth in the crop sector has been negative since 1980. Reversing this trend of declining agricultural productivity would be an important component for a Southeast Asian strategy to cope with the potential risks from the expected changes in climate.

References

Adams, R. M. et al. 1990. Global Climate Change and US Agriculture. *Nature* 345. pp. 219–224.

Ainsworth, E. A. and S. P. Long. 2005. What Have We Learned from 15 Years of Free-Air CO_2 Enrichment (FACE)? A Meta-analytic Review of the Responses of Photosynthesis, Canopy Properties and Plant Production to Rising CO_2. *New Phytol* 165 (2). pp. 351–372.

Anderson, J., P. Pardey, and J. Roseboom. 1994. Sustaining Growth in Agriculture: A Quantitative Review of Agricultural Research Investments. *Agricultural Economics.* 10 (2). pp. 107–123.

Anderson, K., W. Martin, and D. van der Mensbrugghe. 2006. Market and Welfare Implications of Doha Reform Scenarios. In W. Martin and K. Anderson, eds. *Agricultural Trade Reform and the Doha Development Agenda*, pp. 333–399. New York: Palgrave Macmillan and the World Bank.

Armington, P. S. 1969. A Theory of Demand for Products Distinguished by Place of Production. *International Monetary Fund Staff Papers* 16 (1) (March). pp. 159–176.

Bosworth, B. and G. Chodorow-Reich. 2006. Saving and Demographic Change: The Global Dimension. Prepared for the 8th Annual Joint Conference of the Retirement Research Consortium: Pathways to a Secure Retirement, 10–11 August, The Brookings Institution, Washington, DC.

Bruinsma, J., ed. 2003. *World Agriculture: Towards 2015/2030—An FAO Perspective.* UK: Earthscan.

Challinor, A. J., T. R. Wheeler, J. M. Slingo, and D. Hemming. 2005. Quantification of Physical and Biological Uncertainty in the Simulation of the Yield of a Tropical Crop Using Present-day and Doubled CO2 Climates. *Philosophical Transactions of the Royal Society B.* 360. pp. 2085–2094.

Cline, W. 1996. The Impact of Climate Change on Agriculture: Comment. *American Economic Review* 86 (5). pp. 1309–1311.

———. 2007. *Global Warming and Agriculture: Impact Estimates by Country.* Washington, DC: Center for Global Development and Peterson Institute for International Economics.

Darwin, R., M. Tsigas, J. Lewabdrowski, and A. Raneses. 1995. *World Agriculture and Climate Change: Agricultural Economic Report.* No. 703, June. Washington, DC: US Department of Agricultural, Economic Research Service.

Fischer, G., M. Shah, F. N. Tubiello, and H. van Velthuizen. 2005. Socio-economic and Climate Change Impacts on Agriculture: An Integrated Assessment, 1990–2080. *Philosophical Transactions of the Royal Society B.* 360. pp. 2067–2083.

Hertel, T. W., ed. 1997. *Global Trade Analysis: Modeling and Applications.* Cambridge: Cambridge University Press.

Hertel, T. W., C. E. Ludena, and A. A. Golub. 2009. Economic Growth, Technological Change, and the Patterns of Food and Agricultural Trade in Asia. In F. Zhai, ed. *From Growth to Convergence: Asia's Next Two Decades*, pp. 175–204. London: Palgrave Macmillan and the Asian Development Bank.

IPCC. 2007a. *Climate Change 2007: The Physical Science Basis—Contribution of Work Group I to the Fourth Assessment Report of the Intergovernmental Panel on Climate Change.* Cambridge, UK: Cambridge University Press.

———. 2007b. *Climate Change 2007: Impacts, Adaptation and Vulnerability—Contribution of Work Group II to the Fourth Assessment Report of the Intergovernmental Panel on Climate Change.* Cambridge, UK: Cambridge University Press.

Long, S. P., E. A. Ainsworth, A. D B. Leakey, J. Nosberger, and D. R. Ort. 2006. Food For Thought: Lower Than Expected Crop Yield Stimulation with Rising CO_2 Concentrations. *Science.* 312 (5782). pp. 1918–1921.

Ludena, C. E., T. W. Hertel, P. V. Preckel, K. Foster, and A. Nin. 2007. Productivity Growth and Convergence in Crop, Ruminant and Non-ruminant Production: Measurement and Forecasts. *Agricultural Economics.* 37 (1). pp. 1–17.

Mendelsohn, R., and A. Dinar. 1999. Climate Change, Agriculture, and Developing Countries: Does Adaptation Matter? *World Bank Research Observer.* 14 (2). pp. 277–293.

Porter, J. R., and M. A. Semenov. 2005. Crop Responses to Climatic Variability. *Philosophical Transactions of the Royal Society B.* 360. pp. 2021–2035.

Rimmer, M. T., and A. A. Powell. 1996. An Implicitly Additive Demand System. *Applied Economics* 28 (12). pp. 1613–1622.

Rosenberg, N. J., ed. 1993. *Towards and Integrated Assessment of Climate Change: The MINK Study.* Dordrecht: Kluwer Academic Publisher.

Rosenzweig, C., and M. L. Parry. 1994. Potential Impact of Climate Change on World Food Supply. *Nature* 367. pp. 133–138.

Shoven, J. B., and J. Whalley. 1992. *Applying General Equilibrium.* Cambridge: Cambridge University Press.

Van der Mensbrugghe, D. 2005. LINKAGE Technical Reference Document: Version 6.0. Mimeo. Washington, DC: World Bank.

Yu, W., T. Hertel, P. Preckel, and J. Eales. 2004. Projecting World Food Demand Using Alternative Demand Systems. *Economic Modelling.* 21. pp. 99–129.

Zilberman, D., X. Liu, D. Roland-Holst, and D. Sunding. 2004. The Economics of Climate Change in Agriculture. *Mitigation and Adaptation Strategies for Global Change.* 9 (4). pp. 365–382.

Chapter 3

Monitoring the Vulnerability and Adaptation Planning for Food Security

Vangimalla R. Reddy, David H. Fleisher, Dennis J. Timlin, Venkatachalam Anbumozhi, K. Raja Reddy, and Yang Yang

3.1 Introduction

Feast and famine, breakfast and basket case, hope and despair—these are some of the epithets commonly used to describe the Asian food security scenario prior to 1970. But, scientific and technological innovations have opened up new frontiers through the new fields of food production such as genetic engineering, farm-machinery development, water productivity, biomass utilization, climate-database management, and land informatics through remote sensing.

As a result of these advances, by the end of 2007, the proportion of undernourished people in developing countries reached an all-time low since 1961 (Food and Agriculture Organization [FAO] 2010). Numerous attempts have been made in recent years to grapple with the questions of whether Asian food production can keep pace with population growth and reduce malnutrition in the coming decades. International Food Policy Research Institute (IFPRI 2000) estimates that the world's population is expected to increase to 8.3 billion by 2015, a 239% increase from 1950. It is anticipated that 70% of this growth will be in the developing countries of Asia. While the rate of global population growth is expected to decline, the anticipated fall from 1.85% per year to 1.70% is marginal (FAO 1999). The same FAO study also projected an increase in cereal crop production by 90% from the present value of 1.8 billion tons during the next two decades as well as a dramatic rise in per capita food consumption, particularly meat and poultry production that would balance the production gains. Since these projections assume normal conditions in agricultural productivity and investment, this would not necessarily reflect the effects of major changes in crop production or degradation of land, water, and other environmental resources, or abnormal climate events. Moreover, because the potential for increasing the available land area for cultivation is limited, most of the increase in crop production will have to be met by more intensive cultivation of land already cropped leading to the possibility of serious damage to land and water systems. This may also require greater use of nonrenewable and potentially polluting farm inputs such as fertilizers and pesticides whose production and use may contribute to climate change and cause further environmental pollution.

3.2 Climate Change Impacts and Mitigation Efforts

Global climate change is not a new phenomenon. The planet's climate has changed tremendously over geologic time, and these natural changes are still occurring and will continue to do so. But what does appear to be different is the possibility of a new anthropogenic driving force affecting the rate of the climate change. Changes normally observed over a geologic time period may be happening over a shorter time span, particularly since the start of the Industrial Revolution. Apparently, human activities are causing climate change. The concentration of key anthropogenic greenhouse gases (GHGs) such as carbon dioxide (CO_2), methane (CH_4), nitrous oxide (N_2O), and tropospheric ozone have reached their highest levels ever, primarily due to the combustion of fossil fuels, industrial activities, agriculture, and land-use changes (Korner and Bazzaz 1996; Rosenzweig and Hillel 1998; Wittwer 1995). Pre-industrial concentrations of CO_2, CH_4, and N_2O were about 280 parts per million (ppm), 700 parts per billion (ppb), and 270 ppb, respectively (IPCC 2001). Ozone-depleting chemicals, such as chlorofluorocarbon (CFC-11) and hydrofluorocarbon (HFC-23), did not exist during that period, and perfluromethane was about 40 parts per trillion (ppt) (IPCC 2007). The current CO_2 concentration is about 386 ppm (increasing at the rate of 1.9 ppm/year), CH_4 is about 1,745 ppb (7.0 ppb/year), and N_2O is about 314 ppb (0.8 ppb/year) (IPCC 2007). Even if we curtail emissions today, these gases will stay in the atmosphere for a long time as the atmospheric lifetimes for these chemicals vary (5 to 200 years for CO_2, 12 years for CH_4, and 114 years for N_2O). These changes in the atmospheric chemistry are causing the so-called greenhouse effect. If current GHG emission rates continue, agriculture and crop production will face enormous pressure from the stress caused by these heat-trapping gases. Past changes have presumably resulted in about 0.6°C increase in global temperature over the last century. Climate models projected that there will be even greater warming during the 21st century. The CO_2 concentration is projected to reach 405 to 460 ppm by 2025, 445 to 640 ppm by 2050, and 540 to 970 ppm by 2100 (IPCC 2001). The projected global mean temperatures for those CO_2 stabilization scenarios are 0.4°C–1.1°C by 2025, 0.8°C–2.6°C by 2050, and 1.4°C–5.8°C by 2100 above the values of 1990 (IPCC 2001). It is also projected that all land areas will warm more rapidly than the global average, particularly at high northern latitudes in the winter season. In addition, these projections indicate that there will be more hot days, fewer cold days, cold waves and frost days, and reduced diurnal temperature range with higher night temperatures. As the world becomes warmer, the hydrological cycle will also become more intense resulting in more uneven and intense precipitation. This will result in increased summer drying and associated risk of both droughts and floods. The projected climate change will have beneficial and adverse effects on both environmental and socioeconomic systems, but the larger and more abrupt changes in climate will cause more adverse effects on crop production and thus affect regional food security.

3.3 Application of Models in Designing Adaptation Strategies

There are several options for reducing GHG emissions and mitigating climate change impacts on human, animal, agricultural, and natural ecosystems (Figure 3.1). Technological and other methods of limiting GHG emissions and developing alternative energy sources should be given higher priority in all countries. As far as agriculture is concerned, we should focus on developing efficient sustainable crop management strategies to minimize future agricultural contribution to climate changes (Reddy and Hodges 2000). Other options include developing cultivars tolerant to climate stresses either through traditional breeding (Hall and Ziska 2000) or biotechnology methodology (Cheikh et al. 2000; Penfield 2008), or strategies that minimize impacts of drought stress by devising water saving irrigation technologies.

Preserving crop genetic resources should also be given utmost importance as they may provide the genetic materials for developing stress-resistant crop cultivars. This is very important for countries closer to the equator where high temperatures are currently limiting crop yields, and in a changing climate, these regions will have fewer options to change management strategies to optimize crop production. The crops in these regions are already experiencing yield losses because of higher than optimum temperatures and prolonged drought during the cropping seasons.

To assess the degree of sustainability of a particular agricultural production system and its impact on environmental resources as well as to develop adaptation measures, we need to understand quantitatively the processes determining crop production and how these are influenced by climate characteristics, environmental conditions, and management practices. Field studies and long-term experimentation can be used to acquire this information. Field studies, replicated across locations and years, are laborious, resource consuming, and take considerable time to generate outputs for use in the decision-making process. Moreover, they are site-specific in nature; variability in environmental conditions makes them difficult to duplicate in other places. On the other hand, crop models offer a less expensive and quicker complementary approach and can easily evaluate a number of alternative strategies and risks for agricultural decision making. Model usage and outcome can be used to indicate future trends and prescribe appropriate remedial actions, such as suitable crops, best water management practices, and changes in agronomic practices that maximize profit and minimize negative environmental impacts.

A model is a mathematical representation of a real crop-production system, and usually describes the structure or function of that particular system. Simulation is a means of using mathematical equations written in computer code to predict how a particular crop would grow in a natural environment of soil, water, and weather

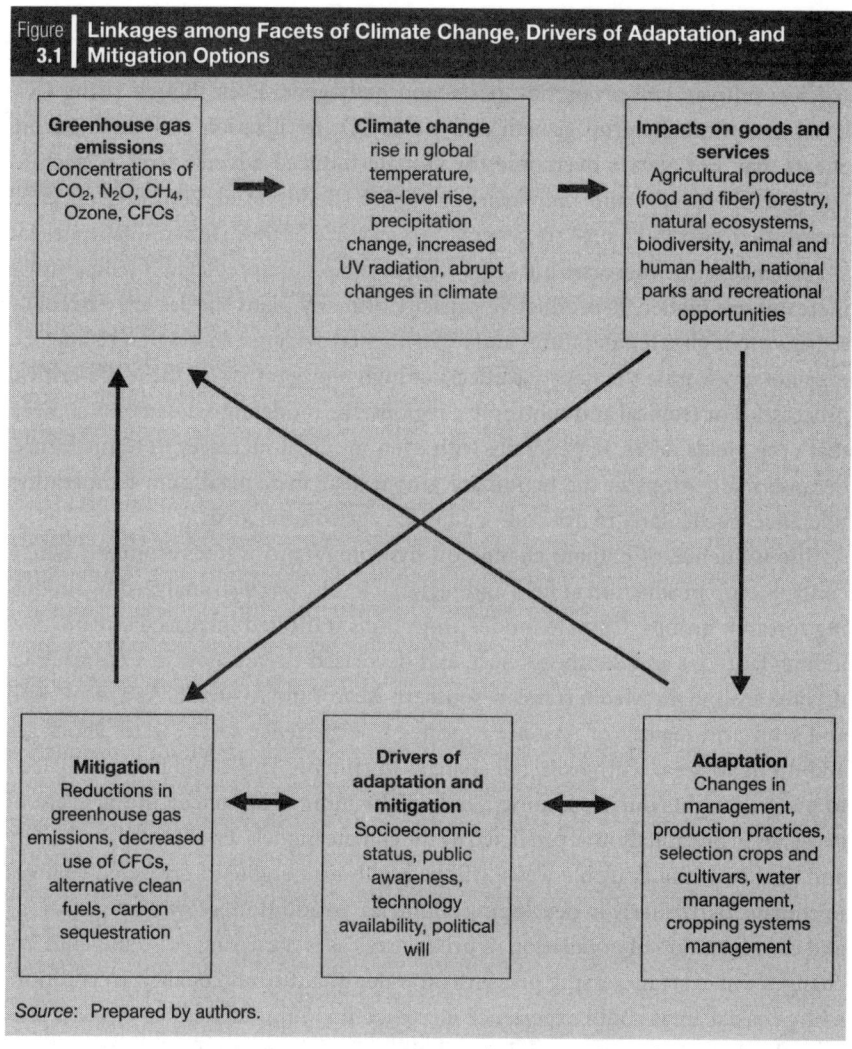

Figure 3.1 | Linkages among Facets of Climate Change, Drivers of Adaptation, and Mitigation Options

Source: Prepared by authors.

(Whisler et al. 1986). In its simplest form, crop simulation involves implementing a mathematical model to produce data that are used to study the relationship between various interlinked components of the soil–plant–atmosphere continuum and natural-resources environment. Such tools are widely used as a component of the decision-making process. Over the past decades the United States Department of Agriculture–Agricultural Research Service (USDA–ARS) Crop Systems and Global Change Laboratory at Beltsville, Maryland, US, in collaboration with state universities has developed, applied, and disseminated several crop-simulation models as decision-support tools within the mandate of USDA–ARS.

Based on model predictions and supported by experiments, it was shown that crop yield responses to projected climate change vary widely, depending upon species, cultivar, soil properties, pests, and pathogens. Even though rising CO_2 levels can stimulate crop growth and yield, CO_2 fertilization and water saving effects may not always overcome the climate-induced adverse stresses such as rising temperatures and increasing droughts (Reddy et al. 2002). Crop yield responses to climate change are generally projected to be positive at midlatitudes for small temperature increases but large yield decreases are predicted if temperature increases are higher. Reproductive capacity of many plant species are extremely sensitive to higher temperatures and experimental evidence shows that rising CO_2 may not ameliorate the negative effects of high temperature on these life-critical processes. For tropical and subtropical regions, the modeling assessments showed that crop yields decrease generally with even minimal increases in temperature, because many crops in the region are grown near their maximum temperature tolerance, particularly in dry and/or rain-fed agricultural areas.

The influence of climate change on hydrology and water resources, which controls crop production at field and regional levels, was also analyzed by modeling research groups. Climate model projections indicated increased stream-flow in high latitudes and Southeast Asia, and decreased stream-flow in Central Asia, the area around the Mediterranean, southern Africa, and Australia. As a result, arid and semi-arid regions of Asia are expected to experience severe water shortages and drought stress with projected changes in climate. This could potentially lead to more droughts during summer, and will be more pronounced in the event of extremes in precipitation as predicted by the climate models. Fresh water availability and quality of the available water will be highly vulnerable to projected changes in climate, particularly in developing countries. In addition, growing populations and concentration of population in urban areas as is occurring now, and land use changes will exert increasing pressure on water quantity and quality. In addition, many coastal areas could experience increased flooding, accelerated erosion, loss of wetlands and mangroves, and inundation of seawater into fresh water sources due to projected changes in climate.

3.4 Case Study on Use of Crop Models in Decision Making

The challenge for crop modelers is to develop user-friendly, economically viable technology that farmers and policymakers can readily adopt. This requires more interactive communication between the model developers and other stakeholders such as extension staff, crop consultants, farmers, researchers, and policymakers for improving the transferability of models from laboratories to various modes of applications. In past modeling projects with USDA–ARS, it was found that crop modeling could best be used as an aid for on-farm decisions. The GOSSYM (Baker et al. 1983) model is a cotton-crop growth simulator that incorporated important physiological processes and addressed various physical processes such as water and nutrient uptake, sunlight, and temperature, into one package that would correctly predict growth and yield under varying environmental conditions. The details of model building, structure, and application characteristics of this model are available in Reddy, Hodges, and McKinion (1997). Almost immediately after GOSSYM was tested and validated, farmers indicated the need for less complicated decision-support systems (DSS) for practical use. So, USDA–ARS developed a model-based reasoning system, COMAX, to simplify information input and provide a user-friendly output format for crop management decisions such as when to irrigate and how much fertilizer to apply, the extent of land to lease and equipment leasing. There had been extensive consultation with farmers during testing and validation of this model before it was released for public use (Whisler et al. 1986). Positive feedback from many farmers in the mid-south area of the US during various stages of model building led to the adoption in various locations and wide use by farmers in the entire cotton belt of the US. The cotton model, integrated with climate change scenarios, has been used in the US for national-level applications for a cotton-belt analysis in changed climate (Reddy, Pachepsky, and Marani 1997; Reddy et al. 2002; Doherty et al. 2003).

Since development, testing, and validation of the GOSSYM/COMAX models, a similar model, GLYCIM was developed for soybeans. Along with the soybean model, GLYCIM, a user-friendly interface called GUICS for Windows-based computers was developed (Acock et al. 1999). Substantial efforts have been made by the model builders in making model development a participatory research project involving farmers, consultants, and extension staff as well as being resilient enough to support decision makers at various levels.

3.5 Other Applications of Crop Models in Natural Resource Management Decision Making

Crop-simulation models are increasingly available as a part of DSS. DSS are used to extrapolate results for strategic decision-making tasks such as regional planning, policy analysis, and poverty alleviation. In the last two decades, it has been recognized that crop-system models have a useful role to play in agricultural planning, on a commercial level in developed countries like the US, France, and Australia, and as a part of internationally funded projects in developing countries. Regional or national planning involves analyzing information that covers different crop-production systems to make decisions like best land use to meet specific development goals. Coupling of crop-simulation models with geographic information systems (GIS) containing land and water characteristics of a region, was found to be a successful approach. In India, Selvarajan et al. (1997) used the ORYZA1 and WTGROW models for analyzing tradeoffs between water use, farm income, and adoption risks at the district level. Jansen (2001) developed a methodology called Sustainable Options for Land Use (SOLUS) in which the crop-simulation model MACROS is coupled with a GIS and a linear program (LP) model to define crop options and associated management practices in Costa Rica. The same approach was later used to evaluate policy issues such as taxing chemicals to reduce environmental contamination and maintenance of forests through subsidies (Schipper et al. 2001). The SARP and SYSNET projects of the International Rice Research Institute (IRRI) involved scientists from India, Malaysia, the Philippines, and Viet Nam to develop and evaluate methodologies and tools for land-use analysis. These methodologies were applied at regional levels to support agricultural and environmental policy making. For example, in this study in India, land units were defined by AEZ based on soil and weather characteristics. WTGROW and CERES-RICE models were used to explore possible combinations of integrated farming to achieve the goals of maximizing food, minimizing water use, and controlling environmental degradation and soil salinity (Aggarwal et al. 2000).

Researchers and decision makers have begun to apply the results of crop-simulation models to strategic policy analysis. Integration of crop-simulation models with GIS and expert systems facilitated models as useful tools for investment decisions. Beinroth et al. (1998) described the development and use of agricultural and environmental GIS (AEGIS) for application with DSSAT-type models in land-use analysis in Colombia. Beinroth et al. (1998) also evaluated land-resource utilization in Colombia using crop models and AEGIS as a part of rural development policy. In this study, outputs from crop models were used as inputs to other models of second-order effects. Parry et al. (1988) used CERES models to study the first-order effects of climate on cereal yields. Farm-level profitability was then investigated as a second-order effect of the climate-induced yield change by balancing the gross return per unit of production. Finally, the implications of changes in crop yields and production for agricultural policy were examined both at the national and international levels. Crop models in combination with spatial analysis tools have the potential to develop rural poverty reduction strategies and evaluate changes in government-support programs in different agro-climatic zones. An example of how the use of crop-simulation models can help to reduce poverty in Kenya is presented by McCown et al. (1994). They described the development in the region as a spiraling "poverty trap" due to increasing population pressure, nutrient depletion, soil degradation, low crop yields, and income reduction. Use of a crop-simulation model, complemented by a small set of on-farm trials showed that using small amounts of fertilizer was an efficient strategy to break the poverty cycle. Extensive field research conducted during the past several years did not consider this as an option as the farmers in that area never used fertilizer for crop production.

The above examples indicate that there are many potential uses of crop-simulation models to support strategic policy decisions at the regional and/or national levels (Anbumozhi et al. 2003). Policymakers and aid agencies like the World Bank and the Asian Development Bank can greatly benefit from the use of crop-simulation models in evaluating the type of interventions they must make for conserving the natural resource base. Some specific areas where systems analysis will help in this endeavor are: *(i)* to assess the changes in the natural resource base because of new policies, *(ii)* to evaluate advantages and disadvantages of different policy packages such as changing cropping patterns and shifting production basins, *(iii)* to analyze farmer responses to policy changes, and *(iv)* to design new policies based on sustainable rural development by determining acceptable levels of tradeoff between development and natural resource depletion. Well-informed decisions for natural resource management have the potential to reduce rural poverty and in this sense crop-simulation models will be highly relevant and useful for development-assistance programs.

3.6 Conclusion

Achieving food security through increased crop production while maintaining environmental integrity is a challenge posed not only to farmers but also to the research community and decision makers. Policymakers need more accurate and detailed information to make informed decisions and also to understand the impacts of their decisions on the natural resource base, farmers' livelihoods, and the national economy. Moreover, global environmental issues like climate change cannot be viewed in isolation, but rather must be considered in a broad context of other human and naturally occurring stresses on ecosystems. These include air and water pollution, habitat destruction, fragmentation and land-use changes, biodiversity, and invasive or species dominance, which are closely linked to regional food security issues. The agricultural research community must develop tools that support decision makers. Crop-simulation models enhance natural resource management decisions at different levels. At the farm level, crop-simulation models are used to investigate long-term changes in the environmental quality of soil and water, and yield stability. Model applications at the farm level include selection of new cropping systems that adapt to micro-climate change, socioeconomic viability, and analysis of yield gaps between experimental stations and field production. Regional-level use of crop-simulation models allows the aggregation of crop-production responses to various environmental changes such as climate change and soil erosion, for use by researchers and policymakers. National- and global-level studies using crop-simulation models are already underway, notably to investigate the large-scale impacts of CO_2 and temperature change on crop production.

Improvement in models has been continuing and more scientific understanding is needed to deal with sensitivity of crop production to dynamic changes taking place in the natural resource base. It is recognized that there are large numbers of national agricultural organizations and university departments investigating various aspects of crop production through modeling approaches in different parts of the world with their own information. However, the developing countries in Asia are rarely fully involved in model development, data acquisition, validation, and application. These countries can benefit from formal and informal networks, particularly in cases where data about their region are used in developing models. Global, regional, and local information sharing can be highly complementary but more efforts are needed in this area. The information generated from such modeling efforts will serve as a sound basis to make new policy interventions to attain food security at the regional level.

References

Acock, B., Y. A. Pachepsky, E. V. Mironenko, F. D. Whisler, and V. R. Reddy. 1999. GUICS: A Generic User Interface for On-farm Crop Simulations. *Agronomy Journal.* 91. pp. 657–665.

Aggarwal, P. K., N. Kalra, S. Kumar, H. Pathak, S. K. Bandopadhayay, A. K. Vasissht, R. P. Roette, and C. T. Hoanh. 2000. Haryana State Case Study: Trade-off Between Cereal Production and Environmental Impact. In R. P. Roetter, H. Van Keulen, and H. H. Van Laar, eds. *Synthesis of Methodology Development and Case Studies.* SysNet Research Paper Series No.3, pp. 11–18. Los Banos, Philippines: International Rice Research Institute.

Anbumozhi, V., V. R. Reddy, F. Y. Lu, and E. Yamaji. 2003. The Role of Simulation Models in Agricultural Research and Rural Development: A Review. *International Agricultural Engineering Journal.* 12 (1–2). pp. 1–18.

Baker, D. N., J. R. Lambert, and J. M. McKinion. 1983. GOSSYM: *A Simulator of Cotton Growth and Yield.* South Carolina Agricultural Experiment Station Technical Bulletin No. 1089.

Beinroth, F. H., J. M. Jones, E. B. Knapp, P. Papajorgji, and J. Luyten. 1998. Evaluation of Land Resources Using Crop Models and a GIS. In G. J. Tsuji, G. Hoogenboom, and P. K. Thornton, eds. *Understanding Options for Agricultural Production. Systems Approaches for Sustainable Agricultural Development*, pp. 293–312. Dordrecht, Germany: Kluwer.

Cheikh, N., P. W. Miller, and G. Kishore. 2000. Role of Biotechnology in Crop Productivity in a Changing Environment. In K. R. Reddy and H. F. Hodges, eds. *Climate Change and Global Crop Productivity*, pp. 425–436. Wallingford, Oxon, UK: CABI.

Doherty, R. M., L. O. Mearns, K. R. Reddy, M. Downton, and L. M. Daniel. 2003. Spatial Scale Effects of Climate Scenarios on Simulated Cotton Production in the Southeastern U.S.A. *Climatic Change.* 60 (1–2). pp. 99–129.

Food and Agriculture Organization (FAO). 1999. *Food Outlook: Global Information and Early Warning System on Food and Agriculture*, No. 2. Rome: FAO.

———. 2010. *The State of Food Insecurity in the World: Addressing Food Insecurity in Protracted Crises.* Rome: FAO.

Hall, A. E., and L. H. Ziska. 2000. Crop Breeding Strategies for the 21st Century. In K. R. Reddy and H. F. Hodges, eds. *Climate Change and Global Crop Productivity*, pp. 407–424. Wallingford, Oxon, UK: CABI.

International Food Policy Research Institute (IFPRI). 2000. *Annual Report*. Washington DC: IFPRI.

Intergovernmental Panel on Climate Change (IPCC). 2001. *Climate Change 2001: Mitigation, Intergovernmental Panel on Climate Change*. Cambridge, UK: Cambridge University Press.

———. 2007. *Climate Change 2007: The Physical Science Basis—Contribution of Working Group I to the Fourth Assessment Report of the Intergovernmental Panel on Climate Change*. Cambridge and New York: Cambridge University Press.

Jansen, H. G. P. 2001. A Decade of Interdisciplinary Land Use Research in Costa Rica by the Research Program on Sustainability in Agriculture (REPOSA): Achievements and Lessons. Proceedings: Third International Symposium on Systems Approaches for Agricultural Development, November 8–9, 1999, Lima.

Korner, C., and F. A. Bazzaz, eds. 1996. *Carbon Dioxide, Populations, and Communities*. San Diego, California, US: Academic Press.

McCown R. L., P. G. Coc, B. A. Keating, G. L. Hammer, P. S. Carberry, M. E. Probert, and D. M. Freebairn. 1994. The Development of Strategies for Improved Agricultural System and Land Use Management. In P. Goldsworthy and F. W. T. Penning de Vries, eds. *Opportunities, Use and Transfer of Systems Research Methods to Developing Countries: Systems Approaches for Sustainable Agricultural Development*, pp. 81–96. Dordrecht, Germany: Kluwer.

Parry, M. L., T. R. Carter, and N. T. Konjin, eds. 1988. The Impact of Climatic Variations on Agriculture. Volume I (Assessments in Coll Temperate and Cold Regions), p. 876. Dordrecht, The Netherlands: Kluver.

Penfield, S. 2008. Temperature Perception and Signal Transduction in Plants. *New Phytology*. 179. pp. 615–628.

Reddy, K. R., and H. F. Hodges, eds. 2000. *Climate Change and Global Crop Productivity*. Wallingford, Oxon, UK: CABI.

Reddy, K. R., H. F. Hodges, and J. M. McKinion. 1997. Crop Modeling and Applications: A Cotton Example. *Advances in Agronomy*. 59. pp. 225–290.

Reddy, K. R., P. R. Doma, L. O. Mearns, H. F. Hodges, A. G. Richardson, M. Y. L. Boone, and V. G. Kakani. 2002. Simulating the Impacts of Climate Change on Cotton Production in the Mississippi Delta. *Climate Research*. 22 (3). pp. 271–281.

Reddy, V. R., Y. A. Pachepsky, and A. Marani. 1997. Carbon Portioning in Cotton and Soybean Crops in Southern US under Climate Change Conditions. *World Resources Review*. 7. pp. 359–371.

Rosenzweig, C., and D. Hillel. 1998. *Climate Change and the Global Harvest: Potential Impacts of the Greenhouse Effect on Agriculture*. New York, US: Oxford University Press.

Schipper, R., H. G. P. Jansen, B. A. M. Bouman, H. Hengsdijk, A. Nieuenhuyse, and F. Sáenz. 2001. Integrated Bioeconomic Land-use Models: An Analysis of Policy Issues in the Atlantic Zone of Costa Rica. In D. R. Lee and C. B. Barrett, eds. *Tradeoffs or Synergies? Agricultural Intensification, Economic Development, and the Environment*, pp. 267–284. Wallingford, Oxon, UK: CABI.

Selvarajan, S., P. K. Aggarwal, S. Pandey, F. P. Lansigan, and S. K. Bandyopadhyay. 1997. Systems Approach for Analyzing Tradeoffs between Income, Risk, and Water Use in Rice–Wheat Production in Northern India. *Field Crops Research.* 51. pp. 147–161.

Whisler, F. D., B. Acock, D. N. Baker, R. E. Fye, H. F. Hodges, J. R. Lambert, H. E. Lemmon, J. M. McKinion, and V. R. Reddy. 1986. Crop Simulation Models in Agronomic Systems. *Advances in Agronomy.* 40. pp. 141–208.

Wittwer, S. H. 1995. *Food, Climate, and Carbon Dioxide: The Global Environment and World Food Production.* Boca Raton, Florida, US: CRC-Lewis Press.

Chapter 4

Framework Conditions for Integrating Climate Change Adaptation into Natural Resource Planning

Venkatachalam Anbumozhi

4.1 Introduction

Global warming is posing a massive threat to development. The Fourth Assessment Report of the Intergovernmental Panel on Climate Change (IPCC) stated that "warming of the climate system is unequivocal and carbon dependency of the world economy is the cause of it" (IPCC 2007). Among the simulations compiled to predict future climate change, even the most conservative scenario foresees that by the end of the 21st century, the average global temperature will rise by approximately 1.8°C–6.0°C from the average seen at the end of the 20th century (Figure 4.1).

An increase in temperature has the potential to disrupt weather patterns, cause a rise in sea levels, and produce significant changes in the amount of precipitation. Other expected effects include change in agricultural yields, modification of trade routes, glacier retreat, species extinction, and an increase in disease vectors. These events will destroy lives, force population migration, and contribute to food and water shortages. Across the world, about 40 million people are exposed to regular coastal flooding events and by the 2050s this number could rise to 150 million (Agrawala and Fankhauser 2008). Collectively, these challenges will severely constrain the ability of Asia and the Pacific to sustain its recent economic prosperity. The Stern Review (2008) and the IPCC Fourth Assessment Report (IPCC 2007) have both confirmed that not only is the cost of taking action now far smaller than the cost of inaction, but that even when the most aggressive action to address climate change is taken, it would have an almost imperceptible impact on the 150% growth in the global economy by 2050. The Stern Review stated that reductions of 75% or more from the 2000 level of global emissions will be required by 2050.

But even now, given the current level of GHGs in the atmosphere, it is too late to avert some serious consequences. Therefore, not only must Asian economies cut future global emissions but we must also learn to adapt to a warmer and more unpredictable world. The poorest populations of poor countries—the least developed countries and small island developing states—face the challenge of tackling the worst of the impacts and with the least capacity to do so. For this group, enhancing adaptation to climate impacts must become a crucial priority. However, for

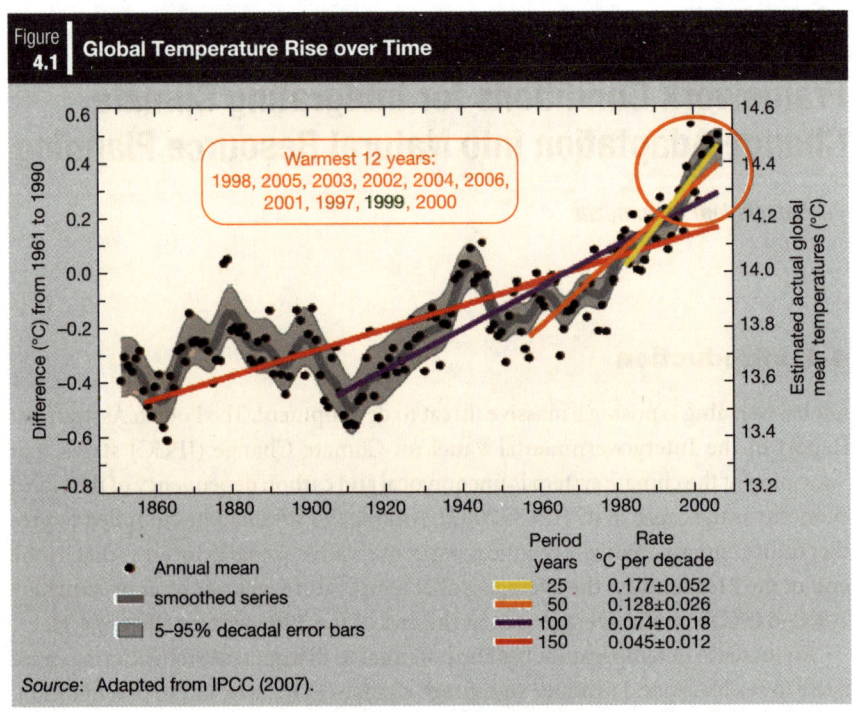

Figure 4.1   Global Temperature Rise over Time

Source: Adapted from IPCC (2007).

most of these countries economic development is often the priority, not climate change adaptation. Development and adaptation need to be considered in tandem, otherwise development will result in actions that do not succeed in reducing vulnerability of systems and social groups, which is also called maladaptation. One way of ensuring an integrated approach is through mainstreaming adaptation into development strategies, that is, integration of adaptation objectives and measures into national and regional strategies.

4.2 Climate Change Adaptation in a Development Realm

Adaptation to climate is not a new phenomenon, indeed, throughout history, societies have adapted to natural climate variability by altering settlement and agricultural patterns and other facets of their economies and lifestyles. But human-induced climate change lends complex new dimensions to society: economic vulnerability to natural climate variability, its degree of exposure, and its capacity to adapt. Exposure has two principal elements: climatic conditions, and the extent and character of natural systems and human environment (including population, institutions, wealth, and levels of awareness). Adaptive capacity is a society's ability to adapt to changing climatic conditions, whether by reducing harm or their level of sensitivity to climate impacts, or by exploiting beneficial new opportunities, or both.

With population and income growth and with the expansion of human settlements into most climate-vulnerable zones, the number of people and the level of economic wealth exposed have steadily grown. Insurance industry data shows rising losses from extreme weather events over the last decade. While the greatest losses (measured in absolute terms) occur in industrialized countries, when measured relative to wealth, losses from extreme weather are substantially higher in developing countries (Huq 2007). From 1984 to 2003, losses as a percentage of national income were three times higher in low and lower middle income countries (constituting 80% of the world's population) than in higher income countries (Agrawala 2005). Increasingly, these rising losses threaten the very purpose of development. High-risk countries must frequently invest for disaster reconstruction, which may not necessarily contribute to economic growth or poverty reduction (World Bank 2003).

It is against this backdrop of increasing exposure and losses that climate change presents a significant new set of adaptation challenges to the countries in the Asia and Pacific region. Figure 4.2 portrays the likely physical impacts of climate change over multi-dimensions aggregated across several studies. The arrows indicate the approximate temperature at which impacts are projected to occur in Asian and Pacific countries. Risks are severe and will globally affect the natural resources sector

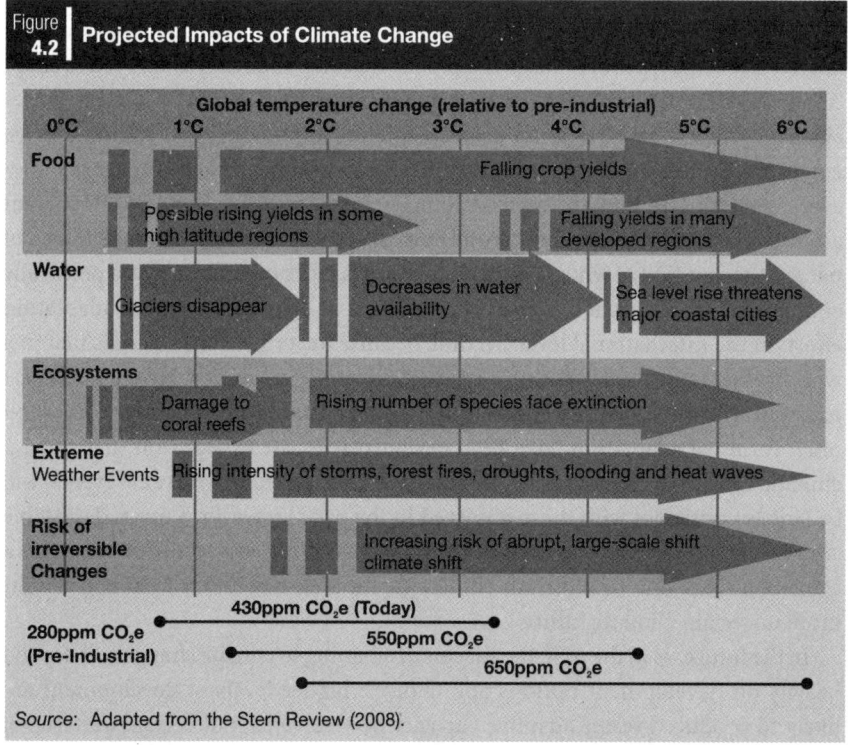

Figure 4.2 | Projected Impacts of Climate Change

Source: Adapted from the Stern Review (2008).

in all Asian countries, in particular in dense coastal areas. Impacts include changes in rainfall patterns, melting of glaciers, increasing risk of floods and intense bursts, increase in sea levels, and decline in crop yields.

The impacts of climate change will be felt across all sectors, but especially in the agriculture and water sectors. While vulnerability to climate change varies from region to region and country to country, it is highest in arid, tropical, and coastal zones. On the one hand, they are more exposed by virtue of being at lower latitudes, where impacts such as increased disease and extreme drought will be more pronounced, and because they derive a larger proportion of their economic output from climate-sensitive sectors such as agriculture, fishing, and tourism. In addition, developing countries generally have low per capita incomes, weaker institutions, and less access to technology, credit, and international markets; hence, lower adaptive capacity.

In assessing vulnerability in developing countries, many studies estimated that as many as 40% of the development programs financed by overseas assistance and concessional loans are sensitive to climate risk (ADB 2009b; Kalirajan et al. 2010; OECD 2008). Without adaptation, vulnerable developing countries could face severe costs from climate change, amounting to a considerable percentage of GDP, or up to US$100 billion, a year (Müller 2007). Absorbing these impacts will hamper the achievement of many United Nations Millennium Development Goals (MDGs).

In some respects, the added challenge posed by climate change is one of degree. The same types of policies and practical strategies already employed to adapt to natural climate variability (such as dams to control flooding, coastal defenses against cyclones, and irrigation projects to endure drought) will continue to be employed, though on a larger scale, in different locations, and at greater cost. However, adaptation to human-induced climate change is different in two important aspects. First, because it results from human activity, rather than pure forces of nature, the question of who pays for adaptation is more complicated and politically contentious. This question is especially relevant in considering future adaptation efforts at the international level. Second, because the critical difference is that in a world subject to climate change, historical climate records that have guided past adaptation have become less reliable. Cropping patterns, engineering works, and other forms of adaptation have been designed with the expectation that general climatic conditions, as well as the frequency and magnitude of extreme events, will be largely consistent with those observed in the past. However, a normal or stable climate can no longer be assumed. The challenge is not to successfully manage a transition from one equilibrium climate to another, but rather to adapt to a far more uncertain climatic future.

In the future, as in the past, the success of adapting to climate change will depend heavily on development options and choices: higher levels of development are likely to produce greater adaptive capacity or vice versa, but certain patterns of development can undermine these advances by exposing populations to ever higher levels of climate risk. Integrating climate change considerations into development becomes no longer an option but a prerequisite for sustainable development.

4.3 Climate Impacts and Adaptive Capacity of Climate Sensitive Sectors

The impacts of climate change are being felt worldwide. The present state of climate modeling allows some macro-level forecasting of the full extent or distribution of impacts at any given magnitude of climate change. Projected impacts in climate-sensitive sectors are presented in the following sections.

4.3.1 Agriculture

Agriculture predominates most Asian economies. The production of food crops is the most climate dependent economic activity. Changes in climate can be expected to have significant impacts on crop yields through changes in both temperature and moisture. As climate patterns shift, changes in the distribution of plant diseases and pests may also have adverse effects on agriculture (Parry et al. 2007). At lower altitudes, especially in seasonally dry and tropical regions, crop production is projected to decrease with even small local temperature increases. At the same time, agriculture has proven to be one of the most adaptable human activities to varying climate conditions. Many investments are relatively short term and crops and cultivars can be quickly changed to suit new conditions. There is flexibility also in farming practices, the application of irrigation water, and other inputs such as drought tolerant seeds. For these reasons—and if given sufficient time—agriculture at the global level can probably adapt to a moderate amount of global warming. There are likely to be considerable regional variations however; crops in low latitudes are more often close to their limits of heat tolerance, while growing conditions are likely to improve in higher latitudes, where agriculture might gain in competitive advantage. As in other sectors, adaptive capacity is likely to be a major factor in determining the relative distribution of adverse impacts.

4.3.2 Water Resources

Climate change is expected to have significant impacts on water supplies creating or exacerbating chronic water shortages and water quality. There is already widespread acceleration of glacial retreat and in many areas stream-flow is shifting from spring to winter peaks. If continued, these shifts could affect the availability of water for agriculture, industry, and domestic use. A rise in sea levels will result in saltwater intrusion into coastal freshwater aquifers, potentially reducing water resource availability. Changes in quantity and intensity of precipitation are likely to result in more floods and droughts and increased demand for irrigation water. Water management often requires costly investment in infrastructure. Given the long economic and physical life of reservoirs, water withdrawal, treatment, delivery, and disposal systems, adaptive responses are generally slower in water management than in agriculture.

4.3.3 Coastal Resources

One of the most certain effects of a warmer climate is a rise in sea levels. Although estimates of future sea-level rise vary, the scientific consensus is that it will be significant and will continue for centuries. Small island states and low-lying coastal areas will be subject to inundation, and risks of flooding and wind damage from coastal storms will increase. Many of the world's largest cities are at or close to sea level, and densely populated agricultural areas are situated on major river deltas. In high income countries and communities, coastal engineering can provide protection against all but the most extreme events and for medium-income countries less costly vegetative buffer zones can help, but elsewhere, evacuation and retreat may be the only option.

4.3.4 Human Health

The causes of adverse health effects are multi-factorial and there are few observed changes that can be confidently attributed to climate change. However, hundreds of millions of people could be at risk from increased morbidity or mortality resulting from climate change. Infectious diseases may become more prevalent as their reach increases and seasonality expands; the frequency and intensity of heat waves and natural hazards such as droughts, floods, and cyclones may increase, causing adverse health effects, and levels of air pollution may increase. Small changes in climate can result in substantial changes in health risks. The increased health risks are likely to be most acute in tropical countries. This is because many climate-related infections or vector-borne diseases are associated with warm or hot weather conditions. Most importantly, public health systems, which can substantially reduce health risks, are relatively weak in many Asian countries. A key factor in reducing future health risks in developing countries is the strengthening of public health systems, including monitoring and surveillance, public health infrastructure, and the development of effective adaptation measures.

4.3.5 Ecosystems and Biodiversity

Changes in natural ecosystems are among the first observable impacts of climate change. Changes in plant flowering dates and bird migration and distribution have already been widely recorded (Both ENDS 2005). Natural ecosystems are highly adapted to specific climatic conditions to migrate rapidly enough in response to changing climate patterns, however many ecosystem components, including many tree species, have much lower mobility. Even where migration is a theoretical possibility, human development has fragmented many ecosystems, weakened them through pollution and other forms of degradation, and in many places limited or cut off migration routes. The combined effects of human development and the slow rate of natural adaptation suggest that considerable ecosystem disruption will take place as the climate changes, and that substantial loss in the diversity and resilience of species is likely to occur.

4.4 Climate Change Adaptation Measures and Policies

Climate change adaptation practices are referred to as actual adjustments, or changes in decision environments, which might ultimately enhance resilience or reduce vulnerability to observed or expected changes in climate. Investments in coastal protection infrastructure to reduce vulnerability to storm surges and an anticipated sea-level rise are examples. An example of change in the policy environment includes the development of climate risk screening guidelines, which might make development projects more resilient to climate risks.

Adaptation practices can be reactive or planned, structural or non-structural, no-regret autonomous or cost effective. Examples of proactive measures include crop diversification, seasonal climate forecasting, community-based disaster risk reduction, insurance, water storage, and supplementary irrigation. Examples of reactive measures include emergency response, disaster recovery, and migration (IPCC 2007). Another classification of measures is with respect to the timing of responses to current variability, to observed medium- and long-term trends in climate, and anticipatory planning in response to model-based scenarios of long-term climate change. But there is no consolidated database on a country's exposure to climate risk at sectoral levels, so that the decision makers can prioritize and allocate resources to implement different response measures in a cost-effective way.

In the context of development programs, the cases of greatest concern are when there is a conflict between policies that promote development and those that protect individuals against the impacts of climate change (Gigli and Agrawala 2007). For example, activities such as shrimp farming and conversion of coastal mangroves may promote livelihoods, but they could also exacerbate vulnerability to rises in the sea level. Table 4.1 provides examples of some adaptation measures implemented by developing countries. Public expenditure is an important part of the adaptation strategy.

The distinction between specific adaptations and enhanced adaptive capacity is not always clear-cut. Some activities may serve both purposes: for instance, resources

Table 4.1	Selected Adaptation Initiatives in Developing Asian Countries	
Country	**Climate impact**	**Adaptation practice**
Bangladesh	Sea-level rise, saltwater intrusion	Consideration of climate change in the national water management plan, building of flow regulators on coastal embankments, use of alternate crops and low-tech water filters, cyclone shelters
Philippines	Droughts, floods	Adjustment of silvicultural treatment schedules to suit climate variations, shift to drought resistant crops, use of shallow tube wells, irrigation-water rotation methods, adoption of soil and water conservation measures for upland farming
Nepal	Glacial lake expansion	The Tsho Rolpa risk-reduction project to address creeping threat
Small island countries	Sea-level rise, drought, saltwater intrusion	Capacity building for shoreline defense systems, participatory risk assessment, grants to rehabilitate coastal infrastructure, cyclone-resistant housing, reforestation of mangroves, review of building codes, rainwater harvesting, hydroponic farming

Source: Author's compilation.

and training to integrate adaptation considerations into development planning; expanded research into alternative crops or cropping patterns; or the strengthening of public health systems. Here again, these are steps with multiple benefits beyond climate adaptation. Many specific adaptations can be effective in reducing certain climate risks also. For example, cyclone shelters in Bangladesh have proven very effective in reducing deaths during cyclones. However, specific adaptations deliver fewer ancillary benefits. In addition, where a daptive capacity is limited, the potential benefits of specific adaptations may be quite limited. For example, a weather warning system is of limited value if the people at risk do not have televisions or radios, or a means of evacuation. Thus, one objective of climate-adaptation policies should be to ensure that specific adaptations are successful and cost-effective by coupling them with corresponding advances in adaptive capacity.

It is also difficult in most instances to distinguish the specific impacts of discrete human-induced climate change from those from natural climate variability. This lack of certainty complicates the policy questions surrounding costs and burden-sharing that invokes competing notions of equity and responsibility. On the ground however, the distinction between climate variability may be completely irrelevant. In some cases, it may be possible to establish with reasonable confidence that a given impact results from climate change. One study, for instance, calculates with 90% confidence that the risk of a heat wave like the one that killed 30,000 people in Europe in 2003 has more than doubled as a result of climate change (Burton and van Aalst 2004). In some cases, an adaptation response may be driven solely by—and protected solely against—a discrete human-induced impact such as sea-level rise. Far more often, however, the impact will not be entirely new and discrete but rather intensify underlying risks such as drought, floods, or storm surges. In these cases, actions to adapt to climate change would invariably address risks arising from natural climate variability as well. From a policy perspective, an overriding objective should be a comprehensive, integrated approach to managing climate risks of all types, regardless of their cause.

4.5 Mainstreaming Climate Adaptation into Development Planning

Mainstreaming refers to the incorporation of initiatives, measures, and strategies to reduce vulnerability to climate change into other existing policies, programs, resource-management structures, and other livelihood-enhancement activities, so that adaptation to climate change becomes part of, or consistent with, other well-established sectoral programs. By mainstreaming into development planning, adaptation could increasingly be seen as key to good development practice. Conversely, development to improve the livelihood conditions of the people most affected by climate change is increasingly seen as essential to successful adaptation. At the same time, there can be tradeoffs between climate change and development, particularly when the two are considered in isolation. For instance, economic-development strategies can increase dependency on climate-sensitive resources, or there could be a mismatch between adaptation activities and the development priorities of donors and recipient countries. Climate change also adds urgency to vulnerability reduction: as new, unstable, and uncertain climatic conditions emerge, resources and research will need to keep pace. Mainstreaming could also be seen as a more sustainable, effective, and efficient use of resources rather than designing and managing separate climate policies (Smith et al. 2003). This could involve the following four stages that could span over a five-to-ten-year period (Figure 4.3).

4.5.1 Stage One: Awareness Building

The relevance of climate change to development processes needs to be highlighted and climate change must be identified as an urgent priority across sectors. This should be presented in the context of existing development concerns. Addressing the underlying causes of vulnerability must be part of the first step. It can help to draw on experiences of climate change at the household, community, and district levels, and make clear how the risks people experience have been affected by climate variability and climate change. The existing tools for climate-change data analysis,

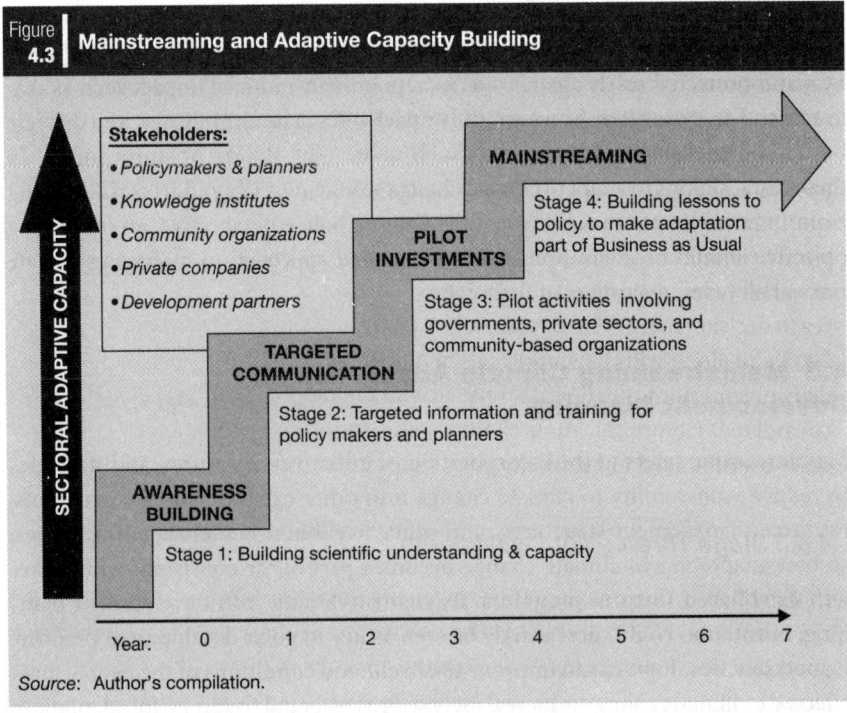

Figure 4.3   **Mainstreaming and Adaptive Capacity Building**

Source: Author's compilation.

such as crop-simulation models, are needed to provide information that is relevant, credible and, in a useful format, to the stakeholders. Investing in scientific and technical capacity on climate change is needed to ensure processed climate information is accurate and can best inform development policies and plans. Part of this is downscaling climate modeling data as far as can be usefully meaningful, as well as sustaining climate observation networks at local and regional levels (McGray et al. 2007). In addition to generating climate information, the existence of this information and its relevance to policymakers must be communicated. Institute for Global Environmental Strategies consultations (IGES 2008) show that climate information that could aid decision makers in making climate-smart decisions does exist; however, this information is seldom used. Improving climate services, raising awareness about available climate information, and providing evidence of its value to decision makers are all essential to aligning development and climate change priorities, and building capacity on climate change.

4.5.2 Stage Two: Targeted Communication

Scientific information will need translating into a format that stakeholders, including policymakers, planners, civil society organizations, and research communities, can use in practice. Incorporating this kind of data will strengthen the links between development processes and the negotiations at the UN Climate Change Conventions. However, the relevant sectors and organizations will need to become receptive to the idea of using scientific information. This demands investment in institutional capacity across all levels, but particularly at the subnational level, by participation from local communities. In parts of Southeast Asia, for instance, a combination of poor local climate data and a failure to communicate and incorporate climate considerations into policy and practice has meant that the benefits of early warning systems and improved climate science are generally failing to reach decision makers (International Institute for Sustainable Development [IISD] 2007). So along with new information and technologies, the processes needed to deliver, communicate, finance, receive, and operationalize adaptation, need to be considered. Communication channels and forums to support information and skills transfer need to be developed.

4.5.3 Stage Three: Pilot Investments

In this stage, action on adaptation, involving governments and nongovernmental organizations (NGO) and the private sector needs to be piloted to demonstrate good practice. Policymakers and planners must be convinced of the relevance of climate change to their work and be able to learn from demonstrable results. For example, in Bangladesh, although research dating back to the late 1990s has shown the implications of climate change, it took three major natural disasters in 2005 and 2007, coupled with evidence of effective systems for dealing with them, to bring the importance of adaptation to policymakers. Project planners and managers will need assistance to align and integrate risk reduction and climate change adaptation information into their development priorities.

4.5.4 Stage Four: Mainstreaming

This stage is where climate change is fully integrated. It demands a shift from business as usual to investments and planning that incorporate climate change information. Further capacity building will be needed at the policy level across sectors to ensure that lessons from stages one to three can be effectively built into the policy process. This capacity building at the national and sectoral levels should start alongside stage one to ensure the targeted stakeholders are fully engaged in the entire process. However, it may take several years before the lessons drawn from stages one to three are fully mainstreamed (Lim et al. 2005). Once climate change awareness and capacity start to grow, full integration into country, sectoral, and local development plans can begin. At the country level, bilateral country programs can support the integration of climate change priorities into country-planning strategies, for example, water management plans, infrastructure planning, and poverty-reduction strategy papers. This should set the stage for the integration of climate change programs at sectoral and local levels, given that all development planning should tie in with country development priorities. One means of more closely integrating adaptation into development decision making, as emerging in some OECD countries would be the systematic application of climate risk assessment to new projects (ADB 2008, 2009a). Proposed investments could be assessed for their own vulnerability to climate variability and human-induced climate change and for any broader effect on climate vulnerability within the host country. As with the environmental impact assessments now performed, these would provide critical information to decision makers. Multilateral institutions like the World Bank have begun to develop screening tools to help project developers assess whether proposed investments face significant climate risk.

For assessments to contribute effectively to risk reduction, the information generated should be formally taken into account in project design, review, and approval. One option would be to make the approval process of projects stringent to a set of vulnerability criteria or indicators. A project that would itself be highly vulnerable to climate risk, or would otherwise contribute to heightened societal vulnerability, would be financed only if modified to reduce projected risks to acceptable levels. For instance, a proposed highway might be rerouted to steer development away from flood-prone areas. Conversely, projects that substantially reduce climate vulnerability could be identified as priorities in country adaptation strategies, and thus, might be given preferential treatment. The criteria for choosing priorities could be established by local governments and private lenders.

Efforts to fully mainstream adaptation into development planning may encounter inherent institutional resistance at different levels of government and increase cost, particularly if they entail new conditionality. Institutions that have overriding objectives of economic and social development may view the introduction of climate concerns as a distraction from their core missions. Objections could be especially strong if new measures are not accompanied by increased assistance so that it appears that existing funds are being diverted to needs other than development. However, with or without additional assistance, routine climate risk assessment could contribute to development objectives, rather than compete with them, by helping to ensure that whatever assistance is available is wisely invested.

4.6 Partnering for Enhanced Adaptive Capacity and Mainstreaming

In principle, adaptation was established as a priority at the very start of the international climate effort. In the United Nations Framework Convention on Climate Change (UNFCCC), all parties committed to undertake national measures and to cooperate in preparing for the impacts of climate change. The UNFCCC also calls for full consideration of the specific needs and concerns of developing countries—especially the least developed—arising from the adverse effects of climate change. More concretely, developed countries committed to help vulnerable countries meet the costs of adaptation. Nearly 15 years after the convention's negotiation, however, the international adaptation effort is more an irregularly funded patchwork of multilateral and bilateral initiatives than a fully conceived and functioning one.

In 1995, the conference of the parties (COP) established a three-stage framework for addressing adaptation. Stage I, to be carried out in the short term, was to focus on identifying the most vulnerable countries or regions and adaptation options. Stage II was to involve measures, including capacity building, to prepare for adaptation. Stage III was to entail implementing measures to facilitate adaptation. The latter two stages were to be implemented over the medium and long term.

The effort to date has centered primarily on stage I and stage II type activities, more often simultaneously than sequentially. Multilateral and bilateral support has focused on building the capacity of developing countries to assess their vulnerability to climate change and examine adaptation needs and options. For example, with assistance provided under the UNFCCC, Bangladesh and small island states in the Pacific have examined their vulnerabilities to climate change and are assessing options for adaptation. The Asian Development Bank, the World Bank, and the UN Environment Programme have worked with 12 countries on in-depth assessments of vulnerability, while the UN Development Programme has assisted many countries in assessing adaptation needs (World Bank 2003). In addition, several countries, including Canada, Germany, Japan, the Netherlands, the United Kingdom, and the United States, have provided bilateral assistance. Bilateral programs have committed US$110 million to more than 50 adaptation projects in 29 countries (PEW 2006).

Recently, the emphasis has shifted to setting priorities among adaptation options (Table 4.2).

More than 40 least-developed countries have received funding under the UNFCCC to prepare national adaptation programs of action (NAPAs) addressing urgent needs. The NAPAs are meant to draw on existing information and community-level input to assess vulnerability to current climate variability and areas where risks will be heightened by climate change, and to identify priority actions. The Global Environment Facility (GEF), which administers adaptation funding under the convention, recently approved the first allocation for implementation projects through a US$50 million strategic priority on adaptation (SPA) initiative (UNEP 2006).

One significant constraint on adaptation efforts to date has been limited funding. In 2001, COP 7 established three GEF-managed funds dedicated to supporting adaptation. However, not all funds pledged by developed countries have been made

Table 4.2	Categorized Estimates of Adaptation Funds (US$ billion)		
Action	**Public**	**Private**	**Total**
Agriculture research and development	15.0	0.0	15.0
Agriculture production and processing	0.0	5.5	5.5
Health	3.0	2.0	5.0
Coastal infrastructure	5.0	0.0	5.0
Water supply infrastructure	9.0	0.0	9.0
Climate proofing new infrastructure (lower bound)	1.5	0.5	2.0
Climate proofing new infrastructure (upper bound)	31.5	9.0	40.5
Source: IPCC (2007).			

available, and some developing countries cite difficulties in accessing what funds are available. The World Bank Group (2006) reported that its support for adaptation had been approximately US$50 million over about five years, mainly through the GEF. To supplement donor country contributions, it was agreed that one of the three new funds would be supported by a levy of 2% on proceeds from emission credits generated through the Kyoto Protocol's Clean Development Mechanism (CDM). Future CDM flows, however, are uncertain. Within the UNFCCC negotiations, administration of the Adaptation Fund remains highly contentious, with many developing countries maintaining that as the funds are not from donor countries, they should be managed by an entity other than the GEF. Improved partnerships and more sustainable funding mechanisms will need to be developed in order to enhance and implement adaptation policies.

4.7 Conclusion

Climate change is a crosscutting challenge. Exposure to climate risk and the capacity to adapt are closely related to sustainable development. The adaptation challenge cuts across key economic sectors, and consequently, a wide range of policy areas. In the absence of an explicit adaptation policy, a country's de facto response to climate risks is a reflection of other policies and priorities. A strategic response to the increased risk of climate change must reach into agriculture, water resources, health, and trade policy, among others. Effective adaptation requires discrete institutions and policies to assess priorities, direct resources, and focus efforts. To be addressed successfully, and as cost-effectively as possible, adaptation concerns and priorities must be integrated across the full breadth of the economic decision-making process. Thus mainstreaming climate change into development planning is necessary.

Climate change adaptation could be a risk management strategy too. When viewed in this context, that is, as not just an environmental problem, the importance of engaging many levels of government, including economic and financial professionals, becomes apparent. Climate change impacts can have debilitating effects on the macroeconomic settings of countries, especially vulnerable developing countries. There can also be upside opportunities for the climate dependent sectors, such as agriculture and water, which may benefit from climate change.

Mainstreaming adaptation into development planning requires a range of inputs and actions, some of which are more open to international assistance than others. If it is to be effective, mainstreaming adaptation requires an understanding of the following elements:

- Decision support systems (DSS) such as vulnerability assessment tools and simulation models that predict the impacts of projected climate changes on key sectors such as agriculture, water, and infrastructure
- Structural measures and nonstructural practices to adapt to expected changes
- Effective policy framework that facilitates forward planning and climate risk management

While the first two elements are open to international assistance, the third element is closely linked to overall governance structures at the national and subnational levels. Without proper governance structures, the provision of international assistance is unlikely to be efficient or effective. Effective mainstreaming of adaptation will also depend to a large extent on domestic policies and the international enabling framework in which adaptation occurs, and the degree to which this framework fosters strategic planning and flexible response measures.

References

Asian Development Bank (ADB). 2008. *Strengthening Mitigation and Adaptation in Asia and the Pacific*. Manila: ADB.

———. 2009a. *Mainstreaming Climate Change in ADB Operations*. Manila: ADB.

———. 2009b. *Under the Weather and the Rising Tide: Adapting to a Changing Climate*. Manila: ADB.

Agrawala, S., ed. 2005. *Bridge over Troubled Waters: Linking Climate Change and Development*. Paris: Organisation for Economic Co-operation and Development.

Agrawala, S., and S. Fankhauser, eds. 2008. *Economic Aspects of Adaptation to Climate Change: Costs, Benefits and Policy Instruments*. Paris: Organisation for Economic Co-operation and Development.

Both ENDS. 2005. *Local Contributions to the Rio Conventions*. http://www.bothends.org/strategic/BE-workingpaper1-RioConventions.pdf

Burton, I., and M. van Aalst. 2004. *Look before You Leap: A Risk Management Approach for Incorporating Climate Change Adaptation in World Bank Operations*. Washington, DC: World Bank.

Gigli, S., and S. Agrawala. 2007. *Stocktaking of Progress on Integrating Adaptation to Climate Change into Development Co-operation Activities*. Paris: Organisation for Economic Co-operation and Development.

Huq, S. 2007. *Climate Change and the Big Question We Have Not Yet Tried to Answer*. London: International Institute for Environment and Development.

Institute for Global Environmental Strategies (IGES). 2008. Reuniting Climate Change and Sustainable Development. White paper on climate change policies in the Asia-Pacific. Kanagawa, Japan.

International Institute for Sustainable Development (IISD). 2007. Community-based Adaptation to Climate Change Bulletin. A summary of the second international workshop on community-based adaptation to climate change. New York, US: IISD Reporting Services, International Institute for Sustainable Development.

Intergovernmental Panel on Climate Change (IPCC). 2007. Summary for Policymakers. In S. Solomon, D. Qin, M. Manning, Z. Chen, M. Marquis, K. B. Averyt, M. Tignor, and H. L. Miller, eds. *Climate Change 2007: The Physical Science Basis—Contribution of Working Group I to the Fourth Assessment Report of the Intergovernmental Panel on Climate Change*, pp. 1–18. Cambridge and New York: Cambridge University Press. http://www.ipcc.ch/pdf/assessment-report/ar4/wg1/ar4-wg1-spm.pdf

Kalirajan, K., K. Singh, S. Thangavelu, A. Venkatachalam, and K. Peera. 2010. Climate Change and Poverty Reduction: Where Does Official Development Assisstance Money Go? ADBI Working Paper No. 318.

Lim, B., E. Spanger-Siegfried, I. Burton, E. Malone, and S. Huq, eds. 2005. *Adaptation Policy Frameworks for Climate Change: Developing Strategies, Policies and Measures*. Cambridge: Cambridge University Press.

McGray, H., A. Hammill, and R. Bradley. 2007. *Weathering the Storm: Options for Framing Adaptation and Development*. Washington, DC: World Resources Institute.

Müller, B. 2007. Nairobi 2006: Trust and the Future of Adaptation Funding. Oxford Institute for Energy Studies EV38. January. http://www.oxfordenergy.org/wpcms/wp-content/uploads/2011/03/EV38-Nairobi2006TrustandtheFutureofAdaptationFunding-BMuller-2007.pdf

Organisation for Economic Co-operation and Development (OECD). 2008. Physical and Socio-economic Trends in Climate-related Risks and Extreme Events, and Their Implications for Sustainable Development. Technical Paper GE.08-64299. Paris: Organization for Economic Co-operation and Development.

Parry, M. L., O. F. Canziani, J. P. Palutikof, P. J. van der Linden, and C. E. Hanson, eds. 2007. *Contribution of Working Group II to the Fourth Assessment Report of the Intergovernmental Panel on Climate Change*. Cambridge: Cambridge University Press.

PEW Centre on Global Climate Change (PEW). 2006. *Adaptation to climate change, international policy options*. http://www.pewclimate.org/docUploads/PEW_Adaptation.pdf

Smith, J. B., R. J. T. Klein, and S. Huq, eds. 2003. *Climate Change, Adaptive Capacity and Development*. London: Imperial College Press.

Stern Review. 2008. *The Economics of Climate Change: Executive Summary, 2008*. http://www.hm-treasury.gov.uk/media/8AC/F7/Executive_Summary.pdf

United Nations Environment Programme (UNEP). 2006. Financing Initiative, Adaptation and Vulnerability to Climate Change: The Role of the Private Sector—CEO Briefing, November. http://www.unepfi.org/fileadmin/documents/CEO_briefing adaptation_vulnerability_2006.pdf

World Bank. 2003. *Poverty and Climate Change: Reducing the Vulnerability of the Poor through Adaptation.* Washington DC: World Bank.

World Bank Group. 2006. *Managing Climate Risk: Integrating Adaptation into World Bank Group Operations.* http://www.siteresources.worldbank.org/

PART II

Evolving Adaptation Measures in the Region

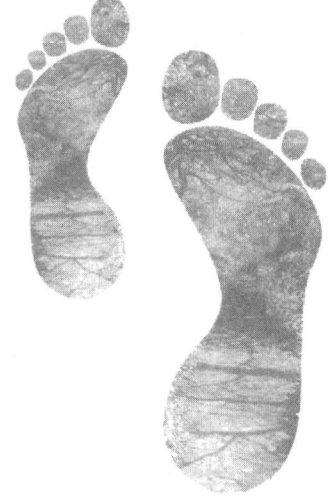

Key Messages

Diversified adaptation measures are being undertaken by Asian and Pacific countries. The effectiveness of a measure depends on the location, sectoral capacity, and socioeconomic situation. Structural and nonstructural measures are used to deal with floods and inundation. Those measures could also be classified into structural or nonstructural measures and building the resilience capacity involves finding a better combination of both. These measures could also be grouped into sectoral, cross sectoral, or multisectoral. Adaptation measures in one sector often involves a strengthening of the policy that already exists, emphasizing the importance of including long-term climate change considerations, such as extreme weather events, existing local coping mechanisms, and integrating these into sectoral development plans.

Evolving multisectoral adaptation measures are related to the management of natural resources that span sectors, for example, integrated water resource management. Linking management measures for adaptation to climate change with those management measures identified as necessary from other sectoral plans is deemed effective in developing no-regret strategies. The United Nations Framework Convention for Climate Change (UNFCCC) compendium on methods to evaluate the impacts of vulnerability, and the FAO applications used for performance evaluation, benchmarking, and modernizing irrigation systems address a range of adaptation choices.

Cross sectoral measures also span several sectors, usually including systematic improvements in risk assessment and developing effective communication mechanisms. For example, countries in the Mekong region provide an example of a cross-sectoral initiative. This initiative involves facilities for forecasting or early warning systems to provide timely information on seasonal risks to agricultural production, which also include early-warning and disaster-management strategies for food security and emergency relief to vulnerable communities during extreme weather events.

Adapting to climate change also demands the prevention and removal of maladaptive practices. Maladaptation refers to measures that do not succeed in reducing vulnerability but instead increase it. Examples of measures that prevent or avoid maladaptation include better management of local irrigation systems, and removal of laws that inadvertently increase vulnerability such as destruction of mangroves in coastal zones. There is a large body of knowledge and experience within local communities in coping with climatic variability, extreme weather events, and health risks, which should become important elements in urban planning and bringing co-benefits.

Chapter 5

Valuing Natural Resource Management: Climate Change Adaptation in the European Union

Meinhard Breiling

5.1 Introduction

We are experiencing global climate change and adverse impacts as a challenge to sustainable development on regional and local scales. The majority of Europeans became aware of climate change in 2007 when the International Panel of Climate Change (IPCC) published its fourth assessment report (Alcamo et al. 2007; IPCC 2007). Two things are obvious: climate change is real and the European Union (EU) will not succeed in completing the Kyoto Protocol. There is broad support by government agencies and the EU institutions for science-based indicative goals for the reduction of greenhouse gases (GHGs) to limit the rise in temperature by 2°C above preindustrial levels. These goals include halving GHG emissions by 2050 and setting a low personal emissions quota for everyone. This has led to the political realization that urgent action is needed, not only for climate change mitigation, but also for adaptation.

Before 2007 climate change research focused primarily on the climate system impacts in general terms, and on mitigation. New challenges are being posed by the emergence of climate change adaptation policies across Europe. Climate policy integration and coherence will be essential to bring together the environmental, economic, and social impacts of both adaptation and mitigation policies. In this context natural resource management will gain increased importance.

Natural resource management includes the two aspects constituting human life: environment and culture. Environment is the entity of all natural resources—soil, water, energy, and materials. Culture—which manifests physically in agriculture, industry, tourism, settlement patterns, and all other land uses—is the way humans use these natural resources to make a living. For millions of years the intensity of natural resources use was modest and the possibilities of humans limited. During a 250-year industrialization period and in particular during the last 50 years, natural resource use has increased tremendously. It might take decades to stabilize resource consumption globally and perhaps hundreds of years to significantly reduce GHG concentrations in the atmosphere.

As a consequence we have to consider different scenarios of climate change and various climate change impacts in our future planning. Under the ongoing gradual climate change and an expected increase in vulnerability to extreme climate events such as floods and droughts, climate change adaptation has become a necessity, even in regions that have not considerably contributed to global resource consumption.

5.2 Conceptual Framework to Evaluate Possibilities and Readiness for Climate Adaptation Measures

In a recent review Wilby et al. (2009) asserted that climate change scenarios can meet some, but not all, of the needs of adaptation planning. The possibilities and readiness to adapt are different around the globe. Often, but not always, these are related to financial resources. In addition, the knowledge of what could be done precisely and how to use available financial resources in an optimal way is missing. More guidance on adapting to the risks of climate variability and change over nearer time horizons, that is, the 2020s, is required. The United Nations Framework Convention on Climate Change (UNFCCC) secretariat published a report based on a workshop using the experience of several countries, sectors, and related impacts (UNFCCC 2010). The report addressed a wide range of climate change impacts and set out main approaches to adaptation planning. These approaches relate to the way different sectors (such as agriculture and tourism) could cope with impacts such as droughts and storms. The vulnerability of certain groups such as farming communities is targeted by community-based approaches that are combined with climate-resilient development projects. Climate-resilient development projects include a step-by-step approach and facilitate adaptation planning in particular ways within specific contexts, involving different stakeholders and with different requirements for technical, institutional, and financial resources. Several steps can be distinguished: provide climate information, screen vulnerability, identify adaptations, conduct analysis, select course of action, implement adaptations, and evaluate adaptations (GIZ 2011; USAID 2007). Several countries in the Asia and Pacific region have developed national climate change adaptation plans considering the range of possible adaptations in a country context: examples are the Kingdom of Cambodia (2006), the People's Republic of China (PRC) (2009), Pakistan Government (2010), and the Philippines Office of the President (2010).

Consequently, efforts of various regional and local government agencies have begun to consider how to integrate these approaches, since no single adaptation planning approach is sufficient to address the array of complex situations where adaptation takes place. There is also a need to avoid isolated adaptation planning. Figure 5.1 is based on the workshop of the UNFCCC (2010) and proposes a conceptual framework to climate adaptation planning. The integration of adaptation planning approaches could be achieved through close coordination and cooperation across administrative levels (that is, vertical integration), across economic sectors (that is, horizontal integration) or through the consideration and reduction of the

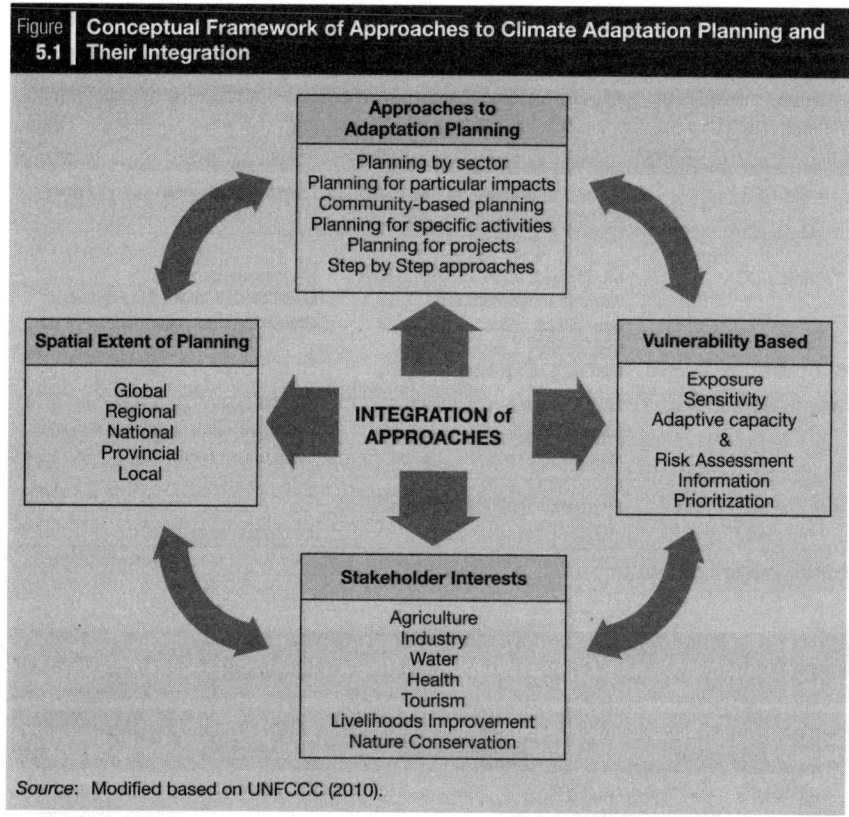

Figure 5.1 | Conceptual Framework of Approaches to Climate Adaptation Planning and Their Integration

Source: Modified based on UNFCCC (2010).

vulnerability of a particular group or system (that is, vulnerability-based integration). Stakeholder engagement, vulnerability and risk assessment, advanced local, regional, and specialist information, and prioritization of adaptive measures are equally important in this process.

The approaches to climate change adaptation refer to social, technical, and physical measures. Social measures are related to institutional and financial means or so called nonstructural measures with the aim of raising awareness and preparing people for adverse impacts in climate conditions. More people are informed through orientations to local residents about what is likely to happen in the short, medium, and long term, and to enable them to plan for their future and to avoid risks.

Appropriate spatial scales (Table 5.1) and time frames (Table 5.2) have to be adjusted to the needs of the region of concern.

Technical measures refer to concrete actions in the landscape and require engineering knowledge. Measures include digging holes to store water and to provide retention areas, planting tree lines or forests to hold soil and water even in the case of strong rains, and building walls and dams against flooding.

Table 5.1	Spatial Assessment for Climate Change Adaptation	
Climate Change Adaptation (CCA) for a particular area	**Possibilities (social, technical, physical measures)**	**Readiness (financial, legal, institutional provisions)**
Social CCA	Raising awareness: Emergency action plans related to particular problems and risks.	Nonstructural measures. Relatively cheap and fast to implement.
Technical CCA	Modify use: Improved soil tillage, water management for agricultural crops. Conversion: Convert to uses with less climate risk.	Structural measures. Local provision of CCA measures. Cheap to expensive.
Physical CCA	Give up hazard zones or use: Migration to better places with less climate risk. Local, regional, national, international refuges for climate victims.	Do we have places to shift/migrate? How many victims can be taken? Expensive measures. Source of conflicts.

Source: Author's compilation.

Table 5.2	Timing of Climate Change Adaptation Related to Selected Measures		
Measure	**Social Climate Change Adaptation (CCA)**	**Technical CCA**	**Physical CCA**
Immediate	Awareness raising to climate change and CCA	Avoided deforestation	Emergency shelters after catastrophes
Short term <2 years	Risk-management plans, emergency-action plans	Avoided soil erosion by building terraces, water-storage facilities	Avoid management of high-risk areas
Medium term <5 years	Educational programs, legal provisions	Improvement of territory with regard to water storage capacity and top soil	National climate refugee centers
Long term >5 years		Forestation, green belts, green nets	

Source: Author's compilation.

Physical climate change adaptation means to switch the location of activities. On a local scale these include choosing wind protected, cooler, or shaded places in the neighborhood and avoiding the most risk-prone hazard zones. On a regional scale, inhabitants may move away from coastal areas to escape risks like hurricanes or move higher up if they live in the mountains to have better access to water. In some cases—in densely populated or flat areas (for example, in Bangladesh or island states like Maldives) even regional movement will not be possible and other regions may be confronted with climate refugees. Climate change aggravated or induced tensions between regions will not be unlikely in future.

Some measures such as avoiding deforestation can be implemented immediately. Other measures will take different time spans: up to two years to develop risk-management and emergency-action plans, up to five years to develop legal provisions for climate change adaptation and particular educational programs to train target groups, or more than five years for tree planting and forestation measures.

5.3 EU Provisions for Valuing Climate Change Adaptation in Natural Resource Management

It is important to link climate change adaptation measures to programs already in place and then develop particular actions not sufficiently well covered by other programs. In European countries climate change adaptation measures were not on the agenda until the fourth IPCC impact assessment report was published in 2007. Many of the established frameworks had not explicitly considered climate change adaptation until then. This has widely changed in the last years. The European Commission (EC) published a white paper on climate change (EC 2009a) and for Austria a climate change adaptation strategy has been in place since late 2009 (Umweltbundesamt 2009). Austria is a small and rich EU country with 8.4 million people and 84,000 km² and represents 1.7% of the EU population and 1.9% of the EU territory. Two thirds of its area is mountains. While the majority of climate induced changes are perceived adversely, some climate changes might provide new opportunities and are also considered within the strategy. Instead of developing a new and particular climate change adaptation instrument, European and Austrian authorities preferred to integrate needed measures into frameworks that already existed. This work is now in progress.

With regard to global problems like climate change the EU acts as one single player and the national policies within the EU are adjusted to each other. Individual countries can always make more strict regulations, but the frameworks are a minimum consensus often negotiated over many years. Natural resource management is widely covered by sector directives and guidelines. Standards are the same, but not all countries in the EU have had them for the same period. For example, the new EU members—Bulgaria and Romania—only adopted them in 2007 when they entered EU.

Sector programs are provided on national levels in conformity with EU regulations and/or directives. Intra-national comparison and control became a main policy in the EU. The final aim is to harmonize the development in all EU countries with regard to natural resource management and to balance out the differences. Agreed reporting systems, procedures, and measures are resulting from this. This chapter describes some important relevant EU directives.

Water issues are regulated by the water framework directive (EC 2000) and supporting directives: the flooding directive (EC 2007), and the groundwater directive

(EC 2006). River basin management plans covering the entire EU territory had to be developed by 2010 and will be implemented by 2015, and in particular cases by 2027. The aim is to achieve in EU countries a good status of water bodies with regard to identified ecological and chemical parameters. In addition, member states have to produce flood hazard and risk maps. Recently the EU included a technical note with regard to climate change adaptation (EC 2009b) that had not been previously considered. This guidance document describes 62 guiding principles for adaptation, and relates each to steps in the river basin management plans. The principles are purposely broad to be applicable across all member states not considering regional variations in potential impacts. Examples demonstrate how the principles work in practice.

Soil and soil management is not yet covered by any EU directive. Currently there are discussions going on to introduce a soil directive. As the topic is closely linked to water issues a combination with the water framework directive is also possible. An important issue in relation to climate change adaptation is the increase of soil organic matter and the fight against soil erosion. A high content of soil organic matter increases water storage capacity and avoids soil erosion. Agricultural practices can support this process. New practices, not directed to maximum yield, but to best ecological performance our now being introduced. No tillage and low-level plowing are two methods in this line. Retention tunnels and retention areas provide safety in case of flooding. A major problem is also land converted from biologically active to biologically inactive sealed areas. This is primarily the conversion of unsealed agricultural land to sealed urban land with settlement and road areas. Now there are new methods to improve sealed areas to semi-sealed areas, avoiding excessive runoff.

Biodiversity issues or qualitative aspects of natural resource management are covered by the EU habitat directive (EC 1992) that regulates protected areas within the EU. The EU region has 17% of its land area protected as Natura 2000 areas (EEA 2010a). Wetlands are a direct link to the water framework directive and treated within the international Ramsar convention. This is an intergovernmental treaty of 1971—worldwide the only one dealing with a particular ecosystem—that provides the framework for national action and international cooperation for the conservation and wise use of wetlands and their resources (Ramsar Secretariat 2010). The EU is committed to halting the loss of biodiversity and thereby adaptation to climate change is one of the target areas. Invasive species are a major threat. Within a 2°C temperature rise, the so-called pioneer plants are advancing and pushing out the specialized and established species, with the consequence of possible extinction of endangered species.

Biomass as a quantitative aspect of natural resource management is one of the topics of the recent renewable energy directive (EC 2009c). In contrast to other world regions, the use of biomass has decreased in recent decades along with a neglect of forest management in periphery and mountainous regions. Renewable energy systems again increase the demand in biomass and forest use. Additionally a promotion of biomass use is also good for climate change mitigation as a carbon sink.

There are no particular directives for major types of sensitive ecological regions, like mountains, arctic areas, coastal zones or urban settings. These areas are incorporated within each directive. However, climate change and climate change adaptation can be very particular in different ecological regions. Mountains and arctic regions are changing faster than other regions and the degree of warming is twice as fast as the global average. This is due to the retreat of snow and ice. The snow line has moved up in altitude and latitude and less water is stored as snow and ice (EEA 2010b). Thawing and melting patterns are changing with a significant impact on agriculture and tourism. Urban areas experience an additional heat effect that is not caused by climate change, but by the situation in urban areas. Here in particular it is necessary to provide more trees and water areas to cool the places naturally during the hot season. Green belts around and green nets in cities are major ways to adapt to more heat and more frequent storms associated with climate change (Amati 2008).

5.4 European Local-scale Climate Adaptation

All EU directives are sector oriented and implemented on a larger scale with medium-term goals. It is hard to know all the factors of influence dealing with areas over several thousands of square kilometers. To counter this situation, the Council of Europe proposed a European Landscape Convention (Council of Europe 2000), an initiative related to comprehensive landscape planning on a community scale, usually covering areas between tens to hundreds of square kilometers or 10-to-100 times smaller than the smallest scales of the sector approach under the EU directives. Like the European landscape convention, other municipal efforts to combine sectors are currently on a voluntary basis, carried out by nongovernmental organization (NGO) workers who receive limited funding and are not present everywhere in Europe. Because of the voluntary character and the enthusiasm of the people involved, these initiatives work well, but do not necessarily share the same methods of implementation.

The Climate Alliance is an NGO in Europe dealing with climate change and climate change adaptation (Climate Alliance 2007) at the local level and has grown gradually since 1997. Currently 1,600 municipalities in 17 countries are members. Recently a coalition with the European soil alliance—another European NGO working on the same scale—was undertaken (ELSA 2010). The two NGOs have since jointly developed a set of best practices that they have promoted in their member countries. In some European states and municipalities there exist

Agenda 21 plans aimed at sustainable development. These plans were developed after the UN conference in Rio in 1992 with the motivation to stimulate environmental benign actions. These actions include saving water, energy, and other resources or to avoid waste on the local scale and target neighborhoods with smaller groups, households or individuals. These plans can be additionally highlighted for climate change adaptation (ICLEI 2009).

5.5 Climate Adaptation and Natural Resource Management in Austria

The following section describes local approaches of actual—and often inexpensive—measures of climate change adaptation in Austria. Local measures are limited in scale, but are practiced by many people. Adaptation measures are different in less populated rural areas than in densely populated urban areas.

In agriculture, climatic changes will affect crop yields, livestock management, and the location of production. The increased likelihood and severity of extreme weather events—like hail in certain regions—will considerably increase the risk of crop failure. Climate change will also deplete soil organic matter—a major contributor to soil fertility. No tillage and minimum tillage practices including plowing less deep can protect and alter soil organic matter. Austria has the highest share of organic farm units in the EU. This fact is considered very benign, both for soil conservation and for protecting rural biodiversity. Concerned farmers can even profit from climate change. In some parts of Austria—like some regions in England and Sweden where it was previously too cold—vineyards are planted now. Fruit trees like apricots, figs, and kiwis, previously rarely available, can offer interesting alternatives in areas when spring frosts disappear. The effects of climate change on forests are likely to include changes in forest health and productivity. Certain trees are more robust to change and will increase their populations while others will become less common. Water management aims to keep water in the landscape and even to create new water bodies or restore former wetlands converted to agriculture. However, the rural economy might be adversely affected by climate change impacts. Mountain-based winter tourism, the main contributor to rural economic growth, depends on snow. Snow is less frequent in important winter sports resorts (Breiling 2008). Technical climate change adaptation in relation to artificial snowmaking has been going on since the 1980s; long before any European or Austrian climate change strategy was developed. In cases where climate change adaptation gets too expensive, low lying places are given up for winter tourism or have to look for less profitable alternatives. If higher altitude places become more severely affected, few or no alternatives are currently available.

In urban dwellings future problems can be anticipated. The Vienna water supply comes directly from the Alpine mountains nearby collected over an area that is several times the size of the city and is transported over 100 km in aqueducts.

The water quality is considered to be excellent and not usual for a city of this size. Good forest management and wise land use has therefore a direct impact on the health of people that in their majority do not see the place of origin of their water (Kuschnig 2006). Drought has not yet led to water scarcity, but it has made an impact on water quality. Recent storms damaged parts of the forest and reduced the cleaning potential. Forest pests are anticipated to increase with climate change. The combined impact of many stress factors means a higher vulnerability. Human health in urban regions will be favored by improving the urban climate. Vienna—probably the first town in EU to establish a water school for primary school children—very early on explained the connections between water management, land use, and climate change impacts. For this reason the public developed a good understanding for funding appropriate measures in climate change adaptation. Caring for the urban green-structure areas is essential. For more than 100 years, Vienna has been developing and enlarging its green belt continuously (Breiling and Ruland 2008). Vienna city plans to restore the water bodies that were overbuilt in former decades (Goldschmied 2006). It was the first town to plan a biomass electricity plant (Wien Energie 2011) that supplies energy to 100,000 people. While this can be criticized from a simple economic view—the costs of electricity generation are three times higher than conventional energy production—the benefits are numerous. The surrounding forest received many incentives to be better managed and the plant itself has become a tourist attraction.

5.6 Conclusion

Natural resources can be used, harvested, overbuilt, overused, polluted, or destroyed. Depending on the impact we classify natural resource management into sustainable and non-sustainable activities. More uses will become unsustainable than before if states and local communities do not react to climate change and more uses will again become sustainable if they are accompanied with appropriate actions. The awareness of local people is the first significant step in this context. Taking technical adaptation measures is the next important step and will improve the situation for many places. Physical adaptation is the last option in places that cannot be sustained within acceptable risks. Alternatives have to be provided for the people affected. The examples from Europe and Austria exhibit that concerned people can do a valuable job in postponing or even avoiding climate change impacts on local scales.

References

Alcamo, J. et al. 2007. 'Europe: Contribution of Working Group II to the Fourth Assessment Report of the Intergovernmental Panel on Climate Change. In M. L. Parry et al., eds. *Climate Change 2007: Impacts, Adaptation and Vulnerability*, pp. 541–580. Cambridge, UK: Cambridge University Press.

Amati, M. 2008. *Urban Green Belts in the Twenty-first Century.* London: Ashgate.

Breiling, M. 2008. *Snow in Kitzbühel: How Winter Tourism Is Adapting to Climate Change.* On Snow. Catalogue to Exhibition Snow Affairs, Museum Kitzbühel, November 28, 2008–April 11, 2009.

Breiling, M., and G. Ruland. 2008. The Vienna Green Belt: From Local Protection to a Regional Concept. In M. Armati, ed. *Urban Green Belts in the Twenty-first Century.* London: Ashgate.

Climate Alliance. 2007. *Adaptation and Mitigation: An Integrated Climate Policy Approach.* Frankfurt, Germany: European Secretariat.

Council of Europe. 2000. European Landscape Convention. *European Treaty Series.* No. 176. Strasbourg, France: Council of Europe.

European Commission (EC). 1992. *Council Directive 92/43/European Commission EC on the Conservation of Natural Habitats and of Wild Fauna and Flora.* Brussels: European Union.

———. 2000. *Directive 2000/60/EC of the European Parliament and of the Council of 23 October 2000 Establishing a Framework for Community Action in the Field of Water Policy.* Brussels: European Union.

———. 2006. *Directive 2006/118/EC of the European Parliament and of the Council of 12 December 2006 on the Protection of Groundwater against Pollution and Deterioration.* Brussels: European Union.

———. 2007. *Directive 2007/60/EC of the European Parliament and of the Council of 23 October 2007 on the Assessment and Management of Flood Risks.* Brussels: European Union.

———. 2009a. *Adapting to Climate Change: Towards a European Framework for Action.* EU White Paper. Brussels: European Union.

———. 2009b. *River Basin Management in a Changing Climate.* Guidance Document No. 24. Common Implementation Strategy for the Water Framework Directive (2000/60/EC). Technical Report-2009-040. Brussels: European Union.

———. 2009c. Directive 2009/28/EC of the European Parliament and of the Council of 23 April 2009 on the Promotion of the Use of Energy from Renewable Sources and Amending and Subsequently Repealing Directives 2001/77/EC and 2003/30/EC. Brussels: European Union.

European Environment Agency (EEA). 2010a. *The European Environment: State and Outlook 2010—Synthesis.* Copenhagen: EEA.

———. 2010b. *Signals: People and Their Environment.* Copenhagen: EEA.

European Land and Soil Alliance (ELSA). 2010. *Local Land and Soil News.* 32–33. April.

GIZ. 2011. *Adaptation to Climate Change: New Findings, Methods and Solutions.* Bonn, Germany: Deutsche Gesellschaft für Internationale Zusammenarbeit.

Goldschmied, U. 2006. Living River Liesing: A LIFE-project on Rehabilitation of a Heavily Modified Waterbody in Vienna's Urban Environment. In M. Breiling, ed. *The Implementation of the EU Water Framework Directive from International, National and Local Perspectives.* Vienna: Department for Urban Planning and Landscape Architecture, Vienna University of Technology.

Intergovernmental Panel on Climate Change (IPCC). 2007. *Fourth Assessment Report (AR4) Contribution of Working Group II to the Fourth Assessment Report of the Intergovernmental Panel on Climate Change.* Geneva: IPCC.

Kingdom of Cambodia, National Government. 2006. *National Adaptation Programme of Action to Climate Change.* Phnom Penh.

Kuschnig, G. 2006. The Interreg IIIB CADSES Project KATER II (Karst Water Research Program). In M. Breiling, ed. *The Implementation of the EU Water Framework Directive from International, National and Local Perspectives.* Vienna: Department for Urban Planning and Landscape Architecture, Vienna University of Technology.

Local Governments for Sustainability (ICLEI). 2009. *Changing Climate, Changing Communities: Guide and Workbook for Municipal Climate Adaptation.* Toronto, Canada.

Pakistan Government, Planning Commission. 2010. *Final Report: Task Force on Climate Change.* Islamabad.

People's Republic of China, National Development and Reform Commission. 2009. *China's Policies and Actions for Addressing Climate Change.* Beijing.

Philippines Office of the President, Climate Change Commission. 2010. *National Framework Strategy on Climate Change, 2010–2022,* Manila.

Ramsar Secretariat. 2010. *The Ramsar Handbooks for the Wise Use of Wetlands, 4th edition,* Gland, Switzerland: Ramsar Secretariat.

Umweltbundesamt (Austrian Environment Agency). 2009. *Die Anpassung als zweite Säule der Klimapolitik.* http://www.klimawandelanpassung.at/nationale-anpassungsstrategie/

United Nations Framework Convention on Climate Change (UNFCCC). 2010. Report on the Technical Workshop on Advancing the Integration of Approaches to Adaptation Planning. Item 3 of the Provisional Agenda Nairobi Work Programme on Impacts, Vulnerability, and Adaptation to Climate Change. Bangkok. October 2009. http://unfccc.int/resource/docs/2010/sbsta/eng/02.pdf

US Agency for International Development (USAID). 2007. *Adapting to Climate Variability and Change: a Guidance Manual for Development Planning.* Washington DC: USAID.

Wien Energie. 2011. *Europe's Largest Forest Biomass Power Station.* http://www.wienenergie.at/we/ep/programView.do/contentTypeId/1001/channelId/-27957/programId/19444/pageTypeId/19114

Wilby, R. L. et al. 2009. A Review of Climate Risk Information for Adaptation and Development Planning. *International Journal of Climatology.* 29 (9). pp. 1193–1215.

Chapter 6
Monitoring the Vulnerability and Adaptation Planning for Water Security

Zhijun Chen

6.1 Water Security in Asia and the Pacific

The world is facing changes at a faster rate than ever seen before. Population growth, migration, urbanization, land-use changes, and climate variability and/or change will drive the way water resources need to be managed in the future. They also call for concrete contributions from water policies and actions to help the world cope with these changes. The Istanbul Ministerial Statement, issued at the 5th World Water Forum on March 22, 2009, "recognizes the need to achieve water security," and thinks "it is vital to increase adaptation of water management to all global changes and improve cooperation at all levels" (World Water Council 2009a).

Water security involves the sustainable use and protection of water systems, the protection against water-related hazards (floods and droughts), the sustainable development of water resources, and the safeguarding of (access to) water functions and services for humans and the environment (Schultz and Uhlenbrook 2007). Asia and the Pacific, while enjoying fast-paced economic development, is facing water challenges undermining regional water security and hindering progress of sustainable development, poverty reduction, and achieving the Millennium Development Goals (MDGs). These challenges are discussed in the following sections.

6.1.1 Frequent Water Disasters

Water-related disasters, including floods, windstorms, waves and surges, landslides, droughts and epidemics, were the most frequent in the world in the past century, accounting for about 90% of the most hazardous 1,000 disasters during 1900–2006 (Adikari et al. 2008). Asia and the Pacific is the most vulnerable region and recorded the highest number of water-related disasters and fatalities during 1980–2006 (Figures 6.1 and 6.2): over 600,000 people were killed and 4.5 billion people were affected which accounted for over 80% of the world total. From 2004 to 2006, floods and windstorms alone killed over 33,000 people, affected over 360 million people, and caused economic damage worth US$282 billion, which accounted for nearly one-third of the record damages during 1980–2006 (ICHARM 2007).

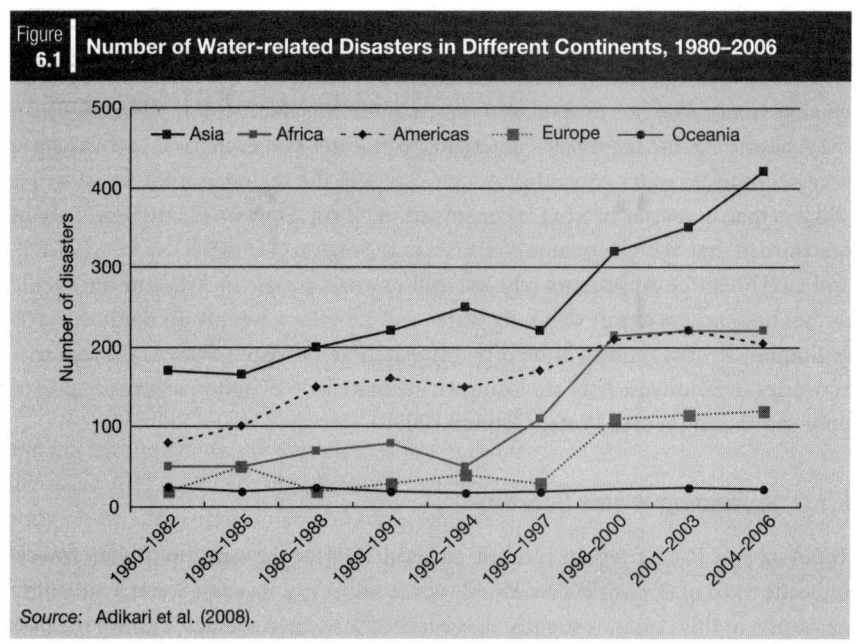

Figure 6.1 | **Number of Water-related Disasters in Different Continents, 1980–2006**

Source: Adikari et al. (2008).

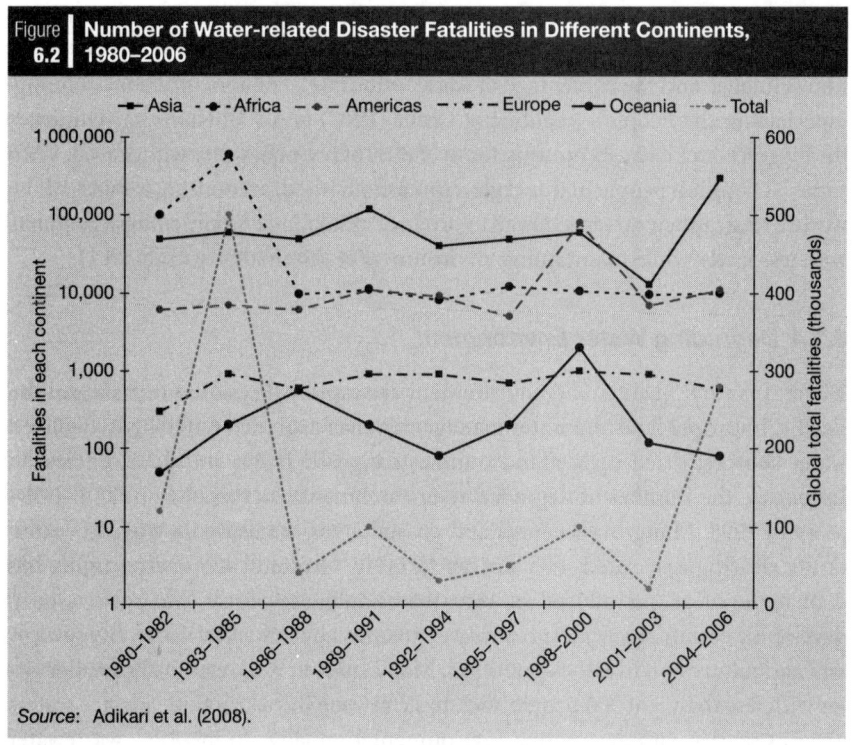

Figure 6.2 | **Number of Water-related Disaster Fatalities in Different Continents, 1980–2006**

Source: Adikari et al. (2008).

6.1.2 Limited Access to Water and Sanitation

According to the UNESCAP, ADB, and UNDP joint report, *The Millennium Development Goals: Progress in Asia and the Pacific 2007*, about 650 million people in the Asia and Pacific region lack access to clean water and even for those who have access, domestic water consumption per capita in the region as a whole (49m³) is still less than a quarter of what is consumed in North America (221m³), and about one third of that of high-income countries in the region (145m³) (UNESCAP, ADB, and UNDP 2007). Approximately 560 million rural people in Asia and the Pacific do not have access to safe drinking water and 1.5 billion people do not have basic sanitation. Recent reports quoted by the Asia Pacific Water Forum showed that countries in Southeast Asia are losing an estimated US$9 billion a year because of poor sanitation (World Water Council 2009b).

6.1.3 Spreading Water Scarcity

The Asia and Pacific region is home to about 61% of the world population with only one third of the world's renewable water resources. Average water availability per capita in this region is slightly above half of the world average. During the past three decades, with rapid population growth and economy development, water use has been increasing at double the rate of population growth; some countries are reaching the limit at which reliable water services can be delivered. Water competition is intense and the percentage of water withdrawal for agriculture is declining, especially in the People's Republic of China (PRC) and Viet Nam. Agriculture is the biggest water user, accounting for over 80% of regional water withdrawal. With some 583 million people in this region undernourished, accounting for 63% of the world's total, agriculture is required to produce more food and agricultural products with less water while maintaining environmental sustainability (Table 6.1).

6.1.4 Degrading Water Environment

Water quantity, quality, and environment are rapidly degrading in Asia and the Pacific. Improper land and water management has degraded watersheds, depleted water sources, dried up wetlands, and endangered fauna and flora species. In Indonesia, the number of degraded river catchments increased from 22 in 1984 to 59 in 1998. Mangrove swamps and coastal areas are being destroyed because of the clearing associated with shrimp farming. Groundwater overpumping has depleted aquifers and induced seawater intrusion in coast areas. An urgent issue in developing countries is widespread water pollution by industrial waste, city sewage, and agriculture and livestock pollution. Many cities in Asia, especially smaller cities with less than 500,000 people, lack basic sewage infrastructure. Surface waters serve as sewers with destructive environmental impacts. In the PRC, wastewater

Table 6.1 Annual Water Withdrawal by Sector, 2004

Subregion	Agriculture km³	Agriculture Percentage of total	Domestic km³	Domestic Percentage of total	Industry km³	Industry Percentage of total	Total Withdrawal km³	Total Withdrawal Percentage of total	Total Withdrawal m³/inhab.	Total Withdrawal Percentage of IRWR
Indian subcontinent	510.7	92.0	27.2	5.0	15.5	3.0	553.4	38.0	500.0	32.0
East Asia	418.3	77.0	26.8	5.0	95.0	18.0	540.1	37.0	428.0	19.0
Far East	73.5	64.0	23.2	20.0	18.4	16.0	115.1	6.0	67.4	23.0
Southeast Asia	82.1	88.0	3.9	4.0	7.0	8.0	93.0	6.0	476.0	5.0
Pacific Islands	127.9	90.0	10.4	7.0	4.3	3.0	142.6	10.0	483.0	3.0
Asia	1212.5	84.0	91.5	6.0	140.2	10.0	1444.2	100.0	476.0	12.0
Oceania	19.0	72.0	5.0	18.0	3.0	10.0	27.0	0.8	564.0	1.5
World	2310.5	71.0	290.6	9.0	652.2	20.0	3253.3	100.0	564.0	8.0
Asia as a percentage of world	52.5		31.5		21.5		44.4			

Source: FAO AQUASTAT Database (2004).

Notes: Asia in the text and Table 6.1 is composed of the following subregions and countries:
Indian subcontinent: Bangladesh, Bhutan, India, Maldives, Nepal, Sri Lanka
East Asia: PRC, Mongolia
Far East: Japan, Republic of Korea
Southeast Asia: Cambodia, Lao People's Democratic Republic, Myanmar, Thailand, Viet Nam
Pacific Islands: Brunei, Indonesia, Malaysia, Papua New Guinea, Philippines
km³ = cubic kilometers; m³ = cubic meters; IRWR = internal renewable water resources

discharge increased from 400 million tons in 1999 to more than 450 million tons in 2003. Irrigation is also responsible for salinization. Today, nearly 40% of irrigated land in dry areas is affected by salinization (UN Water 2009). In the PRC alone, 6.7 million hectares of irrigated land are salinized.

6.1.5 Future Perspectives

The combined effects of social, economic, and natural factors will bring more challenges to water security in the region in coming decades. Continuous population growth and economic development will add pressure to water availability and accelerate water competition. By 2025, 1.8 billion people are expected to be living in countries or regions with less than 500 m^3 of renewable water per year per capita, and two-thirds of the world's population, including one-third of the populations of the PRC and India, could experience "stress" conditions, defined as water consumption levels of between 500 m^3 and 1000 m^3 per year per capita (Seckler et al. 1998). Water shortages are likely to impact most severely on the poorest segments of the population in some areas such as South Asia. To improve food security in the region, irrigation needs to be improved and developed modestly. The number of mega cities with more than 5 million people will increase from 21 in 2000 to 32 in 2015. As most of them are along the coasts, they are vulnerable to floods and tides (UN Department of Social Affairs 2004a, 2004b). Water pollution produced by mid-size and smaller cities in the PRC may increase fivefold (Devan et al. 2008). Pollution pressure is even higher. These are expected to aggravate with the effects of global warming and climate variability, as indicated in the recent findings of the Fourth Intergovernmental Panel on Climate Change (IPCC) Report (IPCC 2007).

6.1.6 The Impacts of Climate Change

Past and present climate trends and variability have already been observed in Asia, including increasing surface air temperature and rainfall variability across the region; decreasing annual mean rainfall in northern PRC, the arid plains of Pakistan, and Northeast India; increasing intensity and frequency of extreme weather events in Southeast Asia; and longer heat wave duration in many Asian countries, which caused the retreat of glaciers and permafrost; intensified drought and water scarcity in many areas; and increased floods, cyclones, landslides, mud flows, and seawater intrusion in coastal areas. Crop yields in some Asian countries have declined. Over 34% of Asian coral reefs, especially in South, Southeast, and East Asia, are reported to be lost. Observed changes in terrestrial and marine ecosystems have become more pronounced (Tables 6.2 and 6.3).

Table 6.2	Summary of Key Observed Past and Present Climate Trends and Variability

Region	Country	Change in Temperature	Change in Precipitation	References
North Asia	Russia	2°C to 3°C rise in past 90 years; more pronounced in spring and winter; highly variable, decrease during 1951 to 1995; increase in last decade	Highly variable, decrease during 1951 to 1995; increase in last decade	Savelieva et al. 2000; Peterson et al.
	Mongolia	1.8°C rise in last 60 years; most pronounced in winter	7.5% decrease in summer; 9% increase in winter	Batima et al. 2005a; Natsagdorj et al. 2005
Central Asia	Regional mean	1°C to 2°C rise in temperature per century	No clear trend during 1900 to 1996	Peterson et al. 2002
	North-west PRC	0.7°C increase in mean annual temperature from 1961 to 2000	Between 22% and 33% increase in rainfall	Shi et al. 2002
Tibetan Plateau	Regional mean	0.16°C and 0.32°C per decade increase in annual and winter temperatures, respectively	Generally increasing in northeast region	Liu et al. 1998; Yao et al. 2000; Liu and Chen 2001; Cai et al. 2003; Du and Ma 2004; Zhao et al. 2004
West Asia (Middle East)	Iran	During 1951 to 2003 several stations in different climatological zones of Iran reported significant decrease in frost days due to rise in surface temperature	Some stations show a decreasing trend in precipitation (Anzali, Tabriz, Zahedan) while others (Mashhad, Shiraz) have reported increasing trends	IRIMO 2006a, b; Rahimzadeh 2006
East Asia	PRC	Warming during last 50 years, more pronounced in winter than summer; rate of increase more pronounced in minimum than in maximum temperature	Annual rain declined in past decade in Northeast and North PRC, increase in Western PRC, Changjiang River, and along southeast coast	Zhai et al. 1999; Hu et al. 2003; Zhai and Pan 2003
	Japan	About 1.0°C rise in 20th century, 2°C to 3°C rise in large cities	No significant trend in the 20th century although fluctuations increased	Ichikawa, 2004; Japan Meteorological Agency 2005
	Republic of Korea	0.23°C rise in annual mean temperature per decade; increase in diurnal range	More frequent heavy rain in recent years	Jung et al. 2002; Ho et al. 2003

(Continued)

(Continued)

Region	Country	Change in Temperature	Change in Precipitation	References
South Asia	India	0.68°C increase per century; increasing trends in annual mean temperature; warming more pronounced during post monsoon and winter	Increase in extreme rains in northwest during summer monsoon in recent decades; lower number of rainy days along east coast	Kripalani et al. 1996; Lal et al. 1996; Lal et al. 2001b; Singh and Sontakke 2002; Lal 2003
	Nepal	For 1977–94: 0.090°C per year in Trans-Himalaya and 0.057°C per year in the Himalaya, more in winter; 0.041°C per year in Terai region	No distinct long-term trends in precipitation records for 1948 to 1994	Shrestha et al. 1999, Shrestha et al. 2000; Bhadra 2002; Shrestha 2004
	Pakistan	0.6°C to 1.0°C rise in mean temperature in coastal areas since early 1900s	10%–15% decrease in coastal belt and hyper arid plains; increase in summer and winter precipitation over the last 40 years in northern Pakistan	Farooq and Khan 2004
	Bangladesh	An increasing trend of about 1°C in May and 0.5°C in November during the 14-year period from 1985 to 1998	Decadal rain anomalies above long-term averages since 1960s	Mirza and Dixit 1997; Khan et al. 2000; Mirza 2002
	Sri Lanka	0.016°C increase per year between 1961 to 1990 over entire country, with regional increases ranging from 0.008°C to 0.025°C per year	Increase trend in February and decrease trend in June	Chandrapala and Fernando 1995; Chandrapala 1996
Southeast Asia	General	0.1°C to 0.3°C increase per decade reported between 1951 to 2000	Decreasing trend between 1961 and 1998; number of rainy days have declined throughout Southeast Asia	Manton et al. 2001
	Indonesia	Homogeneous temperature data were not available	Decline in rainfall in southern and increase in northern region	Manton et al. 2001; Boer and Faqih 2004
	The Philippines	Increase in mean annual, maximum and minimum temperatures by 0.14°C between 1971 to 2000	Increase in annual mean rainfall since 1980s and in number of rainy days since 1990s; increase in inter-annual variability of onset of rainfall	PAGASA 2001; Cruz et al. 2006

Source: IPCC (2007).

Table 6.3	Status of Coral Reefs in Selected Regions of Asia, 2004					
Region	Coral reef area (km²)	Destroyed reefs (%)	Reefs recovered since 1998 (%)	Reefs at critical stage (%)	Reefs at threatened stage (%)	Reefs at low or no threat level (%)
Red Sea	17,640	4.0	2.0	2.0	10.0	84.0
The Gulfs	3,800	65.0	2.0	15.0	15.0	5.0
South Asia	19,210	45.0	13.0	10.0	25.0	20.0
Southeast Asia	91,700	38.0	8.0	28.0	29.0	5.0
East & North Asia	5,400	14.0	3.0	23.0	12.0	51.0
Total	137,750	34.4	7.6	21.6	25.0	19.0
Asia	(48.4%)					

Source: IPCC (2007).
Notes: Destroyed reefs = 90% of the corals lost and unlikely to recover soon; reefs at a critical stage = 50% to 90% of the corals lost or likely to be destroyed in the next 10 to 20 years; reefs at threatened stage = 20% to 50% of the corals lost or likely to be destroyed in the next 20 to 40 years; km² = square kilometers.

Table 6.4 provides a snapshot of the projections on likely increases in area-averaged seasonal surface air temperature and percentage change in area-averaged seasonal precipitation (with respect to the baseline period 1961 to 1990) for Asia. In general, projected warming over all Asian subregions is higher during the northern hemispheric winter than during summer for all time periods. The most pronounced warming is projected at high latitudes in North Asia. Recent modeling experiments suggest that the warming would be significant in the Himalayan highlands including the Tibetan Plateau and arid regions of Asia (Gao et al. 2003). The consensus of the climate models in the IPCC Fourth Assessment Report (AR4) (IPCC 2007) indicates an increase in annual precipitation in most of Asia during this century; the relative increase being largest in North and East Asia. The projected decrease in mean precipitation in Central Asia will be accompanied by an increase in the frequency of very dry spring, summer, and autumn seasons. In South Asia, most of the AR4 models project a decrease of precipitation in December, January, and February. An increase in occurrence of extreme weather events including heat wave and intense precipitation events is also projected in South Asia, East Asia, and Southeast Asia, along with an increase in the inter-annual variability of daily precipitation in the Asian summer monsoon (Lal 2001; May 2004; Giorgi and Bi 2005).

6.1.7 Water Availability

According to the IPCC Fourth Assessment Report (IPCC 2007), freshwater availability in Central, South, East, and Southeast Asia, particularly in large river basins such as Changjiang, is likely to decrease due to climate change. In northern PRC,

Table 6.4 | Projected Changes in Surface Air Temperature and Precipitation in Asia

Subregions	Season	2010 to 2039				2040 to 2069				2070 to 2099			
		Temperature (°C)		Precipitation (%)		Temperature (°C)		Precipitation (%)		Temperature (°C)		Precipitation (%)	
		A1F1	B1	A1F1	B1	A1F1	B1	A1F1	B1	A1F1	B1	A1F1	B1
North Asia (50.0N–67.5N; 40.0E–170.0W)	DJF	2.94	2.69	16	14	6.65	4.25	35	22	10.45	5.99	59	29
	MAM	1.69	2.02	10	10	4.96	3.54	25	19	8.32	4.69	43	25
	JJA	1.69	1.88	4	6	4.20	3.13	9	8	6.94	4.00	15	10
	SON	2.24	2.15	7	7	5.30	3.68	14	11	8.29	4.98	25	15
Central Asia (30.0N–50.0N; 40.0E–75.0E)	DJF	1.82	1.52	5	1	3.93	2.60	8	4	6.22	3.44	10	6
	MAM	1.53	1.52	3	−2	3.71	2.58	0	−2	6.24	3.42	−11	−10
	JJA	1.86	1.89	1	−5	4.42	3.12	−7	−4	7.50	4.10	−13	−7
	SON	1.72	1.54	4	0	3.96	2.74	3	0	6.44	3.72	1	0
West Asia (12.0N–42.0N; 27.0E–63.0E)	DJF	1.26	1.06	−3	−4	3.10	2.00	−3	−5	5.10	2.80	−11	−4
	MAM	1.29	1.24	−2	−8	3.20	2.20	−8	−9	5.60	3.00	−25	−11
	JJA	1.55	1.53	13	5	3.70	2.50	13	20	6.30	2.70	32	13
	SON	1.48	1.35	18	13	3.60	2.20	27	29	5.70	3.20	52	25

Region	Season												
Tibetan Plateau (30.0N–50.0N; 75.0E–100.0E)	DJF	2.05	1.60	14	10	4.44	2.97	21	14	7.62	4.09	31	18
	MAM	2.00	1.71	7	6	4.42	2.92	15	10	7.35	3.95	19	14
	JJA	1.74	1.72	4	4	3.74	2.92	6	8	7.20	3.94	9	7
	SON	1.58	1.49	6	6	3.93	2.74	7	5	6.77	3.73	12	7
East Asia (20.0N–50N; 100.0E–150.0)	DJF	1.82	1.50	6	5	4.18	2.81	13	10	6.95	3.88	21	15
	MAM	1.61	1.50	2	2	3.81	2.67	9	7	6.41	3.69	15	10
	JJA	1.35	1.31	2	3	3.18	2.43	8	5	5.48	3.00	14	8
	SON	1.31	1.24	0	1	3.16	2.24	4	2	5.51	3.04	11	4
South Asia (5.0N–30.0N; 65.0E–100.0E)	DJF	1.17	1.11	-3	4	3.16	1.97	0	0	5.44	2.93	-16	-6
	MAM	1.18	1.07	7	8	2.97	1.81	26	24	5.22	2.71	31	20
	JJA	0.54	0.55	5	7	1.71	0.88	13	11	3.14	1.56	26	15
	SON	0.78	0.83	1	3	2.41	1.49	8	6	4.19	2.17	26	10
Southeast Asia (10.0S–20.0N; 100.0E–150.0E)	DJF	0.86	0.72	-1	1	2.25	1.32	2	4	3.92	2.02	6	4
	MAM	0.92	0.80	0	0	2.32	1.34	3	3	3.83	2.04	12	5
	JJA	0.83	0.74	-1	0	2.13	1.30	0	1	3.61	1.87	7	1
	SON	0.85	0.75	-2	0	1.32	1.32	-1	1	3.72	1.90	7	2

Source: IPCC (2007).

Notes: DJF = December, January, February; MAM = March, April, May; JJA = June, July, August; SON = September, October, November; A1F1 = IPCC emission scenario for a more integrated world under rapid economic growth followed by rapid introductions of new and more efficient technologies with an emphasis on fossil-fuels; B1 = IPCC emission scenario for a more integrated and ecologically friendly world with rapid economic growth and introduction of clean technologies.

irrigation from surface and groundwater sources will meet only 70% of requirements. The maximum monthly flow of the Mekong River is estimated to increase by
35%–41% in the basin and by 16%–19% in the delta while the minimum monthly
flow will decline by 17%–24% in the basin and 26%–29% in the delta, meaning
increased flooding risks in wet seasons and water shortages in dry seasons. Melting of glaciers could seriously affect half a billion people in the Himalaya–Hindu
Kush region and a quarter of a billion people in the PRC. As glaciers melt, river
runoff will initially increase in winter or spring but eventually decrease as a result
of loss of ice resources. The thawing volume and speed of snow cover in spring is
projected to accelerate in northwest PRC and western Mongolia which may lead
to floods in spring and significant water shortage in wintertime by the end of this
century (Table 6.5).

6.1.8 Water Quality

Overexploitation of groundwater has resulted in a drop in its level, leading to
ingress of seawater in coastal areas making the sub-surface water saline. India,
the PRC, and Bangladesh are especially susceptible to increasing salinity of their
groundwater as well as surface water resources, especially along the coast, due to
increases in sea levels as a direct impact of global warming. For example, in the
Zhujiang estuary, a rise in sea level by 0.4 to 1.0 m can induce saltwater intrusion
1 to 3 km further inland (IPCC 2007). Increasing frequency and intensity of droughts

Table 6.5	Vulnerability of Key Sectors to the Impacts of Climate Change in Asia						
Subregions	Food and fiber	Bio-diversity	Water resources	Coastal ecosystem	Human health	Settlements	Land degradation
North Asia	+1/H	−2/M	+1/M	−1/M	−1/M	−1/M	−1/M
Central Asia and West Asia	−2/H	−1/M	−2/VH	−1/L	−2/M	−1/M	−2/H
Tibetan Plateau	+1/L	−2/M	−1/M	Not applicable	No information	No information	−1/L
East Asia	−2/VH	−2/H	−2/H	−2/H	−1/H	−1/H	−2/H
South Asia	−2/H	−2/H	−2/H	−2/H	−2/M	−1/M	−2/H
Southeast Asia	−2/H	−2/H	−1/H	−2/H	−2/H	−1/M	−2/H
Vulnerability:	−2: Highly vulnerable		Level of confidence:	VH: Very high			
	−1: Moderately vulnerable			H: High			
	0: Slightly or not vulnerable			M: Medium			
	+1: Moderately resilient			L: Low			
	+2: Most resilient			VL: Very low			
Source: IPCC (2007).							

in the catchment area will lead to more serious and frequent saltwater intrusion in the estuary and thus a deterioration of surface and groundwater quality.

6.1.9 Droughts and Floods

As a warmer climate is likely to cause variation of precipitation, alter the seasonality and amount of water flows, reduce the capacity of natural systems to store water in the form of snow and glaciers, and induce more evaporation of the land surface, it will not only affect bulk water availability but also worsen the extremes of droughts and floods. At the global level, climate change is expected to account for 20% of the increase in water scarcity. Countries that already suffer from water shortages will be hit hardest. In Asia, it is estimated that under the full range of emission scenarios, 120 million to 1.2 billion people will experience increased water stress by the 2020s, and by the 2050s the number will range from 185 to 981 million (IPCC 2007).

6.1.10 Coastal Ecosystems

Projected rises in sea levels could flood the dwellings of millions of people living in the low lying coast areas of South, Southeast, and East Asia such as in Viet Nam, Bangladesh, India, and the PRC. Even under the most conservative scenario, sea levels will be about 40 cm higher than today by the end of 21st century and this is projected to increase the number of people flooded annually from 13 million to 94 million. Almost 60% of this increase will occur in South Asia (along the coasts from Pakistan to India and Sri Lanka, and from Bangladesh to Burma), while about 20% will occur in Southeast Asia, specifically from Thailand to Viet Nam, including Indonesia and the Philippines. The aquaculture industry and infrastructure, particularly in the heavily populated mega deltas will be seriously affected. Stability of wetlands, mangroves, and coral reefs around Asia is likely to be increasingly threatened. Around 24% and 30% of the reefs in Asia are likely to be lost during the next 10 years and 30 years, respectively (IPCC 2007).

6.1.11 Food Security and Human Health

About 2.5%–10% decrease in crop yield is projected for parts of Asia in 2020s and 5%–30% decrease in 2050s compared with 1990 levels without considering CO_2 effects. An additional 49 million, 132 million, and 266 million people of Asia, projected under A2 scenario without carbon fertilization, could be at risk of hunger by 2020, 2050, and 2080, respectively (IPCC 2007). Increases in coastal water temperatures would exacerbate the abundance and/or toxicity of cholera in South Asia. Natural habitats of vector-borne and water-borne diseases in North Asia are likely to expand in the future (Table 6.6).

Table 6.6 | Regional-scale Impacts of Climate Change by 2080

(millions of people)

	Population living in watersheds with an increase in water resources stress (Arnell 2004)				Increase in average annual number of coastal flood victims (Nicholls 2004)				Additional population at risk of hunger (Parry et al. 2004) Figures in brackets assume maximum direct CO_2-enrichment effect			
					Climate and socioeconomic scenario							
	A1	A2	B1	B2	A1	A2	B1	B2	A1	A2	B1	B2
Europe	270	382–493	233	172–183	1.6	0.3	0.2	0.3	0	0	0	0
Asia	289	812–1197	302	327–608	1.3	14.7	0.5	1.4	78 (6)	266(–21)	7(2)	47(–3)
North America	127	110–145	107	9–63	0.1	0.1	0	0	0	0	0	0
South America	163	430–469	97	130–186	0.6	0.4	0	0.1	27(1)	85(–4)	5(2)	15(–1)
Africa	408	691–909	397	492–559	2.8	12.8	0.6	13.6	157(21)	200(–2)	23(8)	89(–8)
Australasia	0	0	0	0	0	0	0	0	0	0	0	0

Source: IPCC (2007).
Note: A1, A2, B1, B2 = IPCC emission scenarios; A2 = IPCC emission scenario for a world of independently operating, self-reliant nations with continuously increasing population, regionally oriented economic development, slower and more fragmented technological changes, and improvements to per capita income.

The Pacific island countries are especially vulnerable to the effects of climate change. As most of their population and infrastructure are located along the coasts, a sea-level rise will exacerbate inundation, erosion, and other coastal hazards; threaten vital infrastructure, settlements and facilities; and thus compromise the socioeconomic well-being of the people. Most small islands have a limited water supply, and water resources are especially vulnerable to a future variation of rainfall. It its projected that a 10% reduction in average rainfall (by 2050) would lead to a 20% reduction in the size of the freshwater lens on Tarawa Atoll, Kiribati. Climate change is also likely to impact coral reefs, fisheries, and other marine-based resources, and increase the transmission of water-borne diseases. In most cases Pacific island countries have low-adaptive capacity, and adaptation costs are high relative to GDP (IPCC 2007).

6.2 Coping with Climate Change in the Water Sector

Climate change, food security, and sustainable development are major global concerns. What needs to be highlighted is the way these are related to, impacted by, and affected by water. A general consensus on the principles and approaches for improving water security under changed climate has been reached among the international community, national governments, and academic institutes in the Asia and Pacific region. These principles and approaches are discussed in the following sections.

6.2.1 Integrated Approach

It is important to develop water adaptation plans at national, river basin, and local levels, and integrate them into sustainable development strategies, policies, plans, and programs at relevant levels. Adaptation plans should be based on vulnerability assessments, and include integrated options on both autonomous and planned adaptation. Priority should be given to integrated options that can address adaptation and mitigation simultaneously, such as improving water management in high emitting, irrigated rice systems through mid-season drainage or alternative wetting and drying. Rice cultivation mitigation strategies have a technical potential of reducing emissions by 300 million ton CO_2 equivalent per year, and the highest economic potential in developing countries, especially in the PRC, South Asia, and Southeast Asia (Pathak et al. 2006).

6.2.2 Comprehensive Strategy

Adaptation plans at different levels should adopt a comprehensive strategy, with short-term and long-term scenarios; combining considerations from within the "water box" and outside the "water box"; structural measures, that is, physical construction or application of engineering techniques; and nonstructural measures, that is, knowledge, practice, or agreements—especially policies, laws, public awareness raising, training and education—to address various water issues holistically and systematically.

6.2.3 Typological Classification

The IPCC identified areas most vulnerable to the impacts of climate change. These areas include the least developed countries and small island developing states, low-lying densely populated coastal areas, areas affected by glacier melt, and arid areas with fragile populations, economies, and environments. Typological classification will help identify and implement the most suitable mitigation and adaptation options for different areas. Table 6.7 provides an example of typological classification of agricultural water management systems based on different agro-ecosystems.

6.2.4 Multidisciplinary Teamwork

Water is a crosscutting issue, relating to nature, society, economy, and environment. Adaptation measures in the water sector involve multiple options, including water storage development, water productivity improvement, public investment policies, integrated water resources management at river basin level, water quality control, integrated drought and flood management, water environment, and ecosystem protection. It expects technical cooperation and contributions from areas of hydrology, meteorology, engineering, management, economy, sociology, health, resources management, disaster management, and environment. Proper management and institutional innovations should be adopted by relevant stakeholders to enable and facilitate multidisciplinary teamwork.

6.2.5 Cooperation Networks

Cooperation networks should be established at international, regional, national, and local levels with regard to river basins to bring together efforts from development partners, governments, NGOs, and the private sector, as well as to facilitate cooperation and share information on policy innovation, planning formulation, project implementation, technology development, dissemination, and capacity building. As developing countries need additional external financial resources to cope with climate change in the water sector, networks should also work on resources mobilization to close the financial gap for adaptation.

Table 6.7 Typology of Agricultural Water Management Systems

System	Current Status	Climate Change Drivers	Vulnerability	Adaptability	Response Options
1. Snow melt systems					
Indus system	Highly developed, water scarcity emerging. Sediment and salinity constraints.	20-year increasing flows followed by substantial reductions in surface water and groundwater recharge. Changed seasonality of runoff and peak flows. More rainfall in place of snow. Increased peak flows and flooding. Increased salinity. Declining productivity in places.	Run river: very high; Dams: medium high.	Limited room for maneuver (all infrastructure already built)	Water supply management: increased water storage and drainage; improved reservoir operation; change in crop and land use; improved soil management. Water demand management: including groundwater management and salinity control.
Ganges Brahamaputra	High potential for groundwater, established water quality problems. Low productivity.		Falling groundwater tables: high	Still possibilities of groundwater development: medium	
North Western PRC	Extreme water scarcity and high productivity.		Global implications, high food demand with great influence on prices: high	Adaptability is increasing due to increasing wealth: medium	
Red and Mekong	High productivity, high flood risk, water quality.		Medium	Medium	
Colorado	Water scarcity, salinity.		Low	Excessive pressure on resources: medium	

(Continued)

(Continued)

System	Current status	Climate change drivers	Vulnerability	Adaptability	Response options
2. Deltas					
Ganges Brahamaputra	Densely populated. Shallow groundwater, extensively used. Flood adaptation possible. Low productivity.	Rising sea level. Storm surges, and infrastructure damage. Higher frequency of cyclones (E/SE Asia); saline intrusion in groundwater and risers; increased flood frequency. Potential increase in groundwater recharge.	Flood, cyclones: very high	Poor except salinity	Minimize infrastructure development; conjunctive use of surface water and groundwater; manage coastal areas.
Nile River	Delta highly dependent on runoff and Aswan Storage—possible to upstream development.		Population pleasure: high	Medium	
Yellow River	Severe water scarcity.		High	Low	
Red river	Currently adapted but expensive pumped irrigation and drainage.		Medium	High except salinity	
Mekong	Adapted groundwater use in delta—sensitive to upstream development.		High	Medium	
3. Semi-arid/arid tropics					
Monsoonal: Indian subcontinent	Low productivity; Overdeveloped basin (surface water and groundwater);	Increased rainfall. Increased rainfall variability. Increase drought and flooding. Higher temperature.	High	Surface irrigation: low; groundwater irrigation: medium	Storage dilemma; Increase groundwater recharge and use; higher value agriculture (Australia)
Non-monsoonal: sub-Sahara Africa	poor soils; flashy systems; over allocation of water and population pleasure in places; widespread food insecurity.	Increased rainfall variability; increase frequency of drought and flooding. Lower rainfall, higher temperature. Decreasing runoff.	Declining yields in rainfed systems; increased volatility of production: very high	Low	
Non-monsoonal: Southern and Western Australia			High	Low	

4. Humid tropics

Rice: Southern Asia	Surface irrigation; high productivity but stagnating	Increased rainfall. Marginally increased temperatures. Increased rainfall variability and occurrence of droughts and floods	High	Medium	Increased storage for second and third season. Drought and flood insurances, crop diversification
Rice: Southern PRC	Conjunctive use of surface water and groundwater; low output compared to Northern PRC		High	Medium	
Rice: Northern Australia	Fragile ecology		Low	High	
Non-rice – surface irrigation			Low	Medium	
Non-rice – groundwater irrigation			Medium	Medium	

5. Temperate (supplementary irrigation)

Northern Europe	High value agriculture and pasture	Increased rainfall; longer growing seasons; increased productivity.	Low: surface irrigation: medium; groundwater irrigation	Low: surface irrigation; high: groundwater irrigation	Potential for new development. Storage development; Drainage
Northern America	Cereal cropping, groundwater irrigation		Medium	Medium	Increased productivity and outputs; limited options for storage

6. Mediterranean

Southern Europe	Italy, Spain, Greece	Significantly lower rainfall and higher temperatures, increased water stress, decreased runoff	Medium	Low	Localized irrigation, transfer to other sectors
Northern Africa	Morocco, Tunisia, High water scarcity		High	Low	Localized irrigation, supplementary irrigation
West Asia	Fertile crescent	Loss of groundwater reserves	Low	Low	

7. Small islands

Small islands	Fragile ecosystems, groundwater depletion	Sea water rise; saltwater intrusion; increased frequency of cyclones and hurricanes	High	Variable	Groundwater depletion control; water demand management

Source: FAO (2008).

6.3 Vulnerability Assessment and Adaptation Planning

Vulnerability assessment and adaptation planning are two crucial steps in climate change adaptation in the water sector. The reliability of vulnerability assessment relies on the quality of reporting data and the pertinence of methodologies and tools applied. Adaptation planning and its integration into national development plans were high priority areas identified by the first Asia Pacific Water Forum in 2007. A strong international platform for cooperation on development and dissemination of methods and tools for assessing vulnerability and adaptation is indispensable. The UNFCCC Secretariat took a first step in 1999 when it produced the report, *Compendium of Decision Tools to Evaluate Strategies for Adaptation to Climate Change.* The latest version published in February 2008, *Compendium on Methods and Tools to Evaluate Impacts of, and Vulnerability and Adaptation to, Climate Change* includes chapters on complete frameworks and supporting toolkits; crosscutting issues and multi-sector approaches; and sector specific tools.

The Framework and associated toolkits collected a broad range of approaches (Table 6.8). The IPCC Technical Guidelines, the UNEP Handbook, and the United States Country Studies Program represent examples of first-generation approaches to the assessment of vulnerability and adaptation. They have an analytical thrust, and focus on an approach that emphasizes the identification and quantification of impacts. The UNDP Adaptation Policy Framework is a second-generation assessment and places the assessment of vulnerability at the center of the process. The Assessment of Impacts and Adaptations to Climate Change in Multiple Regions and Sectors (AIACC) approach (technically a collection of projects rather than an explicit framework) incorporates elements of both first- and second-generation assessments. The National Adaptation Programs of Action (NAPA) Guidelines provide conceptual and procedural oversight for producing a document that identifies national priorities for adaptation. The UKCIP report provides guidance to those engaged in decision-making and policy processes. It lays out an approach

Table 6.8	Complete Frameworks and Supporting Toolkits
IPCC Technical Guidelines for assessing Climate Change Impacts and Adaptations	
US Country Study Program (USCSP)	
UNEP Handbook on Methods for Climate Change Impact Assessment and Adaptations Strategies	
UNDP Adaptation Policy Framework (APF)	
Assessments of Impacts and Adaptations to Climate Change (AIACC) in Multiple Regions and Sectors	
Guidelines for the Preparation of National Adaptation Programmes of Action (NAPA)	
United Kingdom Climate Impacts Programme (UKCIP) Climate Adaptation: Risk, Uncertainty and Decision Making	

Source: UNFCCC (2008).

to integrate climate adaptation decisions and climate influenced decisions into the broader context of institutional decision making. The UKCIP framework is distinctive in that it casts the assessment process in risk and decision under uncertainty terms.

The crosscutting issues and multi-sector approaches collected a range of applications, including development and application of scenarios that deal with the use of climate or socioeconomic scenario data-decision tools that provide more detail on tools that might be most applicable to a particular step of the vulnerability and adaptation assessment process, stakeholder approaches that encompass a set of tools, and a partial framework that prescribes a process or an approach to undertaking several steps of a complete assessment. Multi-sector tools provide a general evaluation of adaptation options, are easily adapted to different regions and situations, and are used in conjunction with sector-specific tools to develop a comprehensive analysis or in support of a complete framework.

The sector specific tools collected examples for agriculture, water, coastal resources, and human health sectors. The two FAO irrigation models under the agricultural sector, CROPWAT and AquaCrop, may also be used for the water sector. The 11 tools collected for the water sector (Table 6.9) are mathematical models for assessing water resource adaptations to climate change, focusing on regional water supply and demand. Some are long-range simulation tools such as WEAP and IRAS, others are short-range simulation models like River Ware and Water Ware, or economic optimization models like Aquarius. RIBASIM allows for the assessment of infrastructure, operational, and demand management measures. MIKEBASIN provides basin-scale solutions for optimizing water allocations, conjunctive water use, reservoir operation, and water quality issues, emphasizing results visualization through a GIS interface. CALVIN helps identify integrated water management strategies covering surface water, ground water, water conservation, water market, water reuse, and desalination water management options. OSWRM, which simulates water resource supply and demand, was developed to support a stakeholder dialogue process that focuses on the potential role of climate change in water resource management. The European Flood Alert System is a flood forecasting system that provides medium-range flood forecasting information for transnational river basins across Europe.

Gaps and constraints related to the application of some methods and tools include the lack of capacity and tools for integrated assessments at subnational and national levels; lack of capacity for regional climate modeling; lack of a participatory process to better engage all stakeholders; need for better cooperation and coordination in developing and improving methods and tools to avoid overlap and duplication; lack of expertise in some countries for applying the methods and tools and for interpreting the results; lack of basic data and information for adaptation assessment and planning; lack of financial source for conducting the assessment and planning; and

Table 6.9	Water Sector Tools
Water Ware	
Water Evaluation and Planning Systems (WEAP)	
River Ware	
Interactive River and Aquifer Simulation (IRAS)	
Aquarius	
RIBASIM	
MIKE BASIN	
Spatial Tools for River Basins and Environment and Analysis of Management Options (STREAM)	
CALVIN (CALifornia Value Integrated Network)	
OSWRM (Okanagan Sustainable Water Resources Model)	
European Flood Alert System (EFAS)	
Source: UNFCCC (2008).	

lack of awareness of decision makers. More work needs to be done to enable more suitable and efficient use of the information, methods, and tools.

6.4 Other FAO Tools

Two other FAO tools, the Rapid Appraisal Process (RAP) of Irrigation Projects and Modernizing Irrigation Management and the Mapping System and Services for Canal Operation Techniques (MASSCOTE) approach can be used for performance evaluation, benchmarking, and modernization planning of medium-to-large-scale canal irrigation systems.

RAP was first introduced by a joint FAO/IPTRID/World Bank publication, *FAO Water Reports (19)–Modern Water Control and Management Practices in Irrigation–Impact on Performance* (Burt and Styles 1999). RAP is a one-to-two week long process of data collection and analysis in the office and in the field. The process examines external inputs such as water supplies, and outputs such as water destinations (evapo-transpiration and surface runoff). It provides a systematic examination of the hardware and processes used to distribute water internally to all levels within the project (from the source to the fields). External and internal indicators were developed to provide *(i)* baseline of information for comparison against future performance after modernization; *(ii)* benchmarking for comparison against other irrigation projects; and *(iii)* a basis for making specific recommendations for modernization and improvement of water delivery service. An essential ingredient of the successful application of the RAP method is adequate training of the evaluators. Experience has shown that successful RAP programs require *(i)* evaluators with prior training in irrigation; *(ii)* specific training in the RAP method; and *(iii)* follow-up support and critique when the evaluators begin their field work.

MASSCOTE, first introduced in the FAO *Irrigation and Drainage Paper* (Renault, Facon, and Wahaj 2007) can be applied in association with RAP. Based on the evaluation results of system performance, diagnosis on system constraints, and recommendations on system modernization produced by RAP, MASSCOTE will assist managers to address specific modernization needs, issues, and challenges through an 11-step procedure and by dividing the system into relevant subunits. The entry point is canal operation, but the scope is modernization and the goal is to promote service-oriented management, with specific identified targets in terms of effectiveness in relation to money, to water, and with regard to the environment. The methodology capitalizes on many modernization programs in which FAO has been involved in recent years, in particular through associated RAP and MASSCOTE training courses. In the last decade, FAO has trained more than 500 engineers in Asia. This approach has largely been developed in close collaboration with irrigation managers in the field, who are envisaged to be the main users of this product.

6.5 Conclusion

The Asia and Pacific region is the most vulnerable to water-related disasters and faces challenges in water security. Continued population growth, economic development, and urbanization are expected to worsen the situation. Climate change is causing water availability variation, spreading water scarcity, intensifying water disasters, and degrading water quality and environment, hence challenging water security. Principles on climate change adaptation in the water sector include integrated approaches, comprehensive strategies, typological classification, and multidisciplinary teamwork, and cooperation networks. Vulnerability assessment and adaptation planning are two crucial steps.

References

Adikari, Y., J. Yoshitani, N. Takemoto, and A. Chavoshian. 2008. *Technical Report on the Trends of Global Water-related Disasters: Revised Version of 2005 Report.* Japan: Public Works Research Institute.

Burt, C., and S. Styles.1999. Modern Water Control and Management Practices in Irrigation. Impact on Performance. FAO Water Report No. 19.

Devan, J., S. Negri, and J.R. Woetzel. 2008. Meeting the Challenges of China's Growing Cities. *McKinsey Quarterly.* 3, pp. 107–116. http://www.mckinseyquarterly.com/Meeting_the_challenges_of_Chinas_growing_cities_2152

Food and Agricultural Organisation (FAO). 2004. FAO AQUASTAT Database. http://faostat. fao.org/default.aspx

Food and Agricultural Organisation (FAO). 2008. Climate Change, Water and Food Security. Synthesis Report of FAO Expert Group Meeting. Rome, 26–28 February.

Gao, X.J., D. L. Li, Z. C. Zhao, and F. Giorgi. 2003. Climate Change Due to Greenhouse Effects in Qinghai-Xizang Plateau and Along Qianghai-Tibet Railway. *Plateau Meteorology.* 22. pp. 458–463 (In Chinese with English abstract).

Giorgi, F., and X. Bi. 2005. Regional Changes in Surface Climate Interannual Variability for the 21st Century from Ensembles of Global Model Simulations. *Geophysical Research Letters.* 32 (13). L13701.

Intergovernmental Panel on Climate Change (IPCC). 2007. *Climate Change 2007: Impacts, Adaptation and Vulnerability.* In M. L. Parry, O. F. Canziani, J. P. Palutikof, P. J. van der Linden, and C. E. Hanson, eds. *Contribution of Working Group II to the Fourth Assessment Report of the Intergovernmental Panel on Climate Change.* Cambridge University Press, Cambridge, UK.

International Center for Water Hazard and Risk Management (ICHARM). 2007. *Thematic Recommendation—Theme B: Water-related Disaster Management* (Draft). Tsukuba, Japan: ICHARM.

Lal, M. 2001. Tropical Cyclones in a Warmer World. *Current Science.* 80 (9). pp. 1103–1104.

May, W. 2004. Simulation of the Variability and Extremes of Daily Rainfall during the Indian Summer Monsoon for Present and Future times in a Global Time-slice Experiment. *Climate Dynamics.* 22 (2–3). pp. 183–204.

Pathak, H., C. Li, R. Wassmann, and J.K. Ladha. 2006. Simulation of Nitrogen Balance in Rice-Wheat Systems of the Indo-Gangetic Plains. *Soil Science Society of America Journal.* 70 (September–October). pp. 1612–1622.

Renault, R., T. Facon, and R. Wahaj. 2007. Modernizing Irrigation Management: The MASS-COTE Approach, Mapping System and Services for Canal Operation Techniques. FAO Irrigation and Drainage Paper 63. Rome: FAO.

Schultz, B., and S. Uhlenbrook. 2007. *Water Security: What Does It Mean, What May It Imply?* Delft, the Netherlands: UNESCO-IHE Institute for Water Education.

Seckler, D., U. Amerasinghe, D. Molden, R. De Silva, and R. Barker. 1998. World Water Demand and Supply, 1990 to 2025: Scenarios and Issues. *IWMI Research Report 19.* Colombo: International Water Management Institute.

United Nations (UN) Department of Social Affairs. Population Division. 2004a. *World Population Prospects: The 2002 Revision.* New York, US: UN.

———. 2004b. *World Urbanization Prospects: The 2003 Revision.* New York, US: UN.

UNESCAP, ADB, and UNDP. 2007. *The Millennium Development Goals: Progress in Asia and the Pacific 2007.* Bangkok. Thailand: UNESCAP.

UNESCAP and FAO RAP. 2009. *The Study Report: Sustainable Agriculture and Food Security.* Bangkok, Thailand: UNESCAP.

UNESCO. 2009. *The United Nations World Water Development Report 3. Water in a Changing World: World Water Assessment Programme.* Paris, France: UNESCO.

United Nations Framework Convention on Climate Change (UNFCCC) Secretariat. 2007. Synthesis of Information and Views on Methods and Tools Submitted by Parties and Relevant Organizations. September 18. Bonn, Germany: UNFCCC.

UN Water. 2009. *The United Nations World Water Development Report 3: Water in a Changing World.* UNESCO.

UNFCCC Secretariat. 2008. *Compendium on Methods and Tools to Evaluate Impacts of and Vulnerability and Adaptation to Climate Change.* Bonn, Germany: UNFCCC.

World Water Council. 2009a. Istanbul Ministerial Statement. The 5th World Water Forum. 22 March.

———. 2009b. Press Release: Water Security Initiative Announced for Asia-Pacific Region. 15 March. http://www.watermediacenter.org/

Chapter 7

Water Management Practices and Climate Change Adaptation: South Asian Experiences

Selvarajah Pathmarajah

7.1 Introduction

Climate change issues have been increasingly gaining importance in the daily lives of common people and becoming a subject of deep interest to policymakers internationally. While mitigation measures to arrest greenhouse gas (GHG) emissions and global warming are widely being talked about, awareness and interest on adaptation is still an emerging paradigm. Understanding the global concerns and streamlining them to local political agenda remains a challenge. The uncertainty and inconsistency in predictions lead to less interest on investing in serious adaptation measures. However, due to international efforts, there is an increasing awareness on the need to adapt to climate change among scientists and policymakers in South Asia. The Intergovernmental Panel on Climate Change (IPCC) has played a vital role in providing factual scientific information to policymakers during the past two decades (IPCC 1990, 1995, 2001, 2007).

The failure of the climate summit in Copenhagen in December 2009 emphasized the limitations of GHG mitigation as a singular policy response to climate change and highlighted the urgent need to design effective adaptation strategies (Kavi Kumar et al. 2010). It has been repeatedly quoted that developing countries are the most vulnerable to climate change impacts because they have fewer resources to adapt socially, technologically, and financially. Many developing countries have already started to give adaptation action a high priority. However, adaptive capacity varies between countries depending on social structure, culture, economic capacity, geography, and level of environmental degradation (UNFCCC 2008).

7.2 Effect of Climate Change on Water Resources

There is ample evidence to suggest that South Asia's climate has already changed (Eriyagama et al. 2010; Guhathakurta and Rajeevan 2006; IPCC 2007; Malmgren et al. 2003). Water management faces the challenge of managing predicted water stress due to an anticipated increase in atmospheric temperature and irregularity of rainfall patterns including extreme events such as floods and droughts. Soil moisture stress or water logging, groundwater depletion, and saline water intrusion are some of the anticipated effects of climate change that could bring about serious implications. At present, a lack of resources and population growth make South Asian countries vulnerable to any form of natural disaster. Climate change will undoubtedly compound existing food insecurity and vulnerability patterns in these countries. Communities must be made aware to prepare themselves for the possibility of food shortages and make appropriate use of resources to protect their lives and livelihoods as well as property.

In water resources management, the technique of using past hydrological experiences to forecast the future conditions is challenged by climate change. Consequently, the reliability of current water management systems and water-related infrastructure is challenged. Therefore, modeling approaches have gained popularity. The unavailability of dependable information is a constraint for planning beyond the extension of responses to extreme flood and drought conditions that were encountered in the past.

According to the International Water Management Institute, one-third of the developing world will face severe water shortages in the 21st century even though large amounts of water will continue to annually flood out to sea from water-scarce regions. The problem is that the spatial and temporal distribution of precipitation rarely coincides with demand. Whether the demand is for natural processes or human needs, the only way water supply can match demand is through storage (Andrew et al. 2000).

7.3 Adaptation Strategies

IPCC (2007) suggested that it is important to promote "adaptation" to the impacts of global warming as global warming "mitigation" which is centered around the reduction of GHGs has limitations, and global warming impacts would continue over centuries even when mitigation is implemented. Adaptation strategies to avoid drought risk caused by climate change should be considered to be an emerging priority issue in comprehensive water resources management.

There are societal, institutional, and economic coping mechanisms that could be adapted on a short-term and long-term basis. In the past, these measures have been adapted in the region in some form or other to manage the climate variability resulting from monsoonal rainfall patterns with distinct dry and wet seasons. Intensification and improvement of such adaptive measures is a no-regret strategy that would justify immediate investments on climate change adaptation.

According to the UN International Strategy for Disaster Reduction Secretariat (UN/ISDR 2009), "… any physical construction to reduce or avoid possible impacts of hazards, or application of engineering techniques to achieve hazard resistance and resilience in structures or systems …" is defined as structural measure, and a nonstructural measure is "… any measure not involving physical construction that uses knowledge, practice or agreement to reduce risks and impacts, in particular through policies and laws, public awareness raising, training and education." It is not logical however to consider all these measures in isolation. There are existing structures around which nonstructural aspects are established; thus a holistic approach is needed.

There are four major ways of storing water: in the soil profile, in underground aquifers, in small reservoirs, and in large reservoirs behind large dams. If climate change as a result of global warming manifests, the need for freshwater storage will become even more acute. Increasing storage through a combination of groundwater and large and small surface water facilities is critical to meeting the water of the 21st century. This is especially so in monsoonal Asia and the developing countries in the tropics and semitropics (Andrew et al. 2000).

7.4 Climate Change Adaptation Policies and Actions in South Asia

According to the IPCC (AR4) some of the future impacts of climate in South Asia include: glacier melting in the Himalayas is projected to increase flooding and will affect water resources within the next two to three decades; climate change will compound the pressures on natural resources and the environment due to rapid urbanization, industrialization, and economic development; crop yields could decrease up to 30% by the mid-21st century; mortality due to diarrhea primarily associated with floods and droughts will rise; and sea-level rises will exacerbate inundation, storm surge, erosion, and other coastal hazards (IPCC 2007).

The South Asian Association for Regional Cooperation (SAARC) Action Plan on Climate Change (SAARC 2007), from a ministerial level meeting in Dhaka as far back as 2007, was a well-formulated document emphasizing seven thematic areas: adaptation to climate change, policies, and actions for climate change mitigation, policies and actions for technology transfer, finance and investment, education and awareness, management of impacts and risks due to climate change, and capacity building for international negotiations.

Thematic area one, adaptation to climate change, outlined the following for action:

- adaptation to climate change impacts and risks in vulnerable communities, locations, and ecosystems;
- adaptation in sectors (for example, water, agriculture, fisheries, health, and biodiversity);
- adaptation to extreme climate events (for example, flood, cyclone, glacial lake outburst, droughts and heat, and cold waves);

- adaptation to climate change impact (for example, sea-level rise, salinity intrusion, glacial melt, and coastal and soil erosion); and
- adaptation suited to urban settlements, coastal structures and mountain terrain.

The plan that says that it will be initially in action for three years from 2009 to 2011 has not been materialized to the expectation. SAARC did not even participate in the 15th Conference of the Parties (COP 15) in the 2009 United Nations Climate Change Conference that was held in Denmark as a subregional body with a subregional agenda (Kishor Pradhan 2010).

Climate change was the theme of the 16th SAARC summit held in Thimphu, Bhutan in April 2010 (SAARC 2010). The summit arrived at a statement on climate change that focused more on mitigation than on adaptation. There is not much stated about agriculture and water resources management in the context of climate change. However, India has extended its full support in setting up of a fund that would help South Asia effectively meet urgent adaptation and capacity building needs posed by climate change (*The Hindu* 2010).

7.5 Climate Change Adaptation Policies and Actions in South Asian Countries

7.5.1 Bangladesh[1]

Bangladesh has a land area of 144,000 square kilometers (km^2) and a population of over 156 million as per 2010 estimate, with a density of little more than 1,000 persons/km^2, which is one of the highest in the world. The population has grown rapidly during the last three decades reducing the land-man ratio to mere 0.05 ha per capita arable land (Hussain and Asaduzzaman 2010). Thus, the non-climatic pressure exerted by population growth on food security makes the country more vulnerable to the compounded effect of climate change.

Further, Bangladesh's geography makes it one of the most vulnerable countries to climate change. It is dominated by flood plains (80%), with most of the land less than 12 m above sea level. Bangladesh is also affected by two very different ecosystems (the Himalayas to the north and the Bay of Bengal to the south). The overall result is that it has too much water during the monsoon season and too little in the winter. In an average year, approximately one quarter of the country is inundated (UNDP 2009).

Though the agricultural sector contributes only 20% to gross domestic product (GDP), it employs about 48% of the workforce. The people of Bangladesh have

---

[1] The discussion is drawn from UNDP (2009) unless otherwise stated.

been adapting to the risks of floods, droughts, and cyclones for centuries. The high population density, very frequent occurrences of natural disasters, poor infrastructure, and fragile economic resilience to shocks make the country especially vulnerable to climatic risks. Heavy reliance of rural people on agriculture and natural resources increases their vulnerability to climate change. Therefore, supporting rural and urban communities to strengthen their resilience and to adaptation to climate change will remain a high priority in coming decades (Hussain and Asaduzzaman 2010).

The climate change vulnerability of Bangladesh's poor is well recognized, with 70 million people likely to be affected by floods annually by 2050. While there is still considerable uncertainty about the timing and severity of climate change impacts, existing research (Ahmed 2006; Ali 1999; UNDP 2009) suggests the following:

- sea-level rise leading to inundation of up to 7% of Bangladesh's land area by 2050
- increased frequency and intensity of cyclones
- increased salinity in inland areas
- increased severity of river floods
- heavier and more erratic rainfall during the monsoon season, and lower and more erratic rainfall in drier northern and western regions of the country (resulting in increasing droughts)
- higher inundation risks because of the melting of ice caps in the Himalayas

If the sea-level rise is higher than currently expected and coastal embankments are not strengthened or new ones built, 6 to 8 million people could be displaced by 2050. Increased salinity is likely to result in the decline in rice and wheat production, and a more pronounced shortage of drinking water.

Current embankments in Bangladesh are not designed to deal with 50- and 100-year storms. Embankment failure is therefore a serious possibility. Increased salinity will also lead to migration into cities from rural areas, making the management of employment and services for large numbers of migrants another challenge. Increasing water salinity looms as a major problem for agricultural production in Bangladesh. Currently, more than 170,000 hectares (ha) of agricultural land is affected by salt and this problem will only grow bigger (UNDP 2009).

Local institutions in collaboration with international institutions are instrumental in developing salinity and submergence-tolerant rice varieties, and heat- and drought-resistant cereals. For example, rice variety BRRI dhan 47 can withstand a salinity level of 12 dSm^{-1} at seedling stage and up to 6 dSm^{-1} for rest of the growing period. Similarly, new breeds of rice surviving prolonged periods of submergence developed by International Rice Research Institute (IRRI) are already helping farmers in the country (IRRI 2009). A national adaptation program of action (NAPA) document prepared in November 2005 identified 15 immediate and urgent priority projects. The first priority project, "Community Based Afforestation in Coastal Areas" is being implemented by the Forestry Department. As NAPA addressed immediate and urgent adaptation needs it did not cover the major concerns of all the sectors (Hussain and Asaduzzaman 2010). Thus, UNDP (2009) refers to this US$77.4 million worth of climate change adaptation projects as a "wish list."

The Bangladesh Climate Change Strategy and Action Plan 2009 (Government of Bangladesh 2009), a comprehensive document—drawing from experts in all the respective fields—proposed six thematic areas to be implemented by respective line ministries and departments: food security, social protection and health; comprehensive disaster management; infrastructure; research and knowledge management; mitigation and low carbon development; and capacity building and institutional strengthening. However, financial support to address adaptation activities in the country is a crucial issue.

7.5.2 India

India is a vast country covering 3.28 million km^2 with diverse surface features. The climate regimes of the country vary from humid to arid. India, with 17% of the world's population, contributes only 4% of the total global GHG emissions. Agriculture is important in India's economy as it feeds a large and growing population, employs a large labor force, and provides raw material to agro-based industries. Climate change may alter the distribution and quality of natural resources such as fresh water, arable land, and coastal and marine resources. With an economy closely tied to its natural resource base and climate-sensitive sectors such as agriculture, water and forestry, India faces a major threat due to projected changes in the climate (Government of India 2009).

The Ministry of Environment and Forest submitted its initial National Communication (NATCOM) to the UNFCCC[2] in 2004. The NATCOM states that India is faced with the challenge of sustaining its rapid economic growth while dealing with the global threat of climate change. It is concerned about the impacts of climate change as its large population depends on climate-sensitive sectors like agriculture and forestry for livelihoods. Any adverse impact on water availability due to receding glaciers, a decrease in rainfall, and increased flooding in certain pockets would threaten food security, cause dieback of natural ecosystems including species that sustain the livelihoods of rural households, and adversely impact

---

[2] A treaty signed at the 1992 Earth Summit in Rio de Janeiro that calls for the "stabilization of greenhouse gas concentrations in the atmosphere at a level that would prevent dangerous anthropogenic interference with the climate system."

the coastal system due to sea-level rise and increased frequency of extreme events. Apart from these, achievement of vital national development goals related to other systems such as habitats, health, energy demand, and infrastructure investments would be adversely affected (Government of India 2004).

NATCOM further states that India's development plans are crafted with a balanced emphasis on economic development and environment. The planning process, while targeting an accelerated economic growth, is guided by the principles of sustainable development with a commitment to a cleaner and greener environment. Planning in India seeks to increase wealth and human welfare, while simultaneously conserving the environment. It emphasizes the promotion of people's participatory institutions and social mobilization, particularly through the empowerment of women, for ensuring environmental sustainability of the development process. However, strategies to translate plans into actions are not detailed.

Prime Minister Manmohan Singh released India's National Action Plan on Climate Change (NAPCC) on June 30, 2008. It recognizes, "without a careful long-term strategy, climate change may undermine the development efforts, with adverse consequences on the people's livelihood, the environment in which they live and work and their personal health and welfare."

The National Water Mission of NAPCC (Government of India 2008) states that out of the 4,000 billion cubic meters (m^3) of precipitation that India receives annually, only 1,000 billion m^3 are available for use, which comes to approximately 1,000 m^3 per capita per annum. Further, by 2050 it states that India is likely to be water scarce. The National Water Mission thus aims at conserving water, minimizing wastage, and ensuring more equitable distribution through integrated water resource management. It also aims to optimize water use efficiency by 20% by developing a framework of regulatory mechanisms having differential entitlements and pricing. In addition, the Water Mission calls for strategies to tackle variability in rainfall and river flows such as enhancing surface and underground water storage, rainwater harvesting, and more efficient irrigation systems like sprinklers or drip irrigation.

The National Mission for Sustainable Agriculture of the NAPCC (Government of India 2008) aims at making Indian agriculture more resilient to climate change by identifying new varieties of crops, especially thermal resistant ones and alternative cropping patterns. This is to be supported by integrating traditional knowledge and practical systems, information technology and biotechnology, as well as new credit and insurance mechanisms.

In particular the mission focuses on rain-fed agricultural zones and suggests:

- development of drought and pest resistant crop varieties;
- improving methods to conserve soil and water;

- stakeholder consultations, training workshops and demonstration exercises for farming communities, for agroclimatic information sharing and dissemination; and
- financial support to enable farmers to invest in and adopt relevant technologies to overcome climate-related stresses.

In addition, the mission makes suggestions for safeguarding farmers against increased risks due to climate change. These suggestions include strengthening agricultural and weather insurance; creation of Web-enabled, regional language based services for facilitation of weather-based insurance; development of geographic information systems (GIS) and remote-sensing methodologies; mapping vulnerable regions and disease hotspots; and developing and implementing region-specific, vulnerability-based contingency plans. Finally, it suggests greater access to information and use of biotechnology. In 2009, the Ministry of Water Resources released a revised comprehensive mission document detailing the strategy of the National Water Mission under the National Action Plan on Climate Change (Government of India, MoWR 2009).

Climate change impact assessment and adaptation studies require predictions from climate models. To plan for adaptation in a country like India some important changes are required in the inputs provided by current climate models. Climate predictions are needed at finer spatial resolutions than are currently available from the global climate models. This is beginning to happen—for example, the Indian Institute of Tropical Meteorology has developed high resolution (50 × 50 km) regional climate change scenarios for India using the Hadley Centre regional climate model (PRECIS) that is forced by the state-of-the-art coupled general circulation model (HadCM3). Similar exercises are underway to develop suitable high resolution future climate scenarios for India by running three regional climate models (PRECIS, WRF, and RegCM3) using the lateral boundary conditions from five IPCC-AR5 coupled models. As locally downscaled climate information becomes available, advances in policy responses and examination of climate implications will become more feasible (Kavi Kumar, Shyamsundar, and Nambi 2010).

Increases in CO_2 concentration to 550 parts per million (ppm) could increase yields of rice, wheat, legumes, and oilseeds by 10%–20%. However, a degree increase in temperature may reduce yields of wheat, soybean, mustard, groundnut, and potato by 3%–7%. Yield losses are likely to be much higher at higher temperatures. Studies assessing the economic impacts of climate change on agriculture have focused mostly on impacts on cereal crops like rice and wheat. New research findings from crop models on non-cereal and commercial crops have not been integrated yet into economic modeling. (Kavi Kumar, Shyamsundar, and Nambi 2010).

Agriculture and water sectors face considerable non-climatic pressures. Hence the challenge is to integrate responses to non-climatic stresses with those that can minimize potential climate change impacts. Responses to climatic and non-climatic pressures have largely been ad hoc and hence could be inadequate and unsustainable in the long term. Based on fieldwork carried out in Andhra Pradesh and Rajasthan, MSSRF (M. S. Swaminathan Research Foundation 2008) suggests that there are effective ways to make farmers more adaptive to climate changes. They recommend specific changes in traditional water management practices, establishing smart farmer networks that enable farmers to share knowledge on farm management practices, utilizing weather data from simple agro-met stations operated by the farmers, and use of some new rice farming techniques such as "system of rice intensification."

Many storage reservoirs, which were previously used as irrigation tanks in the arid and semiarid tracts of India, have now been converted to recharge ponds, and tube wells have taken the place of irrigation canals. These successful experiments indicate that combinations of big and small reservoirs along with effective aquifer management can provide efficient solutions for conserving water and increasing its productivity. This concept has not been effectively put into practice from the planning stage, although it has been practiced in many areas of the world. With water becoming scarce, use of such integrated planning for conserving water could lead to higher water productivity while maintaining environmental and ecological balances. Combinations of small and large storage and surface water and groundwater recharge are generally the best systems where they are feasible. In monsoonal Asia, research and development are needed on how to manage water under monsoonal conditions (Andrew, Sakthivadivel, and Seckler 2000).

7.5.3 Maldives[3]

Maldives is a small nation with a land area of approximately 235 km². This land is divided over some 1,200 coral islands, of which 96% are less than 1 km² in area. The 194 inhabited islands of the country are scattered over an archipelago. At present, 42% of the population and 47% of all housing structures are within 100 m of the coastline. Over the last six years more than 90 inhabited islands have been flooded at least once, and 37 islands have been flooded regularly or at least once a year. The dispersed nature of the population, combined with poorly developed transportation systems, result in reduced adaptive capacity and increased vulnerability to climate change.

---

[3] This section summarizes the country paper presented by Aishath Shafina (2009) at the Regional Workshop on Mainstreaming Climate Change Adaptations into Developmental Planning.

The observed long-term trend in relative sea level for Hulhulé island where the Malé international airport is located is 1.7 mm per year. For Hulhulé an hourly sea level of 70 cm above mean sea level is currently a 100-year event. It will likely be at least an annual event by 2050. Presently, the maximum storm surge height is reported to be 1.32 m with a return period of 500 years in the Maldives. When storm surges coupled with a high tide, it can generate a storm tide of 2.30 m that is disastrous to islands that have a mean elevation of 1–1.5 m above the mean sea level.

An extreme daily rainfall of 180 mm is currently a 100-year event. It will likely occur twice as often, on average, by 2050. The annual maximum daily temperature is projected to increase by around 1.5°C by 2100. A maximum temperature of 33.5°C is currently a 20-year event. It will likely have a return period of three years by 2025.

Significant investments have been made to develop the country's infrastructure, which is highly vulnerable to sea-level rise and storm conditions. The transport infrastructure includes three major commercial seaports, more than 128 island harbors, and five airports, of which two are international. The infrastructure of the two international airports is within 50 m of the coastline.

The average width of inhabited islands is 566 m, and on both inhabited islands and resort islands more than 75% of the critical infrastructures like powerhouses, communications facilities, and waste disposal sites are located within 100 m of coastline. If appropriate adaptation measures are not taken, frequent inundations could virtually obliterate the critical infrastructure, severely damaging the economy, and threaten the safety and security of the people.

The freshwater aquifer lying beneath the islands is a shallow lens, 1–1.5 m below the surface and no more than a few meters thick. Surface freshwater is lacking throughout the country with the exception of a few swampy areas in some islands. Traditionally people depended on shallow wells to get access to the groundwater lens for drinking water. However, 90% of the atoll households now use rainwater as their principal source of drinking water. In Malé, 100% of the population has access to piped desalinated water. Following the 2004 tsunami, 38 islands have been provided with desalination plants, which are being operated daily or on an emergency basis. Already stressed from overextraction, the freshwater aquifers face the risk of total depletion if dry periods extend. With the predicted sea-level rise and during periods of wave-induced flooding, there is a very high risk of saltwater intrusion into the freshwater lens. Saltwater intrusion would also adversely affect soil and vegetation, damaging agriculture and terrestrial ecosystems.

Rainwater is the main source of drinking water in the atolls. Although the global average precipitation is projected to increase during the 21st century, a marginal decline in precipitation is projected for the Indian Ocean region. The predicted changes have the potential to affect rainwater harvesting across all the atolls and in particular the northern atolls. Even today drinking water shortages during dry periods poses a significant challenge to the atoll population.

Agriculture is vital to the food security, nutritional status, and livelihoods of the atoll population and contributed 2.6% to GDP in 2005. Climate hazards will affect agriculture through heat stress on plants, changes in soil moisture and temperature, loss of soil fertility through erosion of top soil, less water available for crop production, changes in the height of water table, salinization of freshwater aquifer, and loss of land through sea-level rise. The consequences of such impacts are likely to be particularly severe in Maldives because agriculture is already under stress due to poor soil, limited available land, and water scarcity. However, Maldives imports almost all food items other than fresh tuna and coconut.

To address the challenges posed by environmentally vulnerable islands that are currently experiencing severe impacts from climate change and associated sea-level rise, with remote and dispersed population, the government formulated the Population and Development Consolidation program. Under this program the government seeks to resettle populations through incentives to migrate from islands that are environmentally vulnerable. The vulnerability of Maldives was truly exposed following the tsunami of December 2004. While a tsunami of the magnitude experienced in December 2004 is an extremely rare event, with the predicted sea-level rises, flooding may become a more frequent phenomenon.

As one of the key adaptation measures for the predicted climate change, the government formulated a comprehensive NAPA in early 2006 (Government of Maldives 2006). The implementation of NAPA is one of the highest priorities of the government.

The following priority adaptation needs have been identified in the NAPA:

1. Integrate future climate change scenarios in the development of selected population centers or islands to adapt sea-level rise and extreme weather risks associated with climate change.
2. Protect human settlements through innovative coastal protection for development of selected population centers or islands.
3. Improve and promote eco-friendly sustainable housing technology.
4. Build resilience of fisheries.
5. Acquire technologies and appropriate tools to manage water resources.
6. Strengthen capacity for health services.
7. Strengthen agricultural production and increase food security.

Creating awareness among the population on the impact of climate change on water resources, the need to use water efficient devices and promote behaviors that are water conscious are listed as the urgent and immediate adaptation measures required.

7.5.4 Pakistan[4]

Pakistan has a total land area of 880,940 km^2. The country is vulnerable to climate change because it has a warm climate; it lies in a region where the temperature increases are expected to be higher than the global averages; its land area is mostly arid and semi-arid (about 60% of the area receives less than 250 mm of rainfall per year and 24% receives between 250 and 500 mm); its rivers are predominantly fed by the Hindukush–Karakoram–Himalayan glaciers which are reported to be receding rapidly due to global warming; its economy is largely agrarian and hence highly climate sensitive; and because the country faces increasingly larger risks of variability in monsoon rains, large floods and extended droughts.

Pakistan possesses the world's largest irrigation system (commonly known as Indus Basin Irrigation System) that commands an area of about 14.3 million ha representing about 76% of the cultivated area. Besides the river water, large quantities of groundwater are also used for irrigation. The current accountable use of water is: agriculture (92%), industries (3%), and domestic and infrastructure (5%).

Due to silting, the capacities of all the three reservoirs in the Indus Basin have been decreasing with time. The total capacity decreased from their original total capacity of 18.37 million acre-feet (maf)[5] in 1974 to 13.68 maf in 2003 and is projected to decrease to 12.34 maf by 2010. The present reservoir capacity (live storage) corresponds to only 9% of the Indus River System (IRS) average annual flow. At present on average 35 maf of water flows to the sea annually during flood season, while there is need to conserve every drop not required for optimal ecological flow into the sea.

The major climate change related threats to water security are identified as:

- increased variability of river flows due to increase in the variability of monsoon and winter rains and loss of natural reservoirs in the form of glaciers;
- likelihood of increased frequency and severity of extreme events such as floods and droughts;
- increased demand for irrigation water because of higher evaporation rates at elevated temperatures in the wake of reducing per capita availability of water resources and increasing overall water demand;
- increase in sediment flow due to increased incidences of high intensity rains resulting in more rapid loss of reservoir capacity;

---

[4] This section summarizes the country paper presented by Anwar (2010) at the Regional Workshop on Strategic Assessment of Climate Change Adaptation in Natural Resources Management.

[5] 1 acre-foot = 1233.5 m^3.

- changes in the seasonal pattern of river flows due to early start of snow and glacier melting at elevated temperatures and the shrinkage of glacier volumes (this will have serious implications for storage of irrigation water and its supply for *kharif* and *rabi* crops);
- possible drastic shift in weather pattern, both on temporal and spatial scales; and
- increased incidences of high altitude snow avalanches and glacial lake outburst floods (GLOFs) generated by surging tributary glaciers blocking main unglaciated valleys.

So far most of the adaptation efforts undertaken in Pakistan are unplanned and carried out irrespective of the knowledge of climate change. It is now planned to construct a series of large hydropower projects to add 18 maf of new storage capacity by 2030 to the existing 12.5 maf capacity (which is decreasing by 0.2 maf annually due to silting). The construction of a 4,500 MW hydropower plant at Bhasha with 6.4 maf water storage capacity will be complemented by small and medium dams as well as measures for recharging underground reservoirs, and using groundwater aquifers as water storage facilities. According to the *Daily Times*, construction work will start by mid-2012 (Chaudhry 2011). A major program is underway for lining the water channels and plans to monitor continuously the movement of glaciers in northern Pakistan.

It is recommended to add sufficient reservoir capacity on the IRS so that even during high flood years no water flows down in excess of what is necessary for environmental reasons. Other adaptation measures proposed by the government are local rain harvesting and building of surface and subsurface storages for agriculture and other local needs, adoption of stringent demand management and efficiency improvement measures in all water-use sectors, particularly in the supply, distribution and use of irrigation water, and reuse of marginal quality irrigation effluent.

Not much attention has so far been paid to mitigate or adapt to climate change related issues in the agriculture and livestock sectors. It is planned to: *(i)* develop through biotechnology, heat-stress resistant, drought- and flood-tolerant, and water-use efficient high yielding crop varieties; *(ii)* increase irrigation water availability by reducing losses in the irrigation water supply network; *(iii)* implement a "more crop per drop" strategy through improved irrigation methods and practices, water saving techniques in combination with the use of high yielding and water-efficient crop varieties; and *(iv)* increase milk and meat production by developing animals breeds which are less vulnerable to climatic changes, and by improving animal feedstock.

The government has proposed to attend to institutional capacity building for addressing climate change by: *(i)* enhancing capacity of all relevant organizations; *(ii)* introducing climate change–related scientific disciplines in Pakistan's leading universities so as to ensure a regular supply of trained workers; and *(iii)* establishing a national data bank for climatological, hydrological, agro-meteorological, and other climate change–related data to cater for the needs of all relevant institutions.

Pakistan lacks technical capacity and financial resources to address climate change-related mitigation, adaptation, and capacity building needs. International cooperation is needed to develop capacity to deal with disasters like floods, droughts and cyclones and to train young scientists in the fields such as regional climate modeling, watershed modeling, and crop growth simulation modeling; forecasting of seasonal and interannual climatic changes and extreme events; monitoring of temporal changes in glacier volumes and land cover using satellite imagery and GIS techniques.

7.5.5 Sri Lanka

Sri Lanka is a small island nation, with a land area of 65,610 km², in the Indian Ocean off the southern tip of India. The territory includes 1,660 km of coastal areas and 2,905 km² of inland water bodies. The total population in Sri Lanka is 20.03 million and the population growth rate was 1.31% in 2008. Present per capita land availability of the country is 0.3 ha. Arable land extent is nearly 45% of the total land area of the country. The main agricultural land uses include paddy (27%) and plantation crops (tea, rubber, and coconut, 24%) (Wickramasinghe and Punyawardena 2009).

Topographic influences and monsoonal rainfall patterns result in diverse weather patterns ranging from an annual average rainfall of 5,500 mm in the central highlands to between 800 and 1,200 mm in the southeast and northwest coastal areas (Malmgren et al. 2003). Similarly, annual average temperatures range from 15°C in the central highlands to 27°C in the lowlands.

Sri Lanka ratified UNFCCC in 1993 (Government of Sri Lanka 2000) and took its first GHG inventory in 1995. Also, it established the Centre for Climate Change Studies in 1999 at the Department of Meteorology that regularly monitors climatic parameters including the temperature and the rainfall through a well-established island-wide network of monitoring facilities. It ratified the Kyoto Protocol on September 3, 2002 and established two national Clean Development Mechanism (CDM) study centers at the University of Peradeniya and University of Moraruwa.

Sri Lanka falls into the UNFCCC and IPCC category of "vulnerable" small island nations under serious threat from various climate change impacts, such as sea-level rise and severe floods and droughts. These threats could have significant negative consequences on various sectors in Sri Lanka.

Analysis of 130 years of precipitation data from 1870 to 2000 for 15 stations distributed all over the country indicated statistically significant temporal changes in southwest-monsoon (May–September) related precipitation at five of the stations, with three stations showing enhanced rainfall and two stations a decrease in rainfall with time. In addition, one station experienced a decrease of both first (March–April) and second (October–November) inter-monsoon rainfall over time. The stations showing loss of rainfall are confined to higher elevation areas and those exhibiting enhanced rainfall are located in the lowlands in the south-western sector of Sri Lanka. None of the stations show any significant change in northeast-monsoon (December–February) precipitation through time (Malmgren et al. 2003). The analysis shows that the impact of climate change on rainfall may not be spatially coherent and could result in water shortages in certain areas and increased rainfall in other areas. In the country as a whole, the number of consecutive dry days increased while the number of consecutive wet days reduced (Eriyagama et al. 2010).

Seo, Mendelsohn, and Munasinghe (2005) reported that Sri Lanka's dry zone agricultural output will decline significantly in the next 20 to 30 years because of reduced rainfall and hotter weather. In the last six years, droughts hit Hambantota in 2001 and Anuradhapura and Kurunegala districts in 2004, affecting farmers and residents.

Water resource adaptation strategies for climate change in Sri Lanka as observed by Yamane (2003) are:

- Encourage minor storage water reservoirs
- Investigate feasibility of trans-basin diversion schemes
- Conserve seasonal water
- Rehabilitate irrigation water tanks networks
- Promote micro-watershed management
- Prepare groundwater extraction regulation policy
- Introduce permit/monitoring systems for groundwater extraction and water quality assessment in vulnerable areas

Reservoirs with a command area less than 80 ha are considered as minor or small tanks in Sri Lanka. A total of around 15,000 small tanks, both operational and abandoned, are distributed across 70 well-defined river basins in the dry zone. The construction of and settlement around these small village tanks took place over a long period up to around AD 1200–1300. They are an important traditional system of harvesting rainwater runoff and storing it for multiple purposes. Hydrological features of these small tanks suggest that the capture of water in these reservoirs was intended less for irrigated paddy, but more for the needs of human settlement and survival through occasional periods of sustained drought (Panabokke, Sakthivadivel, and Weerasinghe 2002). Tanks were used to sustain various livelihood activities, like agriculture, fisheries, and livestock rearing. Traditionally, community-based institutions took an active role in the management of the tanks. They ensured adequate maintenance, oversaw the distribution of tank water, and provided a forum for all stakeholders to express their interests with regards to the tanks. Considering their agricultural, socioeconomic, and environmental benefits, small tanks are highly relevant to the rural development programs.

Small tanks with an irrigation potential of about 100,000 ha will have a major role to play in addressing the climate change related food security issues in the dry zone of Sri Lanka where ancient minor irrigation tanks are located. These tanks form a series of water bodies along small watercourses in a cascading system. They were designed so that water is repeatedly put into use to counteract irregularities of rainfall, nonavailability of large catchment areas, and the difficulty in constructing larger reservoirs (Gunasena 2001). Furthermore, the seepage and percolation losses from small tanks in Sri Lanka account for 20% of reservoir volume against 5% of reservoir volume in large dams. Therefore, these small reservoirs act as percolation tanks, recharging aquifers and retarding runoff (Andrew, Sakthivadivel, and Seckler 2000).

Modernizing the agricultural production systems within a tank cascade is a need of the hour. However, the management interventions in the minor tank areas during the past three decades have lead to conjunctive use of surface water and groundwater in a haphazard manner. It is noted that agro-wells (large diameter irrigation wells) have been constructed under small tank command areas without considering the hydrological basis of the cascade system. Unplanned exploitation of the shallow aquifer system that exists in the dry zone could lead to serious economic and environmental consequences. Therefore, there is a need to have clear policies based on technical feasibility to utilize the small tanks on a sustainable basis (Pathmarajah 2003). In southern Sri Lanka, construction and linking of a large storage reservoir at Lunugamvehera with five small, existing, cascading reservoirs resulted in a 400% increase in crop production. In fact, cascading small reservoirs can significantly increase crop water use by capturing drainage, return flow, and surpluses from upstream reservoirs (Andrew, Sakthivadivel, and Seckler 2000).

"Dahasak Vew" (tank rehabilitation), a flagship project highlighted in the government's Economic Policy Framework recognizes the augmentation of water supply in basins—where water stress exists—by harnessing rainwater and storing it in existing and abandoned village tank systems. The project proposes rehabilitating 10,000 minor tanks. In addition, diversion of perennial water by means of anicuts (small dams) to farm lands through supply canals for cultivation in the upcountry region is also recognized. It is proposed to develop identified tanks in an integrated manner (Government of Sri Lanka 2005).

Apart from the government's initiatives, during the past 20 years international and national NGOs have undertaken minor tank rehabilitation and agro-well promotion programs. Though their objective was not adaptation to climate change, it had an element of sustainable livelihood under extreme weather conditions.

Furthermore, community participation—women's participation in particular—was effectively incorporated into the tank rehabilitation process by the NGOs. This approach built a sense of ownership by the community.

An encouraging recent addition to community adaptation to water stress is the increasing number of farmer designed and managed micro-irrigation systems in the coastal sand areas that lead to efficient water management and reduced groundwater pollution.

7.6 Conclusion

Climate change and global warming awareness is on the increase at all levels from the general public to policymakers. However, while mitigation measures to arrest GHG emissions and global warming are widely talked about, adaptation actions have not been given similar priority. In addition to a lack of political will, the major constraints for any serious adaptation actions are the lack of resources and ambiguity in the emerging information on climate change impacts and vulnerability. On the other hand, countries whose contribution to global warming (which leads to climate change) is insignificant have no option than investing in adaptation.

The technique of using past hydrological experiences to forecast the future conditions is challenged by climate change. Consequently, the reliability of current water management systems and water-related infrastructure is also challenged. The unavailability of dependable information constrains planning beyond the extension of responses to extreme flood and drought conditions that were encountered in the past.

Water management faces the challenge of managing predicted water stress due to an anticipated increase in atmospheric temperature and irregularity of rainfall patterns including extreme events such as floods and droughts. There are societal, institutional, and economic coping mechanisms that could be adapted on a short- and long-term basis. In the past, all these measures have been adapted in the region in some form or other to manage climate variability resulting from monsoonal rainfall patterns. Therefore, intensifying and improving such adaptive measures is a no-regret strategy that would justify immediate investments on climate change adaptation. However, long-term strategies are yet to evolve to conserve and utilize the large quantity of water that flows into the sea that is not required for optimal ecological purposes. Strategies are also required to arrest degradation of surface water quality resulting from anthropogenic activities.

Though the expected impacts on water resources and the agricultural sector may result in serious impacts on food production, livelihoods, and the economy, there is no incentive for governments to implement adaptation measures that will not guarantee short-term political benefits. Therefore, a national agenda on climate issues has to evolve irrespective of political interests.

Impact and vulnerability assessment, and adaptation planning need to be given national priority beyond academic exercises and interests. Researchers should support policymakers by providing the necessary information on cost of adaptation and non-adaptation and the implications of both adaptation and non-adaptation.

To discuss adaptation strategies at community level, the impacts imposed on society by water-related disasters should be presented appropriately. The selection of appropriate adaptation strategies such as altering the dates of planting, spacing and input management, cultivating alternate crops or cultivars, changing irrigational practices, adapting soil moisture conservation practices, harvesting rainwater, and safe use of groundwater will be possible only after people sufficiently understand those vulnerabilities. This suggests that information flows may be an important mechanism in facilitating climate change adaptation at community level.

Adaptation to environmental change in the form of social programs such as crop insurance, subsidies, and pricing policies related to water and energy are essential. Effective drought relief programs by governments and the promotion of private- and community-managed insurance schemes are options that will encourage farmers to take risks. Governments must support alternative livelihood options that are less dependent on agriculture.

Countries in South Asia lack technical capacity and financial resources to address climate change related issues. Therefore, international and regional cooperation and support are needed to address mitigation, adaptation, and capacity building. Greater support is needed for monitoring and forecasting using satellite imagery, GIS, and other modern tools and techniques. It is essential to introduce climate change related scientific disciplines into high school and university curriculum to build awareness among young people and to ensure a regular supply of trained workers.

The challenge ahead for governments is to mainstream the adaptation measures into the development priorities that are emerging from population growth, urbanization, industrialization, globalization, and local political dynamics.

References

Ahmed, A. U. 2006. *Bangladesh Climate Change Impacts and Vulnerability: A Synthesis.* Dhaka: Climate Change Cell, Department of Environment, Bangladesh.

Ali, A. 1999. Climate Change Impacts and Adaptation Assessment in Bangladesh. *Climate Research.* 12. pp. 109–116. http://v3.weadapt.org/placemarks/files/227/c012p109.pdf

Andrew, K., R. Sakthivadivel, and D. Seckler. 2000. *Water Scarcity and the Role of Storage in Development. Research Report 39.* Colombo, Sri Lanka: International Water Management Institute.

Anwar, M. T. 2010. *Country Paper: Pakistan.* Paper Presented in a Regional Workshop on Strategic Assessment of Climate Change Adaptation in Natural Resources Management. Colombo, Sri Lanka, June, 8–11 2010. http://www.adbi.org/files/2009.06.08.cpp.day1.sess2.4.country.paper.pakistan.pdf

Chaudhry, S. 2011. Work on Diamer-Bhasha Dam to Start by Mid Next Year. *Daily Times.* Pakistan edition, June 11. http://www.dailytimes.com.pk/default.asp?page=2011%5C 06%5C11%5Cstory_11-6-2011_pg5_7

Eriyagama, N., V. Smakhtin, L. Chandrapala, and K. Fernando. 2010. *Impacts of Climate Change on Water Resources and Agriculture in Sri Lanka: A Review and Preliminary Vulnerability Mapping.* Research report 135. Colombo: International Water Management. http://www.iwmi.cgiar.org/Publications/IWMI_Research_Reports/PDF/PUB135/ RR135.pdf

Government of Bangladesh, Ministry of Environment and Forests (MoEF). 2009. *Bangladesh Climate Change Strategy and Action Plan 2009.* Dhaka: Government of the People's Republic of Bangladesh. http://www.moef.gov.bd/climate_change_strategy2009.pdf

Government of India, Ministry of Environment and Forests (MoEF). 2004. *India's Initial National Communication.* Delhi: Government of India. http://www.natcomindia. org/natcomreport.htm

Government of India, MoEF. 2009. *India's National Capacity Needs Self-assessment Report and Action Plan.* Delhi: Government of India. http://www.undp.org/mainstreaming/ docs/ncsa/ncsa-reports/finalreportsandplan/ncsa-india-fr-ap.pdf

Government of India. 2008. *India's National Action Plan on Climate Change.* http://www. climate-leaders.org/climate-change-resources/india-and-climate-change/indias-national-action-plan-on-climate-change

Government of India, Ministry of Water Resources (MoWR). 2009. *National Water Mission under National Action Plan on Climate Change: Revised Comprehensive Mission Documents. Vols I and II.* Delhi: Government of India. http://www.indiawaterportal. org/node/10614

Government of Maldives, Ministry of Environment, Energy, and Water. 2006. *National Adaptation Programme of Action (NAPA).* Republic of Maldives. http://unfccc.int/ resource/docs/napa/mdv01.pdf

Government of Sri Lanka. 2000. *Initial National Communication under the United Nations Framework Convention on Climate Change.* http://unfccc.int/resource/docs/natc/srinc1. pdf

Government of Sri Lanka, Ministry of Finance and Planning. 2005. *Sri Lanka New Development Strategy: Framework for Economic Growth and Poverty Reduction.* Government of Sri Lanka. http://www.recoverlanka.net/data/SLDF05/SLNDS.pdf

Guhathakurta, P., and M. Rajeevan. 2006. *Trends in the Rainfall Pattern over India.* Research Report No: 2/2006. Pune, India: National Climate Centre, India Meteorological Department. http://www.imdpune.gov.in/ncc_rept/RESEARCH%20REPORT%202.pdf

Gunasena, H. P. M., ed. 2001. *Food Security and Small Tank Systems in Sri Lanka: Proceedings of the Workshop Organized by the Working Committee on Agricultural Science & Forestry.* Sri Lanka: National Science Foundation, September 9, 2000.

Hussain, G., and M. Asaduzzaman. 2010. *Country Paper: Bangladesh.* Paper presented in a regional workshop on Strategic Assessment of Climate Change Adaptation in Natural Resources Management. Colombo, Sri Lanka. 8–11 June. http://www.adbi.org/ files/2009.06.08.cpp.day1.sess2.4.country.paper.bangladesh.pdf

Intergovernmental Panel on Climate Change (IPCC). 1990. *First Assessment Report (FAR)*. http://www.ipcc.ch/publications_and_data/publications_and_data_reports.htm#1

———. 1995. *Second Assessment Report (SAR)*. http://www.ipcc.ch/publications_and_data/ publications_and_data_reports.htm#1

———. 2001. *Third Assessment Report (TAR)*. http://www.ipcc.ch/publications_and_data/ publications_and_data_reports.htm#1

———. 2007. *Fourth Assessment (AR4)*. http://www.ipcc.ch/publications_and_data/publica-tions_and_data_reports.htm#1 http://www.ipcc.ch/pdf/technical-papers/climate-change-water-en.pdf

IRRI. 2009. *Rice Today*. 8 (2) April–June. http://beta.irri.org/news/images/stories/riceto-day/8-2/RT%208-2a.pdf

Kavi Kumar, K. S., Priya Shyamsundar, and A. Arivudai Nambi. 2010. *The Economics of Climate Change Adaptation in India: Research and Policy Challenges Ahead*. http:// blogs.dfid.gov.uk/wp-content/uploads/2010/03/adaptation-workshop-MSE-policy-note-March2010.pdf

Kishor Pradhan. 2010. SAARC's Climate Change Review. *The New Nation*, April 26.

Malmgren, B. A., R. Hulugalla, Y. Hayashi, and T. Mikami. 2003. Precipitation Trends in Sri Lanka Since the 1870s and Relationships to El Nino–Southern Oscillation. *International Journal of Climatology*. 23 (10). pp. 1235–1252.

M. S. Swaminathan Research Foundation (MSSRF). 2008. *A Road Map for Policy Develop-ment: Community Level Adaptation to Climate Change and Its Relevance to National Action Plan*. Chennai, India: M S Swaminathan Research Foundation.

Panabokke, C. R., R. Sakthivadivel, and A. D. Weerasinghe. 2002. *Evolution, Present Status and Issues Concerning Small Tank Systems in Sri Lanka*. Colombo, Sri Lanka: Interna-tional Water Management Institute.

Pathmarajah, S. 2003. Use of Groundwater for Agriculture in Sri Lanka: A Synthesis of the Past, Present, and the Future. In *Proceedings of the Symposium on the Use of Ground-water for Agriculture in Sri Lanka*, ed. S. Pathmarajah. Paper presented at a seminar at Peradeniya, Sri Lanka. September 30, 2002.

SAARC. 2007. *Action Plan on Climate Change*. http://www.nset.org.np/nset/climatechange/ pdf/SAARC_Action_Plan.pdf

———. 2010. *Thimphu Statement on Climate Change*., http://www.saarc-sec.org/userfiles/ ThimphuStatementonClimateChange-29April2010.pdf

Seo, N. S., R. Mendelsohn, and M. Munasinghe. 2005. Climate Change and Agriculture in Sri Lanka: A Ricardian valuation. *Environment and Development Economics*. 10. pp. 581–596.

Shafina, A. 2009. *Country Paper: Maldives*. Paper Presented in a regional workshop on Mainstreaming Climate Change Adaptations into Developmental Planning, April 4–7, in Tokyo. http://www.adbi.org/conf-seminar-papers/2009/07/03/3107.country.paper. maldives/

The Hindu. 2010. India Announces SAARC Climate Change Fund. April 28. http://www. thehindu.com/news/national/article413918.ece

United Nations Development Programme (UNDP). 2009. *Climate Change Economics in Bangladesh: A Policy Note.* http://www.unpei.org/PDF/bangladesh-workshop-oct09policy-note.pdf

United Nations Framework Convention on Climate Change (UNFCCC). 2008. *Climate Change: Impacts, Vulnerabilities, and Adaptation in Developing Countries.* Bonn: UNFCCC. http://unfccc.int/resource/docs/publications/impacts.pdf

United Nations Strategy for Disaster Reduction Secretariat (UN/ISDR). 2009. Structural and Non-structural Measures. UN International Strategy for Disaster Reduction Sec, January 15. http://www.preventionweb.net/english/professional/terminology/v.php?id=505

Wickramasinghe, W. M. A. D. B., and B. V. R. Punyawardena. 2009. *Country Paper: Sri Lanka.* Paper Presented in a regional workshop on Mainstreaming Climate Change Adaptation into Developmental Planning, April, 4–7, Tokyo. http://www.adbi.org/files/2009.04.14. cpp.sess2.paper.wickramasinghe.country.paper.sri.lanka.pdf

Yamane, A. 2003. Rethinking Vulnerability to Climate Change in Sri Lanka. 9th International Conference on Sri Lanka Studies, November, 28–30, Matara, Sri Lanka. http://www.freewebs.com/slageconr/9thicslsflpprs/fullp004.pdf

Chapter 8

Adaptation Measures for Climate Change in Japan

Toshio Okazumi and Eiji Otsuki

8.1 Introduction

When the River Bureau of the Japanese Ministry of Land, Infrastructure, Transport and Tourism (MLIT) considers climate change from the perspective of managing water-related disasters, the concern is that the growing risk of sea-level rise and increases in rainfall intensity will bring about an extremely hazardous situation. In spite of full recognition of the need for mitigation measures to promote a reduction of greenhouse gas (GHG) emissions, it is necessary to start considering adapting to future climate change as soon as possible, because even if a reduction in GHG emissions becomes successful, climate change will still inevitably occur. This chapter proposes a framework for a specific Japanese risk-based program based on previous work of an interdisciplinary expert committee (MLIT 2008) to counter impacts as explained in the Fourth Assessment Report of the Intergovernmental Panel on Climate Change (IPCC 2007). The rise in temperature from climate change causes melting of glaciers and Antarctic ice sheets, thermal expansion of seawater, and increase in evapotranspiration. Ultimately, it increases the intensity of the impacts of water-related disasters, giving rise to more frequent storm surges, coastal erosion and floods, and aggravating the damage from debris flow and landslides.

8.2 Projections of Future Climate Change

Using the Japan Meteorological Agency (JMA) measurement stations for the past years (JMA 2010) and JMA models named 20 km-mesh Global Climate Model (GCM20) with a grid of 20 km (JMA 2010), we obtained the mean values of the maximum daily precipitation in the year during the past 20 years and during the same period 100 years into the future, based on the IPCC A1B scenario (IPCC 2007). With emphasis on the balance among various energy sources, the results showed that rainfall intensity will increase nationwide, and that the future rainfall intensity in northern Japan, in particular, will increase by more than 20% (MLIT 2008).

8.3 Adaptation Strategies against Water-related Disasters

In 2008, the MLIT commissioned a team of experts that published a report where major adaptation measures are described. Adaptation measures against water-related disasters include the use of dams, levees, and other structures; measures tied in with community development; measures focused on emergency response; and improved monitoring of the adverse effects of climate change. It is necessary to couple adaptation measures with risk assessment and to arrive at the optimum combination that offers the greatest efficiency and effectiveness and comprises different subsystems such as: *(i)* flood prevention facility management; *(ii)* provision of planning and building regulation; *(iii)* crisis management and emergency plans; and *(iv)* monitoring of river systems, planning efficiency, and preparedness of local people (Okazumi and Otsuki 2010). Japan and MLIT targeted their concern to avoid a 150 years event of flooding in some major rivers. Based on the projection of future climate change, a river with a target flood control safety level of 1/150 years return period is expected to have its flood control safety level drop to 1/40 in 100 years (Okazumi and Otsuki 2010). In this case, major floods will occur more frequently and higher safety demands are required, but considering the socioeconomic conditions, it is not realistic to raise again the target flood control safety level to 1/150 in the face of future climate change. Therefore, to manage future increases in river discharge, it is necessary to implement structural measures such as channel improvement and construction of flood control facilities in drainage basins. Furthermore, emergency measures should be developed in order to reduce fatalities and damage as well as regulation and guidance on land use and development.

8.4 Proposal for a Framework of New Flood Program Based on Risk Assessment

When considering flood control measures in a river basin, it is important to identify and assess the risk level. We considered a range of foreseeable, measurable damage, and selected four kinds of damage—the number of deaths; the number of stranded people; economic damage; and infrastructure—as being the most important. Among these four items, the top priority is to prevent the loss of lives. Therefore, we need to concentrate on ensuring that there are no rocks in riverbeds as they are a major risk of death in flood events. Second, we need to minimize the economic damage. From a range of measures at our disposal, it is necessary to choose the ones that can efficiently mitigate damage. When structural measures are not sufficient to prevent loss of lives or from people becoming stranded, it may be necessary to implement emergency measures, such as evacuation and advance preparations, as well as to provide regulation and guidance on land use. This new flood management program will utilize risk assessment and explore the optimum solution by effectively combining a range of mitigation measures.

When implementing a risk assessment, we need to take into consideration the increasing risks related to future climate change, and carry out an analysis of floods of various magnitudes that may occur at different precipitations. Risk assessment would then be performed for each index, and risk maps are created from these assessments.

8.5 Arakawa River as an Example

Specifically, we used the data on the projection of future climate change as a reference for the likely impacts on the drainage area of the lower Arakawa River. We calculated floods based on the assumption of 20% increase in rainfall. This is necessary to understand the limitations of the currently implemented measures.

Our flood management projects have focused on setting the target discharge for each river. The next step is to assess the possible damage in the entire basin area that may occur from floods of varying magnitudes. The results indicated that the four priority items (number of deaths, number of stranded people, economic damage, and infrastructure), that are most affected from the increase in rainfall, are the most appropriate indices for evaluation (Okazumi and Otsuki 2010).

In considering adaptation measures, we examined the four priority items and concluded that preventing any loss of life is the most important. In this study, there is a risk of a substantial number of deaths only in few zones. We need to execute measures to remedy this situation. Specifically, we reinforce levees alongside the two zones to prevent damage from the anticipated levee breach. Since these measures are still insufficient, we build dams and reservoirs upstream. We also considered measures for minimizing economic damage. Since there is a concentration of assets in the lower drainage area, the economic damage would be considerable, particularly in urban zones. We examined measures for mitigating damage in these zones. We propose to reinforce levees in these zones and replace bridges to increase discharge capacity.

An assessment on the combined effect of the various structural measures discussed, has shown that the maximum water depth will decrease substantially even in the zone where loss of lives was anticipated. The inundated area on the left bank and the upper drainage area will also decrease markedly.

The content of our study, together with the effectiveness of the measures adapted over time, are summarized in a framework. The framework illustrates the zones requiring particular attention, the priority and non-priority zones, the timeline for specific measures, the effect of those measures, and nonstructural measures in zones that require further action.

When structural measures are not enough to eliminate adverse impacts, we can concentrate on nonstructural measures in specific locations to minimize damage as much as possible. Specifically, by considering the types of anticipated flooding, concrete measures should be implemented. When there is a need to reduce the number of stranded people, for example, we would need to focus on emergency measures and work on reinforcing the evacuation program.

An assessment may also be made on the level of preparedness, which may include the preparation and distribution of hazard maps and other public awareness activities. It is, however, difficult to evaluate such activities in terms of numerical data. Nonstructural measures may include land-use regulation and guidance over the long term.

8.6 Conclusion

We have evaluated the effect of individual flood control projects. However it is also necessary to change this to an approach where we introduce risk assessment and evaluate the effect on mitigating the risks of floods in the entire drainage area.

Our framework shows the staggered implementation of structural measures, identified low-priority zones, and examined whether it was possible for those low-priority zones to share risks. This approach can also be applied to the ongoing projects. It allows us to set well-defined, concrete programs for minimizing damage in various projects by combining structural measures on the rivers and implementing emergency measures.

In areas where structural measures are not sufficient to mitigate risks, we need to assess the vulnerabilities, which should be reflected on hazard maps and in land-use regulations.

Because of the uncertainty of future climate change, it is necessary to conduct continuous monitoring, review climate change projections based on the results of the monitoring, and respond flexibly to any revisions. This new approach can be widely used and applied to other studies in Japan as well as developing countries.

In addition, the Integrated Flood Analysis System (IFAS) (Sugiura, Fukami, and Inomata 2008), developed by the International Centre for Water Hazard and Risk Management (ICHARM), can support existing flood control and more effective risk-based disaster plans, especially for countries that have insufficient rainfall and/or discharge gauges.

References

Intergovernmental Panel on Climate Change (IPCC). 2007. *Climate Change 2007: Synthesis Report, Summary for Policymakers.* IPCC.

Japan Meteorological Agency (JMA). 2010. *The National Meteorological Agency of Japan 2010.* http://www.jma.go.jp/jma/en/Activities/brochure201003.pdf

Ministry of Land, Infrastructure, Transport and Tourism (MLIT). 2008. *Climate Change Adaptation Strategies to Cope with Water-related Disasters Due to Global Warming,* www.mlit.go.jp/river/basic_info/english/pdf/policy_report.pdf

Okazumi, T. and E. Otsuki. 2010. Risk-based Flood Management for Adapting to Climate Change. Paper prepared for the Thai National Mekong Committee Secretariat, Department of Water Resources, Bangkok. http://www.tnmckc.org/upload/document/fmmp/6/6.7/Approved%20papers%20and%20documents/Paper%203-1-3%20Okazumi%20et%20al.pdf

Sugiura, T., K. Fukami, and H. Inomata. 2008. Development of Integrated Flood Analysis System (IFAS) and Its Applications. Proceedings of the World Environmental and Water Resources Congress 2008. May 12–17. Honolulu: American Society of Civil Engineers Publication, pp. 1–10.

Chapter 9
Climate Change Impacts on the Mekong River Delta

Shigeko Haruyama

9.1 Introduction

The Mekong Delta suffers from severe flooding every year because of factors between the natural environment and the human environment. Changing land use caused by rapid population increase, urban sprawl, and the expansion of farmlands is having a strong impact on the Mekong Delta, including an increase in flood damage. Climate change is related to sea-level change, with the coastal plain becoming more vulnerable to sea-level rise. Geomorphologic land classification mapping and flood geomorphology are the most important factors for regional planning against flood mitigation in the mega delta. To achieve appropriate prevention works and achieve sustainability, regional planning toward mitigation and capacity building must be improved. These concepts require an understanding of the effects of fluvial geomorphology and flooding characteristics on the traditional agricultural system in the delta. This chapter is an overview of a study undertaken to describe the impact of climate change in the Mekong Delta.

9.2 Geomorphologic Land Classification Maps

The geomorphologic land classification maps reveal the micro landforms and the combination of geomorphologic elements. The maps also provide information of the previous phases of flooding and the state of estimated future flooding because the fluvial landforms have been formed by repeated floods over a long time. When an estimate of the future conditions of floods on the fluvial plain is made, geomorphologic land classification maps should demonstrate the inundation phases, flood direction, the most vulnerable point for breaking, and safety from floods, among others.

The typical combination of geomorphologic landform units in alluvial and coastal plains is fan + natural levee + delta. However, the Mekong Delta does not contain an alluvial fan, but it has a wide-area system of rivers and canals on the low relief structure in the plain. This is because the landform surface is inclined at a very small angle of 0.01/1,000. The Mekong River does not transport gravel to the lower reaches; it flows through several intermountain basins such as the Vientiane plain in the Lao People's Democratic Republic (Lao PDR) where it deposits gravel.

The geomorphologic landforms are related to the following water-related hazards:

1. delta and estuary—storm surge, high tide, tsunami, web and tide
2. back swamp and back marsh—deep inundation, long period inundation
3. former river course—floodwater route
4. alluvial fan—devastated flood damage with gravels and sands
5. sand dune—safety against river flood, but trivial damage under storm surge and tsunami
6. sand ridge—shifting tendency under sever river flood
7. natural levee—well drained after flooding
8. tidal creek—suffered by web and tide
9. alluvial terrace (formed in Holocene)—suffered by severe river flood
10. residual hill—gully forming under torrential rainfall
11. pleistocene terrace—terrace slope suffered by torrential rainfall

In view of the physical environment, the landform structure has a close connection between flooding and inundation.

Figure 9.1 shows a geomorphologic land classification map of the Mekong Delta in southern Viet Nam. The typical combination of landform elements comprising the geomorphologic landform units in the Mekong Delta is as follows: large natural levee + small natural levee + delta + sand dune. The geomorphologic elements of the Mekong Delta area comprise residual hills, marine terraces, Holocene lower terraces, natural levees, back swamps, sand dunes, sand ridges with saline lowland dismals, tidal flats with ebb and tide levees, mangrove forests with nipa palms, submerging forests or inundation forests, and former river courses.

The relief structure of the Mekong Delta is monotonous; however, the forefront delta along the South China Sea is occupied by high elevation sand dunes while the western part of the delta is occupied by a tidal flat facing the Gulf of Thailand. A higher natural levee is distributed along the upper delta and the relative height between the natural levee and the back swamp is 4–5 meters (m) in the upper delta, while a lower natural levee with a relative height of less than 1 m is distributed in the lower delta. These characteristics display the inundation and flooding processes in this delta.

9.3 Mekong River Delta and Changing Flood Patterns

The Mekong Delta has suffered severe flooding in the last two decades. The 1996 and 2000 floods were memorable with record levels of flood damage. After the

Figure 9.1 Geomorphologic Land Classification Map of Mekong Delta in Viet Nam

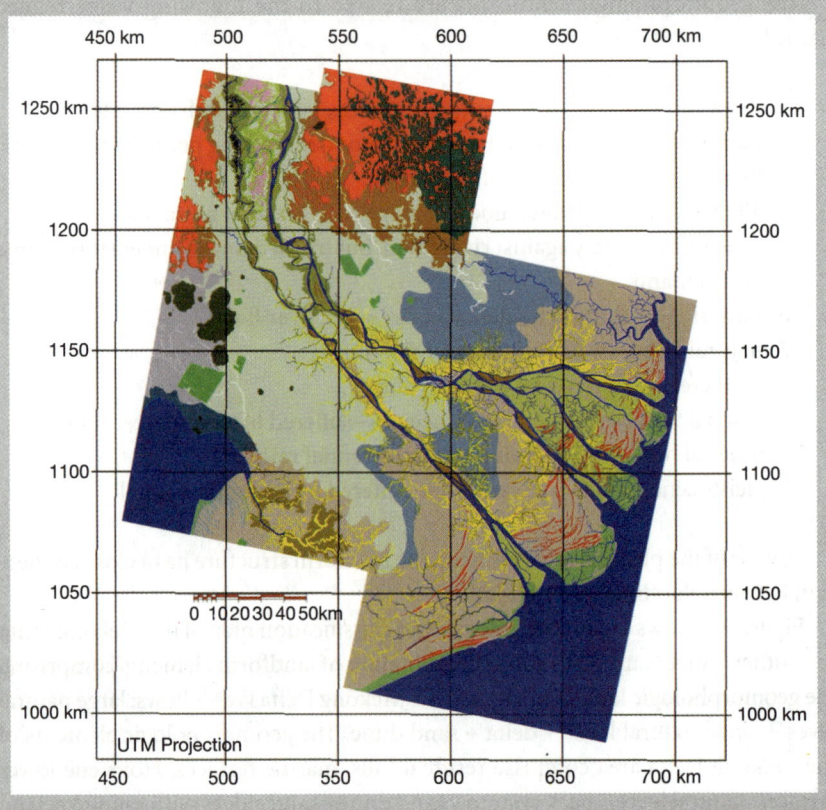

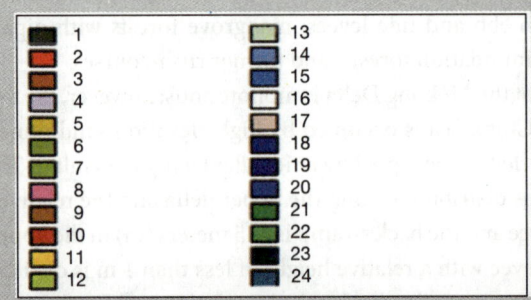

Source: Haruyama and Shida (2008).

Notes: 1 = residual hill; 2 = marine terrace I; 3 = marine terrace II; 4 = alluvial terrace; 5 = natural levee I; 6 = natural levee II; 7 = old natural levee I; 8 = old natural levee II; 9 = sand bar; 10 = sand dune; 11 = tidal canal levee; 12 = barrier; 13 = back swamp I; 14 = back swamp II; 15 = former river course; 16 = unclear former river course; 17 = chenier coastal plain; 18 = main Mekong River course; 19 = tidal creek; 20 = saline swamp; 21 = inundation forest; 22 = mangrove forest; 23 = artificial canal deposit; 24 = permanent swamp.

1996 flood, infrastructure was built on the flood plain in Viet Nam. Following that, flood conditions have changed because of the transformation of hydrologic conditions.

In conducting flood damage analysis in the last 10 years, the study evaluated three types of damage: *(i)* overflow: in the upper reaches—the inundation depth is the greatest and diffusion speed is high; *(ii)* land side water inundation: in the middle reaches—the flooding is gentle and exhibits long stagnation; and *(iii)* tidal effect and saline intrusion: in the lowest reaches—flooding cases such as typhoons and high tides, and the flood depth is low; however, this zone is not greatly affected by river flooding in the Mekong Delta because of the marine process for landform formation.

By comparing the time series of JERS 1 SAR images (satellite data with synthetic aperture radar) in the geomorphologic land classification map of the Mekong Delta, flooding propagation can be monitored (Haruyama and Shida 2008). When flooding usually begins on the flood plain in June it is distributed in the depths of the back swamps, and the flood flowing from Cambodia propagates to the lower reaches around the natural levee in August. The flood spreads and covers the surface of the delta, overflowing the natural levee in September; however, roads and artificial piled lands remain safe. The delta front and the sand dune complex zone are unaffected by the river flood in September. The river flood stops at the central part of the floodplain because the above point is located on a significant boundary that connects the sand dune complex zone and the delta.

The relationship between the micro-geomorphology (Figure 9.1) and the flooding features is as follows:

1. the longest inundation flood zone exists for more than three months in the deeper back swamp, namely, back swamp I
2. the shallow depth inundation area lies in back swamp II and here, the inundation period is less than one month
3. natural level I and II experience overflow but exhibit rapidly reducing flooding in the monsoon season
4. the coastal plain with the sand dunes and the lowland between the sand ridges is unaffected by the overflowing from the main Mekong River; however, it experiences tidal effects, storm surges, and typhoons
5. the length of the inundation period in the lowlands between the sand ridges is less than 1 month, and the inundation depth is less than 50 centimeter (cm)
6. the eastern part of the Mekong Delta, which is a former river course, is a flood flow course in the rainy season
7. the urbanized area experiences flooding; however, the inundation depth is less than 50 cm
8. back swamp II, located in the central part of the delta, experiences inundation for one month

In contrast, human activity such as infrastructure construction has transformed flood characteristics in the Mekong Delta. Hard infrastructure construction includes building facilities of convenience for living, agriculture, and industry. But hard infrastructure has transformed the characteristics of flooding, causing enormous floods and long periods of inundation. The flood flow trend in the Mekong Delta is undergoing extreme variations. A tendency is found in increasing maximum water levels, especially accelerated downstream, but maximum water levels at main observation stations of the Mekong River have decreased (Figure 9.2). The infrastructure development in the lower reaches causes variation in the lower Mekong. The higher embankments and strengthened dikes and roads have become obstructions to flood drainage, thus increasing flood peaks. Human activity has also influenced the inflow to and outflow from the Mekong Delta, by *(i)* decreasing the amount of overbank flow across the border between Cambodia and Viet Nam, *(ii)* increasing the flow in canals, and *(iii)* decreasing drainage to the west Mekong region while increasing the amount drained to the Gulf of Thailand.

In order to understand in greater detail the mechanism of flood propagation and inundation, numerical models should be applied to simulate flood propagation, and comparisons among scenarios must be carried out.

The flood phase of the Mekong River Delta was simulated for future environmental changes such as climate change (Figure 9.3). The low lying fluvial plain will be affected by deep inundation and seawater intrusion as far as 85 kilometers (km) inland.

9.4 Adaptation and Mitigation

What kind of technology is needed for mitigation in a huge low relief fluvial plain and thus in international river basins? The Mekong River Commission has been discussing programs of the Mekong River basin such as development, water utilization, and conservation of the natural environment. Flood management and mitigation are the most important programs in the lower Mekong River Basin (Wicelgartner and Haruyama 2005).

There are very few flood observation data available whereas adaptation and mitigation programs need more exact evidence including: *(i)* transnational sharing of catchment-related information and data; *(ii)* guidance on methods of collecting and analyzing data for monitoring and evaluating risk reduction activities; *(iii)* identification of appropriate indicators for the many different forms that flood reduction can take; *(iv)* implementation of a disaster information system where all relevant data obtained from remote sensing, at gauging stations, in the field, and from statistics or maps is entered; *(v)* ensuring that agricultural practices are in harmony with the geophysical systems to avoid aggravate flood hazards; and *(vi)* preparation of more reliable flood probability information.

Figure
9.2

Historical Records of the Floods at Selected Locations

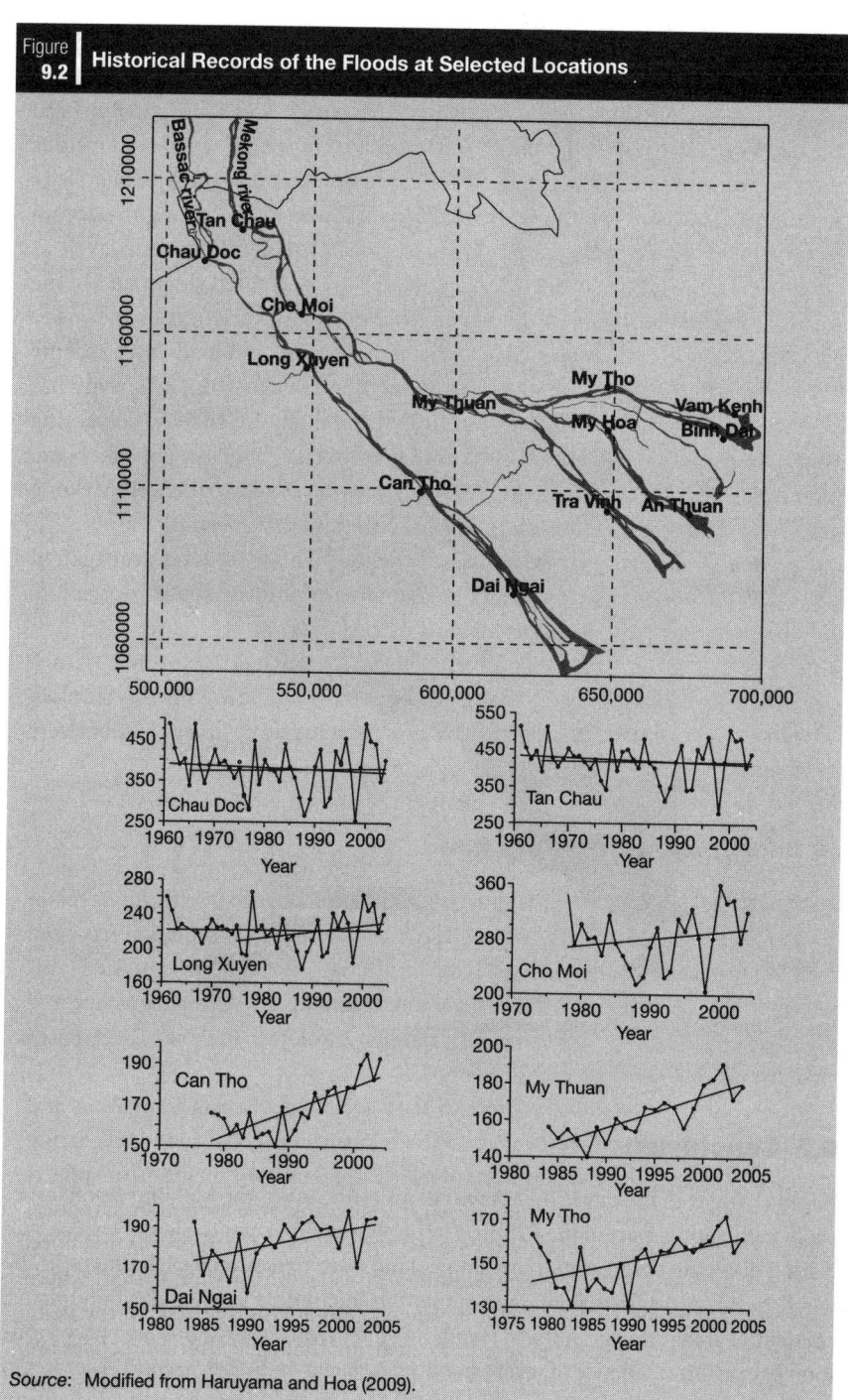

Source: Modified from Haruyama and Hoa (2009).

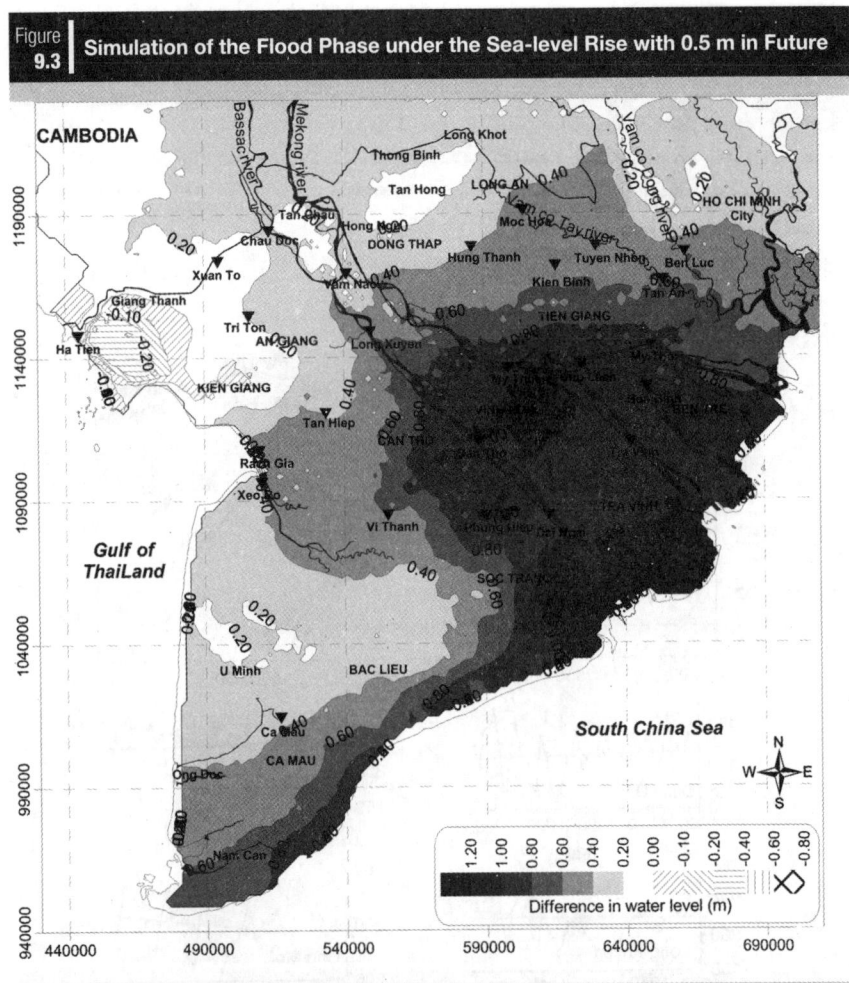

Figure 9.3 Simulation of the Flood Phase under the Sea-level Rise with 0.5 m in Future

Source: Modified from Le Thi Viet Hoa et al. (2008).

9.5 Conclusion

Rapid changes to the land cover have been significant in the Mekong Basin, especially in the upper part in the People's Republic of China and the lower part in Viet Nam. Following deforestation on sloped land, water resource development in the whole basin, and agricultural development in the middle reaches and floodplains, residential areas and agricultural farmland suffer from severe flooding every monsoon season. The evidence of severe flooding is traced in the long period observation data and remote-sensing data analysis in the study area. However, the dimension of flooding now depends on the floodplain landform and landform reformation by human activity. The exact information of the natural environmental structure, natural environmental change, and natural environmental dynamism should be known for better flood forecasting. To decide on the policy of mitigation and adaptation toward future natural hazards for regional planning and for common people living with floods, more exact estimates, simulation, and evaluation for decision making will be required. The future for new projects for flood monitoring of monsoonal Asia, evaluation of nature and extent of risks and vulnerability, organizing knowledge networking, supportive policies, standards, regulations, and design guidance should be considered.

References

Haruyama, S. and K. Shida. 2008. Geomorphologic Land Classification Map of the Mekong Delta Utilizing JERS-1S SAR Images. *Hydrological Processes*. 22 (9). pp. 1373–1381.

Haruyama, S. and Le Thie Viet Hoa. 2009. *HYDRO-GIS: Theory and Lessons from the Vietnamese Delta*. New York: Nova Science Publishers.

Le Thi Viet Hoa, S. Haruyama, Huu Nhan, and Tran Thanh Cong. 2008. Infrastructure Effects on Floods in the Mekong River Delta in Vietnam. *Hydrological Processes*. 22 (9). pp. 1359–1372.

Wicelgartner, J. and S. Haruyama. 2005. Flood Plain Management. *Journal of Rural Planning Association*. 24 (3). pp. 210–214.

Chapter 10

Integrated Approach to Climate Change Impact Assessment on Agricultural Production Systems[1]

Tsugihiro Watanabe

10.1 Introduction

With an expanding and deeper recognition of global climate change, some fundamental questions have been raised, including what impact climate change will have on agricultural production systems, how the systems can adapt to the changes, and what measures should be applied to sustain productivity.

Identifying the direction and dimension of potential impacts on agricultural production systems based on the projection of future regional climate change, is an essential base to adapt to them. The structure and problems of agricultural production systems are elucidated through analyzing the impact of climate change and developing adaptation measures.

As the world population grows and the demand for food increases, agriculture must be more productive—especially in arid and semi-arid areas—while its development is severely restricted by water availability that has been a base for irrigation development resulting in increased food production. On the other hand, in many arid and semi-arid regions, agriculture and irrigation development has caused land degradation and desertification, and created serious problems in the hydrological regime. The changes in agricultural land and water management practices pose serious threats to the sustainability of agriculture itself.

Moreover, future global climate change could provide climatological and hydrological conditions in arid and semi-arid regions with substantial changes in temperature, rainfall, and evapotranspiration, presenting another challenge to the agricultural production system.

Agriculture is a human activity, and works as an interface between human society and nature. To cope with climate and other subsequent changes in natural conditions, humans have adapted to new environments, or taken appropriate measures accordingly. It is timely to reorganize conventional "wisdom" of a region or agriculture so that it is adequate to overcome future global climate change, by utilizing scientific knowledge with state-of-the-art technology.

---

[1] This chapter overviews the research project ICCAP, administered and financially supported by RIHN of Japan and TÜBİTAK of Turkey. This research is also supported financially in part by the JSPS Grant-in-Aid No.16380164 and No. 19208022.

This chapter introduces one challenge in the context above. An integrated assessment system for such ponderous analysis is needed. The Research Institute for Humanity and Nature (RIHN) developed an assessment methodology, aimed at identifying current and future challenges and effective countermeasures against possible climate change impacts.

Agriculture is based on the interaction of human activity with nature including climate change. This relationship is complex and causes various problems if it malfunctions.

The Impact of Climate Change on Agricultural Production System in Arid Areas (ICCAP) research project assessed the impact of global warming on regional climate change and on regional hydrology and agriculture. The direction and dimension of the potential integrated impact of climate change on agricultural production systems have been identified in the Seyhan River Basin, located on the eastern coast of the Mediterranean Sea. The project assessed possible problems by projecting impacts such as rises in temperature, decreases in precipitation, and sea-level rise.

This chapter outlines the project framework and summarizes the outcome, focusing on the challenges of developing a methodology for integrated assessment that is applicable to other agricultural regions in the world.

10.2 Scope and Assessment Framework

The project considered the interaction between human activity and nature by investigating the fundamental structure of land and water management as well as through the projection of abrupt climate changes and impact assessment.

The main objectives of the methodology are:

1. To examine and diagnose the structure of land and water management in agricultural production systems in arid areas, especially to evaluate quantitatively the relationship between cropping systems and the hydrological cycle and water balance in farmland and its environs.
2. To develop a methodology or model for integrated assessment on impacts of climate change and adaptations on agricultural production systems, mainly on the aspect of land and water management.
3. To assist the development and improvement of the Regional Climate Model (RCM) for more accurate prediction with higher resolution of future changes in regional climate.
4. To assess the vulnerability of agricultural production systems from natural changes and to suggest possible and effective measures for enhancing sustainability of agriculture, through integrated impact and adaptive assessment of climate change.

10.3 Validation and Assessment Method

The research was implemented on the east coast of the Mediterranean Sea, mainly in the Seyhan River Basin of Turkey. Firstly, a comprehensive assessment of the basic structure of the agricultural production system was carried out with special reference to regional climate, land and water use, cropping pattern, and irrigation systems. It attempted to predict and evaluate the impacts of future climate change and regional adaptability, and finally through these analyzes, examine in an integrated matter the correlations between changes in nature and human activity.

In this process, regional climate change projections with higher resolution are critical to precise impact assessment. Further, impacts on regional water resources, irrigation and drainage systems, natural vegetation, crop production, farm management, and cropping patterns as well as the effect on the food production and marketing are taken into account. The research aimed at providing suggestions for regional policies and monitoring systems for further climate change impact assessment. The research procedures in the original research plan are shown in Figure 10.1.

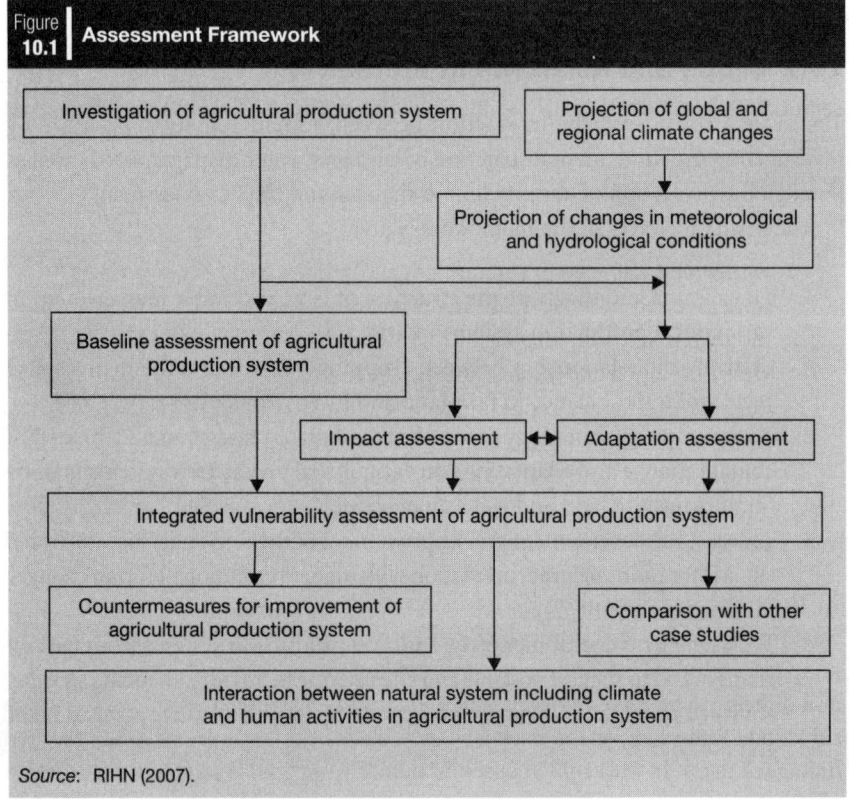

Figure 10.1 | Assessment Framework

Source: RIHN (2007).

10.3.1 Case Study Area

The Seyhan River Basin (Figure 10.2) is dominated by a Mediterranean climate, with most precipitation occurring in winter. Precipitation increases in accordance with elevation but is particularly low in the upper basin, which is dominated by a continental climate. Rain-fed wheat is widespread in the upper hilly areas of the basin. Large-scale irrigated agriculture extends throughout the lower delta, where maize, citrus, cotton, wheat, and vegetables are cultivated. These crops depend on water supplied by reservoirs that store the runoff of winter rain and snow in the upper mountainous areas.

10.4 Approach in Integrated Assessment

Based on the preliminary diagnostic studies on the natural condition of the basin, the study analyzed the present basic structure of the agricultural production system. Simultaneously, future climate change scenarios of the basin in the 2070s were generated by two most advanced general circulation models (GCMs) and RCMs with downscaling methods based on the scenarios A2 of the Special Reports on Emissions Scenarios by the Intergovernmental Panel on Climate Change (IPCC). With generated climate scenarios, particular models developed in this research assessed the impacts of climate change on the regional hydrological regime, natural vegetation, crop productivity, irrigation management, cropping systems, and regional crop production. These assessments proved the basic structure of the present agricultural system and the path of climate change impact on the system, as summarized below:

1. The climate change scenarios for the 2070s of the basin were generated, with which impacts of climate changes on basin hydrology and agriculture were assessed and discussed. The projection of future climate by the climate model has still some uncertainty, while measures for model improvement were developed and applied during the model development stage.
2. The path of climate change impacts on agricultural production in the Seyhan Basin is shown in Figure 10.3 by a framework of associated components, critical factors, and relations. The reciprocal relationships between crops and livestock, pests and diseases, and ecosystem transitions were not direct objects of the assessment.

The combined agricultural production model consisting of some sub-models was not developed and applied, since the basic policy, structure and elements, temporal and spatial resolution of parameters of the sub-models are quite different and difficult to be linked to each other. This research project tried to connect the sub-models implicitly in repeated feedback, avoiding difficult work for the explicit linkage of them. In the implicit combination, the models share the common input

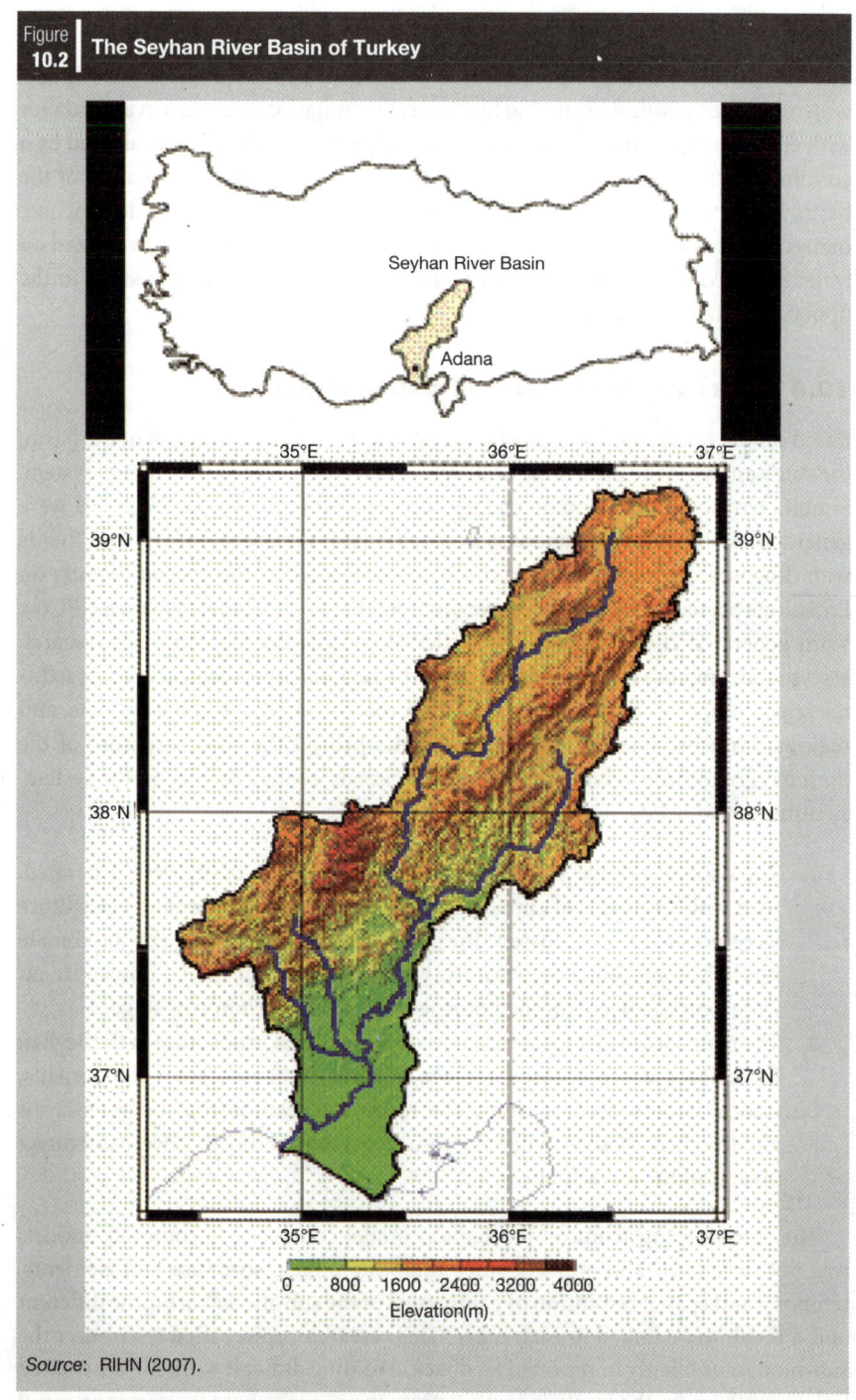

Source: RIHN (2007).

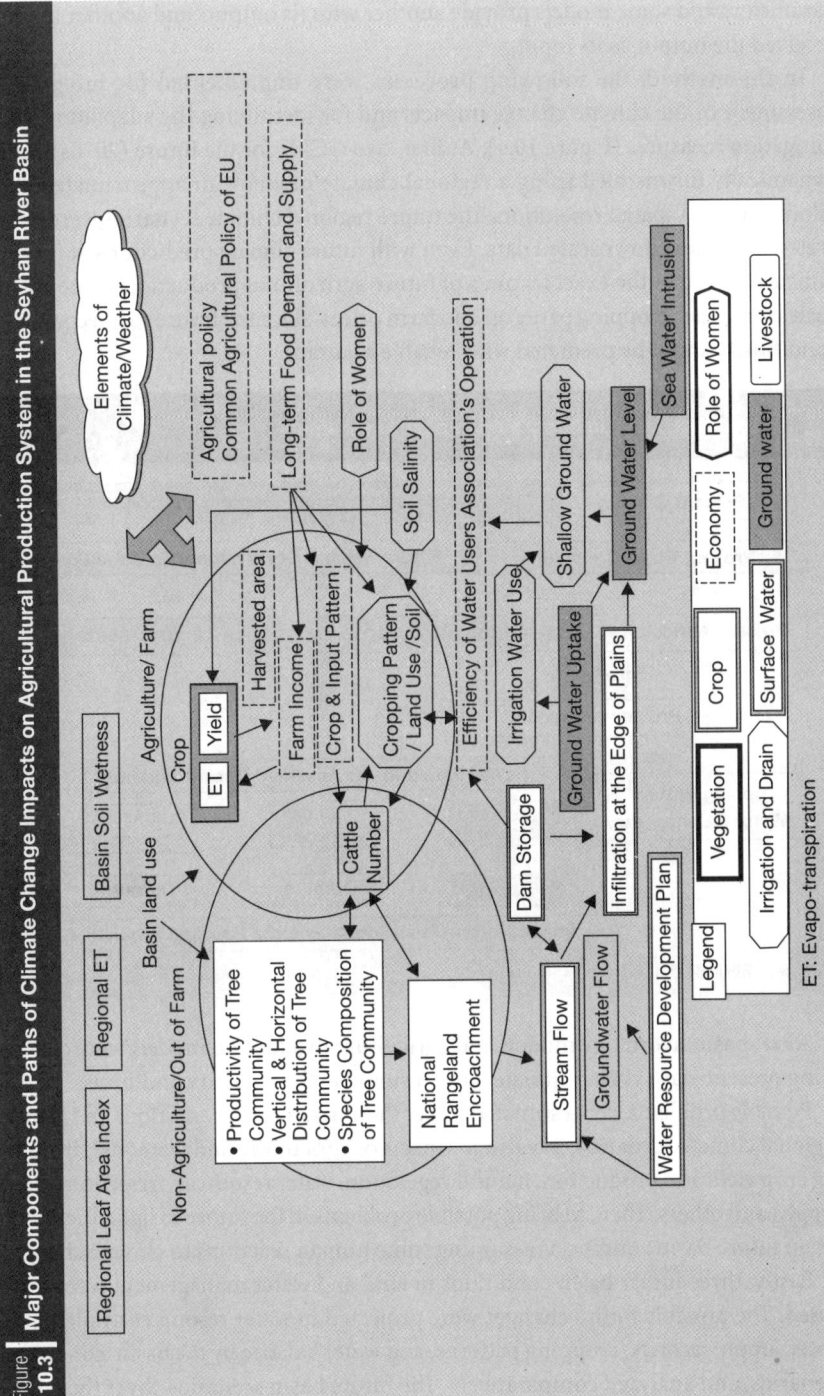

Figure 10.3   Major Components and Paths of Climate Change Impacts on Agricultural Production System in the Seyhan River Basin

Source: RIHN (2007).

parameters and some models provide another with its outputs and another model received the output as its input.

In the method, the following processes were implemented for integrated assessment of the climate change impacts and for identifying the adaptation and mitigation measures (Figure 10.4). At first, two GCMs for the future (2070s) were dynamically downscaled using a regional climate model with approximately 8.3 kilometer (km) spatial resolution. The future regional climate scenarios were generated with these downscaled data. Even with future climate predictions, it is very difficult to predict the exact features of future agricultural production in the basin such as land use, cropping patterns, and farm prices, since the future socioeconomic conditions cannot be predicted with reliable accuracy.

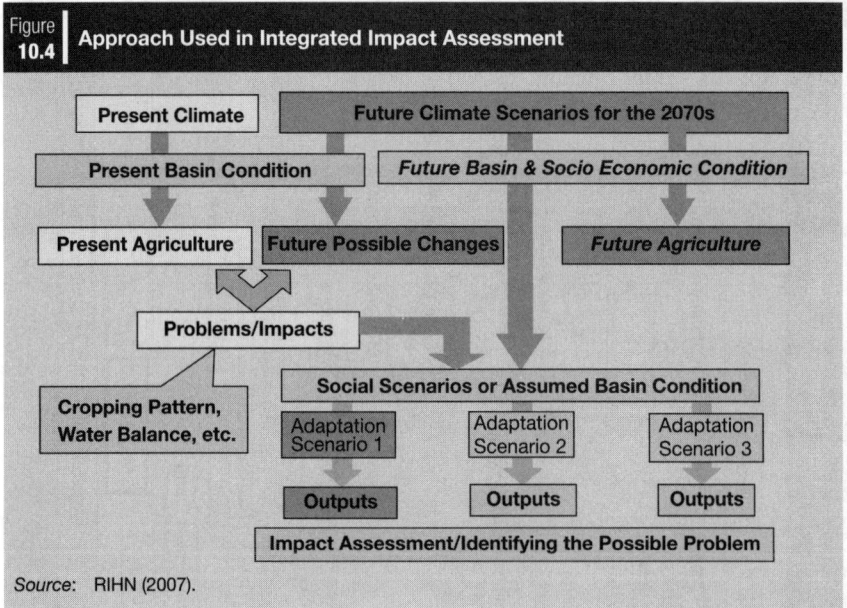

Figure 10.4   Approach Used in Integrated Impact Assessment

Source: RIHN (2007).

After, basin hydrology models and agriculture production models were driven using present-day (1990s) climate data to simulate the current conditions.

Possible problems were estimated under the current basin conditions and future regional climate scenarios, in various aspects of agriculture and water use, including crop yield and production, natural vegetation, water resources, irrigation water supply, and others. Then, to bring possible problems in the future to light, the study set up future basin conditions assuming some human reactions to climate change.

Lastly, three future basin conditions in land and water management were supposed. The possible future changes were projected in water resources availability, water supply security, cropping patterns, and water balance by the basin-condition scenarios, and analyzed comparatively. The future basin scenarios cover the cropping patterns and the land and water use management.

10.5 Implications and Prototype Assessment Method

10.5.1 Future Regional Climate Change

The likely climate change scenarios in the future in Turkey from increased GHGs are estimated for ten years in the 2070s. The outputs of GCMs were downscaled for 10 years' climate during the 2070s in the whole of Turkey with a 25 km grid interval and that in Seyhan with 8.3 km grid interval. The provided climate dataset contains interpolated precipitation, temperature, moisture, and insulation from every observation station in Turkey.

Two independent GCM projections were downscaled by only one RCM. Another GCM with a very high horizontal resolution is used as a reference to assess the reliability of the downscaling done in this research.

Figure 10.5 indicates the downscaling of the model output with an example of the change in winter precipitation of January in the 2070s. The example predicts the possibility of the decreased precipitation in wintertime.

Figure 10.5	Projected Changes in January Precipitation of Turkey in the 2070s

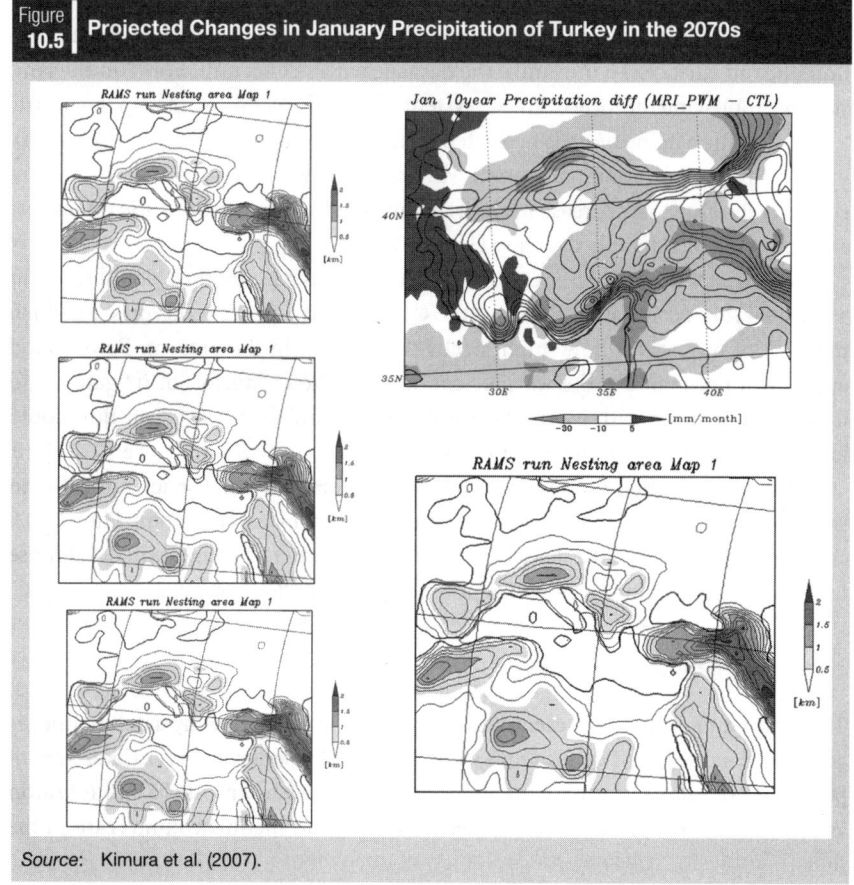

Source: Kimura et al. (2007).

According to the generated scenarios, surface temperature in Turkey may increase by 2.0°C (projected using the GCM developed by the Meteorological Research Institute of Japan [MRI-GCM]) and 3.5°C (projected using the GCM developed by the Center for Climate System Research of University of Tokyo and the National Institute of Environmental Studies of Japan [CCSR/NIES-GCM]). Total precipitation in Turkey may decrease about 20% except in the summer. The projected trend of changes in temperature and precipitation in the Seyhan River Basin is similar to the changes in the whole of Turkey, while there precipitation may decrease about 25% (Figure 10.6).

10.5.2 Hydrology and Water Resources

The direct impact of future sea-level rise on groundwater salinity will not be serious, while increased evaporation and decreased precipitation with sea-level rise could cause significant increase in salinity of the lagoon. Therefore, further groundwater withdrawal may result in saltwater intrusion. Build up of a higher saline zone in the aquifer beneath the lagoon could cause waterlogging on the land surface. Water logging and increased salinity in shallow groundwater may cause salt accumulation on land surface. To minimize the damage with salt accumulation on the land surface, improving the local drainage system is strongly recommended in the future.

Precipitation in the basin is projected to decrease by about 170 millimeter (mm), while evapotranspiration and runoff will decrease by about 40 mm and 110 mm, respectively. Due to a decrease in snowfall and a rise in temperature, the total snow amount will considerably decrease (Figure 10.7).

Compared with the present conditions, decreased precipitation could result in a considerable decrease of inflow to the Çatalan and Seyhan reservoirs, in which the peak of monthly inflow might occur earlier than in the present. Fewer flood events will occur under the warmer conditions. The expansion of irrigated land in the middle basin with increased water demand and decreased river flow could lead to water scarcity for the lower plain of the basin as shown in Figure 10.8. Here, "reliability" is defined as "water supply/water demand," that is an indicator to show how much the demand is satisfied by supply from the reservoirs. Figure 10.8 shows that irrigation in the delta region might face water shortages when water use upstream increases in the case of adaptation scenario no. 1 or no. 2.

10.5.3 Natural Vegetation

The actual and potential present vegetation were estimated using satellite images and field data. Areas of maquis and woodland with broadleaved evergreen trees of potential present vegetation were practically occupied by crop field and *P.bruitai* as secondary forest, respectively. Areas of steppe and maquis will increase in the

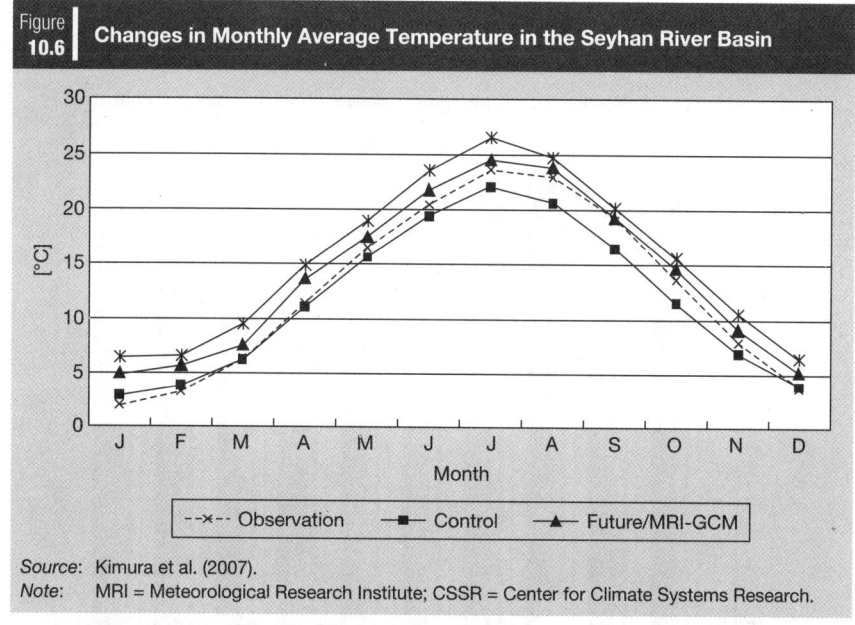

Figure 10.6 | Changes in Monthly Average Temperature in the Seyhan River Basin

Source: Kimura et al. (2007).
Note: MRI = Meteorological Research Institute; CSSR = Center for Climate Systems Research.

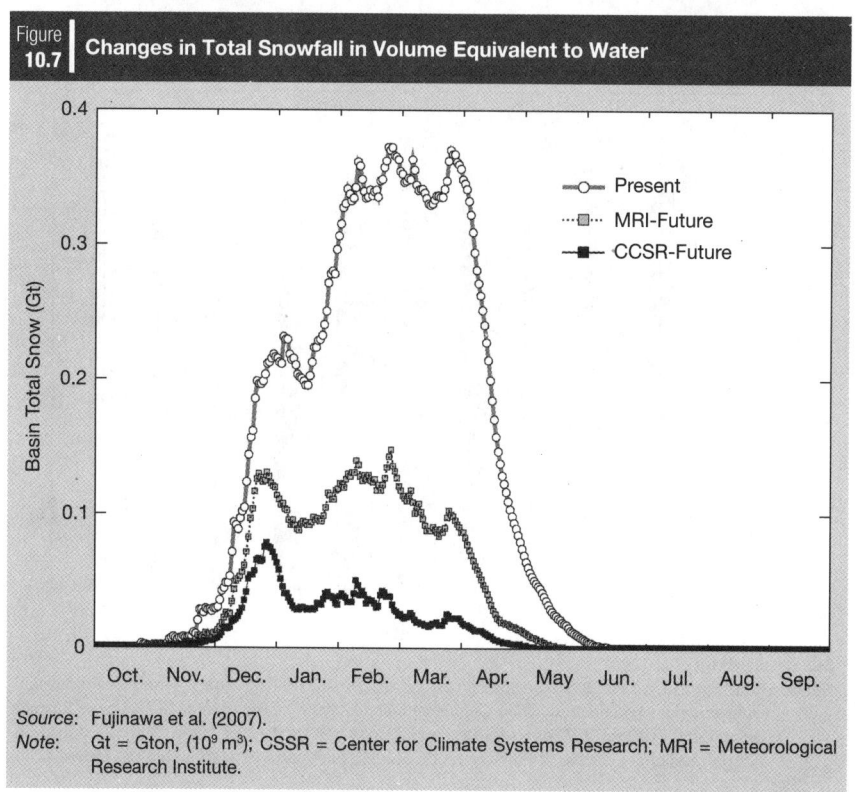

Figure 10.7 | Changes in Total Snowfall in Volume Equivalent to Water

Source: Fujinawa et al. (2007).
Note: Gt = Gton, (10^9 m³); CSSR = Center for Climate Systems Research; MRI = Meteorological
           Research Institute.

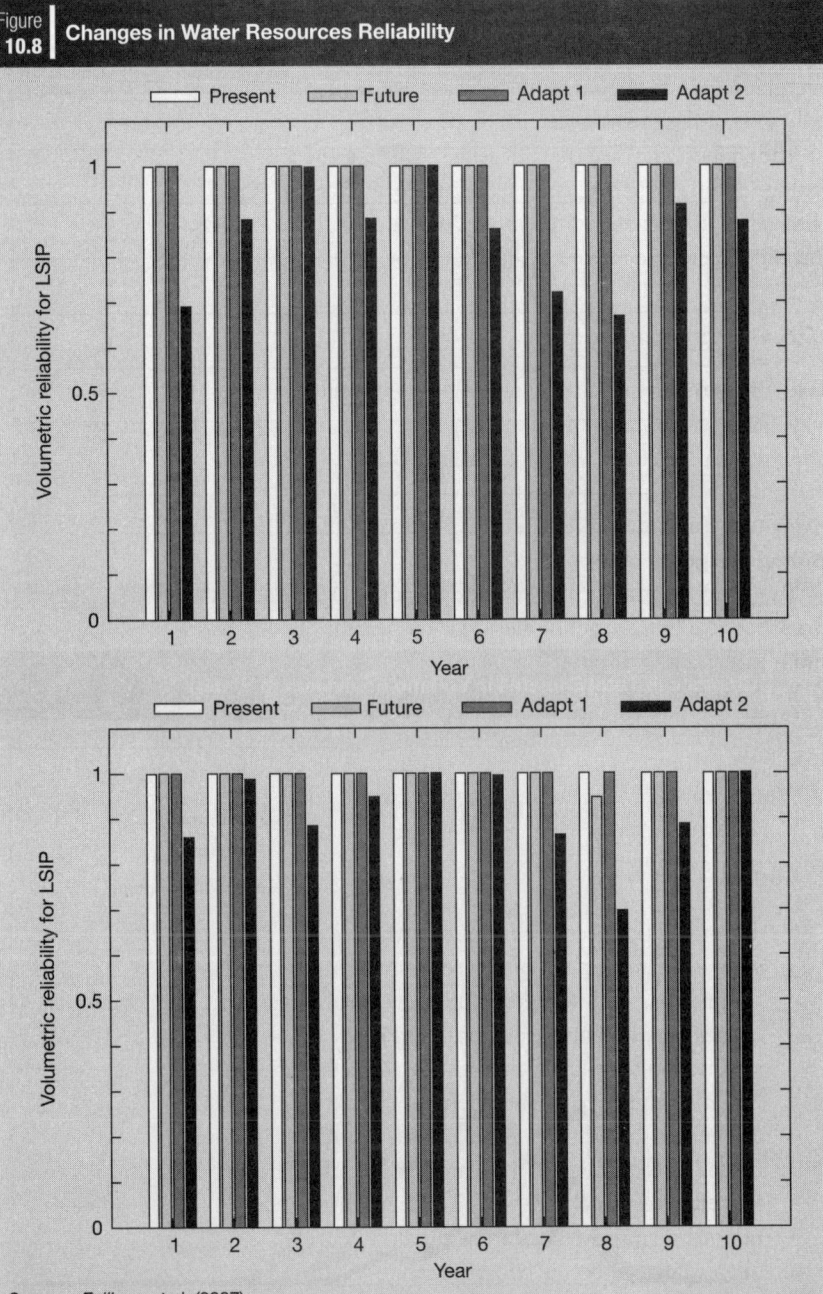

Figure 10.8 | Changes in Water Resources Reliability

Source: Fujihara et al. (2007).
Notes: Top figure by MRI-GCM. Bottom figure by CCSR/NIES-GCM. Adapt 1 and 2: Adaptation scenario no. 1 and no. 2. CCSR = Center for Climate Systems Research; LSIP = Lower Seyhan Irrigation Project; MRI-GCM = Meteorological Research Institute of Japan General Circulation Model; National Institute of Environmental Studies of Japan General Circulation Model.

2070s while those of coniferous evergreen forest will decrease. Biomass of maquis and deciduous broadleaved woodland in the future were increased and coniferous evergreen forest will markedly decrease, while the total biomass in the area will be only 45% of the present one.

The analysis clarified the difference of the area occupied by each vegetation type, and the difference in biomass between the practical and potential present. The future changes in these items are predicted. These outcomes suggest the method to assess the climate change impact on vegetation.

10.5.4 Crop Production

Two crop growth simulation models were developed. The models projected that wheat and maize yields in Adana areas may increase at most by 15% from the current yield in the 2070s with the changed climate in the generated scenarios, while the simplified process model (SimWinc), one of the models, projects that wheat yields that would decrease by 10% if CO_2 concentration is not incorporated in the estimate (Figures 10.9 and 10.10).

The yield estimated by two models suggests that the effect of elevated CO_2 almost offsets the impact of elevated temperature and reduced rainfall on wheat and maize grain yield.

In the future, the models should include accurate estimates for the effects of elevated temperature and water deficit on the harvest index (yield/biomass yield). Further, a sub-model evaluating the effects should be developed.

Wheat growth and yield is one of the main interests attracting public attention in the context of climate change impact on agriculture in Turkey. The findings on this issue are summarized as follows:

1. The projected decrease in precipitation will give negative effects on wheat yield, especially in the plain area of Adana, where total precipitation during the growth period of wheat will decrease to below 500 mm. The amount and intensity of rain at the beginning of the rainy season also may affect wheat production through their effect on the establishment of seedlings.
2. Negative effects of elevated temperature would be expected on wheat yield due to the shortening of the growth duration and some adverse effects on reproductive growth in the plain. On the other hand, increase of temperature will enhance canopy development, resulting in a better yield in the mountainous regions with severe winters.
3. The negative effects of climate change will be at least partly compensated by increased CO_2 in the 2070s.
4. The global warming effects on wheat yield in Adana projected both by the wheat growth model and the economic model are around +13%, while other wheat growth model projects range between +25% and +37%.

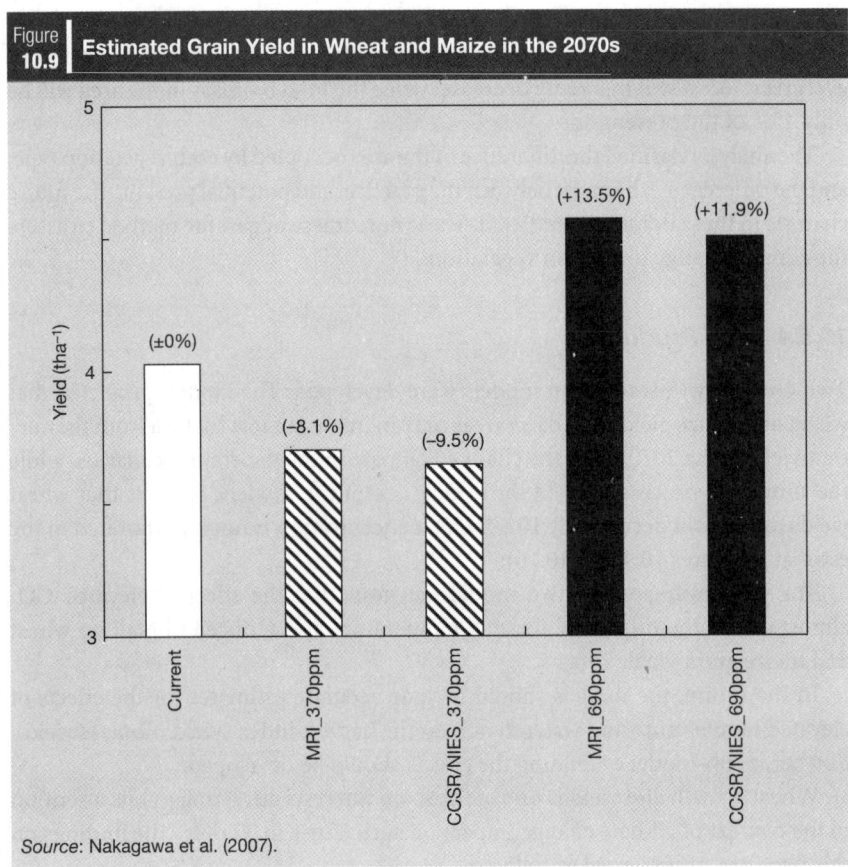

| Figure 10.9 | Estimated Grain Yield in Wheat and Maize in the 2070s |

Source: Nakagawa et al. (2007).

5. Wide spatial variability will be expected in the climate change impacts on wheat yield in Adana Province. Climate change will increase and stabilize the wheat yield in the mountainous area, while it will destabilize the yield in the plain.

10.5.5 Irrigation and Drainage

In the topic of the irrigation and drainage, as the results of the research, typical problems of the present system were identified by visiting and questioning all water users' associations (WUAs) in the area of the Lower Seyhan Irrigation Project (LSIP). Farmers and WUAs had more concerns on recent conflicts over the allocation of water during the peak irrigation season. Through remote sensing, spatial distribution of the land use in the present and in the past were detected and the wheat cultivated area was identified. With field observations, reference water budgets were obtained and the characteristics of the actual irrigation methods were examined.

Figure 10.10	Differences in Changes of Wheat Grain Yield in the 2070s among the Counties in Adana Province

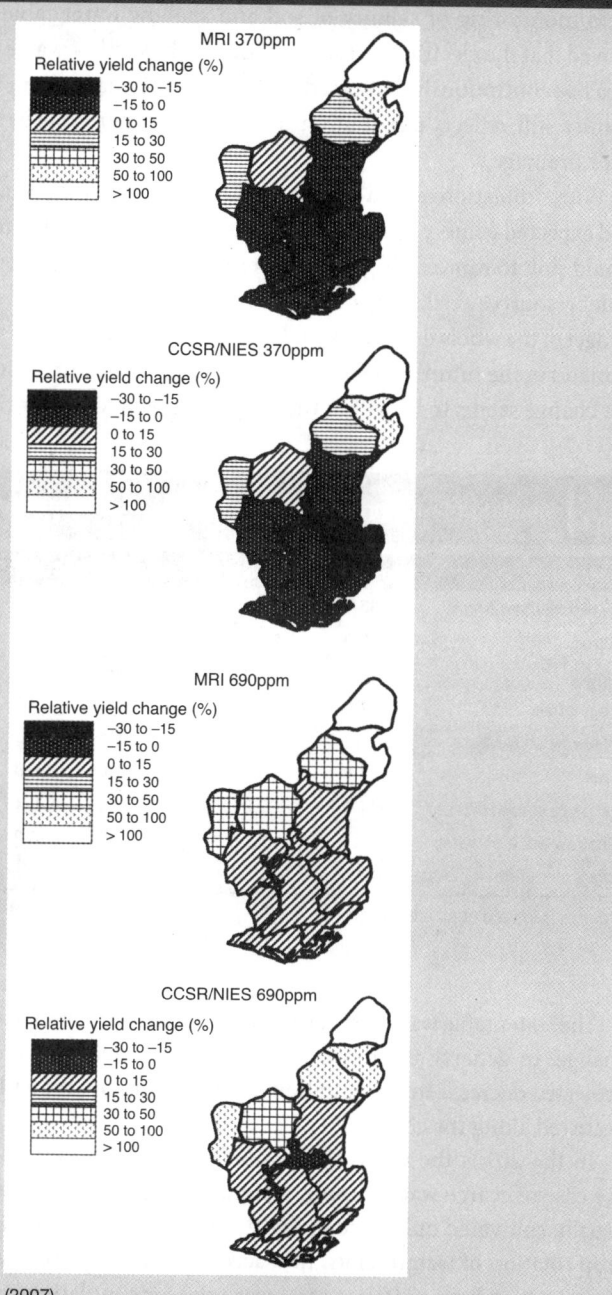

Source: Nakagawa et al. (2007).
Notes: CCSR = Center for Climate Research; MRI-GCM = Meteorological Research Institute of Japan General Circulation Model.

The Irrigation Management Performance Assessment Model (IMPAM) was developed. It was validated by applying it to a small monitored area in the LSIP. Field monitoring of salinity of soil and shallow water table in the coastal area proved that the electric conductivity (EC) of the shallow water table in the irrigated area has continuously decreased over the past 20 years, yet in the coastal area, soil salinity still reflects distribution of a shallow water table as in 1977, suggesting poor drainage.

With simulation of land use changes in the 2070s using pseudo-warming outputs and expected value-variance (E-V) model, it is projected that in the 2070s, land use would shift to more cash-generating crops than the present, even under decreased water resources availability (Table 10.1). Using the IMPAM, crop growth and water budget of the whole delta was simulated, and the results have revealed that irrigation demand in the future will increase due to an extended irrigation period. However, the change seems to be within the range of its adaptive capacity.

Table 10.1	Simulated Cropping Pattern with Climate and Social Scenarios						
Scenario	Base case	MRI-S1	MRI-S2	MRI-S3	CCSR-S1	CCSR-S2	CCSR-S3
Available water (mm)	585	469	429	579	398	330	480
Citrus	22.0	22.1	22.1	21.9	21.9	18.3	21.8
Cotton	59.3	24.0	15.1	48.3	4.3		26.0
Vegetables	7.0	4.4	3.6	6.4	3.0	3.2	4.7
Watermelon & Maize		41.3	51.7	12.9	64.0	78.5	38.8
Fruit	11.6	8.3	7.5	10.4	6.8		8.6
Gross revenue (YTL/da)	717.9	706.9	702.6	715.6	696.4	670.0	707.9
Shadow price of water		0.1	0.1	0.1	0.2	0.1	0.1
Idle water (mm)	23.5						

Source: Umetsu et al. (2007).

The water table was more sensitive to the degree of management than to climate change. In general, the risk of a higher water table seems less possible due to a projected decrease in precipitation and water supply. Water logging only partially occurred along the coast.

In the 2070s the citrus area would remain constant at around 20% and in the case of scarce water supply, watermelons would be grown. Watermelons are usually cultivated only once in five years to avoid replant failure. In order to take crop rotation of watermelons into account, the weighted average of watermelons (one year) and maize (four years) was used for simulation.

10.5.6 Socioeconomics

The econometric analysis estimated the climate change impacts on the production of wheat and barley and the farmers' economy and behavior in the Adana and Konya regions.

Changes in crop yield were predicted with price effect, drought effect, high temperature effect, and CO_2 concentration effect. Changes of the area sown were predicted with price effect and soil moisture effect. According to the predictions, with climate change the wheat and barley yield in Adana will decrease by 18% and 24%, respectively, in the 2070s, while in Konya they will decrease by 48% and 47%, respectively. The larger temperature increase in Konya than Adana may cause this difference. The area sown with wheat and barley in Adana and Konya will decrease slightly. The higher CO_2 will increase the yields of wheat and barley by 18.45%, which was assumed by available references and suggestions of the crop scientists in the research. This assumed value is considerably high for any future changes compared with the predicted extent of changes caused by higher temperatures and decreased rainfall, while this assumption will be modified with further research and information.

Consequently, the total production of wheat in Adana will decrease by 3% and the production of barley will decrease by 13%. On the other hand, in Konya, in the 2070s, the production of both wheat and barley will decrease by 31% and 29%, respectively. The results imply that Turkey could face possible food security problems or food shortages with global warming, since Konya is a representative wheat producing area. Estimates in the case whereby Turkey will gain membership of the European Union (EU) show that the yield decrease of both of wheat and barley will be smaller than the decrease in the case whereby Turkey will not be admitted to the EU.

In the farm surveys, the farm economic situation, rural credit market, rural land tenure problems and their relation to cropping patterns, livestock economy, and other farmers' behavior were studied. The results were used to understand the actual farm situation.

10.6 Conclusion

Although the ICCAP research has made the above preliminary conclusions, predicting future changes caused by global warming is still a difficult undertaking, and in some cases, predicting future agricultural production in a specific place and year, like in one river basin in the 2070s, is considered "impossible." At the moment, future climate change projections are still uncertain and a challenging topic. The response of crops to climate change is also still in the basic study stage, even for a major staple crop like wheat.

If the phenomena or factors associated with climate change and its apparent impact are difficult to appraise, how can humans respond or react? There are problems of natural events that are difficult to simulate or examine quantitatively in the laboratory or by computer. Likewise, the impacts of human activity in a natural system, such as land reclamation or irrigation development, cannot be evaluated precisely in advance even though a substantial knowledge base is available.

One of the more effective and feasible measures for such a dilemma is to take actions incrementally, as in a trial-and-error manner, utilizing the best available current knowledge and past experience, and collecting additional information as needed. In pursuing such an adaptive approach, stakeholders should participate in the decisions and actions taken incrementally. For adaptation and mitigation in agriculture against climate change, farmers and their associations or cooperatives, and other organizations interested in climate, water resources, and agriculture need to be involved jointly.

Agriculture has to adapt to changes in nature including climate. With some uncertainties in the future climate, agriculture should be ready to cope with these unpredictable changes, reorganizing or restoring its fundamental functions to work together with available resources in the region, and much more importantly with others in the same community. To identify and utilize the "available resources" is one of the keys for adapting to climate change.

References

Fujihara, Y., K. Tanaka, T. Watanabe, and T. Kojiri. 2007. *Assessing the Impact of Climate Change on the Water Resources of the Seyhan River Basin, Turkey.* The Final Report of ICCAP. Research Institute for Humanity and Nature and the Scientific and Technological Research Council of Turkey.

Fujinawa, K., K. Tanaka, Y. Fujihara, and T. Kojiri. 2007. *The Impacts of Climate Change on the Hydrology and Water Resources of the Seyhan River Basin, Turkey.* The Final Report of ICCAP. Research Institute for Humanity and Nature and the Scientific and Technological Research Council of Turkey.

Hoshikawa, K., T. Nagano, T. Kume, and T. Watanabe. 2007. *Evaluation of Impact of Climate Changes on the Lower Seyhan Irrigation Project, Turkey.* The Final Report of ICCAP. Research Institute for Humanity and Nature and the Scientific and Technological Research Council of Turkey.

Kimura, F., A. Kitoh, A. Sumi, J. Asanuma, and A. Yatagai. 2007. *Downscaling of the Global Warming Projections to Turkey.* The Final Report of ICCAP. Research Institute for Humanity and Nature and the Scientific and Technological Research Council of Turkey.

Nakagawa, H., T. Kobata, T. Yano, C. Barutçular, M. Koç, K. Tanaka, T. Nagano, Y. Fujihara, K. Hoshikawa, T. Kume, and T. Watanabe. 2007. *Predicting the Impact of Global Warming on Wheat Production in Adana.* The Final Report of ICCAP. Research

Institute for Humanity and Nature and the Scientific and Technological Research Council of Turkey.

Research Institute for Humanity and Nature (RIHN). 2007. *Final Report: The Research Project—Impact of Climate Change on Agricultural Production System in Arid Areas* (ICCAP). Japan: RIHN. http://www.chikyu.ac.jp/iccap/finalreport.htm

Umetsu, C., K. Palanisami, Z. Coşkun, S. Donma, T. Nagano, Y. Fujihara, and K. Tanaka. 2007. *Climate Change and Alternative Cropping Patterns in Lower Seyhan Irrigation Project: A Regional Simulation Analysis with MRI-GCM and CSSR-GCM.* The Final Report of ICCAP. Research Institute for Humanity and Nature and the Scientific and Technological Research Council of Turkey.

Chapter 11

Adaptation in Urban Settings: Asian Experiences

Jostacio M. Lapitan

11.1 Introduction

Climate change will affect the health of half of the world's population living in cities. Irrespective of global warming, cities alter their local climate, particularly from reducing rainfall and increasing night temperatures (World Meteorological Organization [WMO] 1996). The "urban heat island" effect is caused by daytime heat being retained by the concrete fabric of buildings and by a reduction in cooling vegetation. In temperate-latitude cities, this has the effect of raising nighttime temperatures by 1°C–5°C. In tropical cities, the mean monthly urban heat island intensities can reach 1°C by the end of the night, especially during the dry season (WMO 1996).

Cities are a significant source of greenhouse gas (GHG) emissions and have an important role to play in both mitigation and adaptation, where they will also benefit the most. Cities are recognized as major players in "carbon-free economic growth" and paying greater attention to building codes and urban transportation is expected to contribute increasingly to climate change mitigation. The other side of the challenge is adaptation as uncertain climate conditions and the expected rise in sea level put coastal cities at high risk.

The research agenda of the World Health Organization (WHO) Centre for Health Development (WHO Kobe Centre [WKC]) on climate change and health in urban settings is shaped by a World Health Assembly (WHA) resolution (WHA 61.19) and by its geographic position and status within WHO, its historical roots, and its track record as a research center.

11.2 Key Area and Focus

In 2004–2005, the center undertook a process of consultation with its partners and the scientific community to gain perspective on its future work for 2006–2015. An ad hoc Research Advisory Group (RAG) and three sub-groups were convened to delineate the most important research questions related to several driving forces:

1. environmental change and technological innovation;
2. aging and demographic change; and
3. urbanization.

The product of this process was *A Proposed Research Framework for the WKC* (WHO Kobe Centre 2004). This framework served as the basis of the 10-year extension of the Memorandum of Understanding and development of the center's research plans and reflected consequently in the 2006–2007 and 2008–2009 work-plans.

A key message of the center's research framework is that health is essential to development. Without good health, development cannot take place. Aging and demographic change, urbanization, environmental change, and technological innovation can bring positive and negative impacts on health. To achieve development goals, health must first be protected and this is possible through more responsive health and welfare systems. The approach should be holistic and consider the social, political, environmental, and economic conditions where people live, work, learn, and play.

The concept of "health in development" emphasizes that health is central to social and economic development and vice-versa. The interrelatedness of health and development is not limited to developing countries alone: developed countries also face serious challenges to health from advances in social and economic development. The RAG suggested analyzing the complexity and relatedness of driving forces and looking for practical solutions to this complex situation.

With this in mind, the group proposed the following vision statement for the center: *Healthier People in Healthier Environments.* The center adopted this vision statement to highlight the need for healthy populations and communities that are living in healthy social, political, economic, and ecological systems.

11.3 What Are the Health Risks Posed by Climate Change?

Humanity is at a pivotal moment in history. The 21st century is unprecedented in its importance to the future of human life on earth. The year 2007 marked the first time in history when 50% or more of the world's population lives in urban settings. Also, 2007 was the year in which the science on climate change became unequivocal: the earth is warming, as verified by the fourth assessment report of the Intergovernmental Panel on Climate Change (IPCC 2007). Projections for rapid urbanization and increasing health impacts of accelerating climate change point to a perfect storm for public health. Climate change and its health impacts on vulnerable populations in urban settings are burning issues for health, environment, and development workers and disaster risk reducers alike.

As a serious emerging threat to global public health, WHO has emphasized that the health risks and extreme weather events posed by climate change are significant, distributed across the globe and difficult to reverse. Chapter 8 of the IPCC Working Group II report, "Impacts, Adaptation and Vulnerability," has summarized the emerging evidence of climate change effects on human health and the projected trends in climate change related exposures of importance to human health in concise summary tables (Confalonieri et al. 2007). The United Nations Environment Programme (UNEP), WHO, and the World Meteorological Organization (WMO) have underscored that climate change impacts on health include increased health-related mortality and morbidity; decreased cold-related mortality; greater frequency of infectious disease epidemics following floods and storms and substantial health effects following displacement from sea-level rise and increased storm activity (McMichael et al. 2003). Crucially, the effects of global warming are highly inequitable as the greatest risks are to the poorest most vulnerable populations who have contributed least to GHG emissions (Campbell-Lendrum and Corvalan 2007).

Margaret Chan, Director-General, WHO, in her World Health Day statement on April 7, 2008 emphasized, "The core concern is succinctly stated: climate change endangers health in fundamental ways" (WHO 2008c). She announced increased WHO efforts to respond to challenges and a research agenda with partners to obtain better estimates of the scale and nature of health vulnerability and to identify strategies and tools for health protection.

11.4 Who Is at Risk?

Climate change will affect everybody but not in the same way. Geography, health system preparedness, health status, age, social class, and support systems will make a difference. Specific populations differ in vulnerability. As growing and developing beings, children are at risk. In the last few decades, the prevalence of allergies among children has increased in Europe as a consequence of an earlier onset of the spring pollen season. Older persons are more vulnerable to heat stress.

Geographically, populations most at risk are the rural poor and those living in megacities, mountain areas, water-stressed areas, and coastal areas. Knowledge of which groups or geographical areas are most vulnerable to the health impacts of climate change allows health systems to target interventions appropriately, in collaboration with other sectors.

The IPCC has assessed that the global mean temperature is likely to rise by 1.4°C to 5.8°C between 1990 and 2100 (McMichael and Githeko 2001). A WHO quantitative assessment, taking into account only a subset of the possible health impacts, concluded that climate change since the mid-1970s may have caused a net increase of over 150,000 deaths in 2000. This has an important health equity dimension, with the largest health risks posed to children in the poorest communities who have contributed least to emissions (Patz et al. 2005).

The WHO Commission on Social Determinants of Health recognized the urgency of dealing with climate change as disruption and depletion of natural environmental systems, including climate change, has profound implications for the way of life of people globally (WHO 2008a). Adaptation in cities needs the attention of both developing and developed countries.

Ono found out by using ambulance records in an unpublished study that *(i)* the number of heat disorder cases in 18 cities in Japan has been increasing with relatively higher incidence in males and among the elderly and school children; and *(ii)* the significant places for heat disorder cases are homes for the elderly population and sports facilities for schoolchildren and adolescents (Ono 2009).

The WHO Kobe Centre has so far completed two case studies. In one study (WHO 2009a), the impacts of heat on work ability were considered as a chronic condition with a quantifiable link to climate change. The impact on work ability, assuming that working people reduce their work output per hour during hot hours in accordance with the international standard for work in heat (ISO 1989) can be evaluated. When the calculation uses the Wet Bulb Globe Temperature (WBGT) outdoors in the sun, the loss of ability in a heavy labor job (500W) becomes extreme, and in fact during the middle of the day no work of this type can be carried out. For a person working with less physical demand (200W), the work ability losses are less, but still near 0% to 18% in the middle of the day in May in the megacity of New Delhi, India.

The other case study by the WHO Kobe Centre (WHO 2009b) is about city health system preparedness for changes in dengue fever reasonably attributable to climate change. There are no peer-reviewed papers looking specifically at this issue. The objectives of the study were to: *(i)* review what research has been carried out on assessing the preparedness of health plans to changes in dengue fever; and *(ii)* develop a toolkit to assess preparedness. Thus, a toolkit was developed to serve as a prototype in the literature for potential use by other city researchers. The case study conducted in Bangkok attempted to identify national-capital city health system preparedness challenges for effective response to dengue fever.

11.5 What Can Countries and Cities Do?

In May 2008, the 193 member states represented at WHA adopted a new resolution on health protection from climate change, signaling a much higher level of engagement from the health sector. Resolution WHA 61.19 (Appendix 11A) draws attention to further strengthening of the evidence for human-induced climate change, and consequent risks to global health, the achievement of the Millennium Development Goals (MDGs), and health equity.

WHA 61.19 resolution calls on WHO to strengthen its work in raising awareness of the health implications of climate change, collaborating with other agencies within and outside the United Nations (UN), supporting capacity building and research in health protection from climate change, and requesting the organization to consult further on developing WHO support to countries.

WHA 61.19 resolution calls for support for proactive management of the threats that climate change poses to health. In order to support member states, the resolution requested WHO:

> to continue close cooperation with appropriate UN organizations, other agencies and funding bodies, and member states, to develop capacity to assess the risks from climate change for human health and to implement effective response measures, by promoting further research and pilot projects in this area, including work on:
>
> - health vulnerability to climate change and the scale and nature thereof;
> - health protection strategies and measures relating to climate change and their effectiveness, including cost-effectiveness;
> - the health impacts of potential adaptation and mitigation measures in other sectors such as water resources, land use, and transport, in particular where these could have positive benefits for health protection;
> - decision-support and other tools, such as surveillance and monitoring, for assessing vulnerability and health impacts and targeting measures appropriately; and
> - assessment of the likely financial costs and other resources necessary for health protection from climate change.

Urban settings are considered strategic as cities and municipalities are key to slowing climate change: most emissions are generated from producing the goods and services used by urban consumers. Moreover, urban areas in low- and middle-income countries have a large and growing proportion of the world's population most at risk from emergencies and disasters brought about or intensified by climate change related storms, floods, heat waves, and freshwater shortages. A recent paper by David Satterthwaite reviewed data drawn from the fourth assessment report of IPCC, and suggested that the contribution of cities to global anthropogenic GHG emissions is likely to be less than half of all anthropogenic GHG emissions (Satterthwaite 2008).

One of the main entry points for engaging cities on climate change and health is disaster risk reduction. Local climate change impacts will progressively be felt through an increase in severity and frequency of emergencies and disasters in the wake of storms, floods, sea-level rise, and heat waves.

Interventions will also need to consider climate change risks that vary across cities depending on factors such as economic development, geographical location, and/or population density. A worldwide study funded by the United States National Science Foundation shows that approximately 600 million people currently live in low elevation coastal zones, less than 10 meters above sea level (McGranahan, Balk, and Anderson 2007). To adapt to climate change impacts, there is a need to project where demographic growth will occur, estimate climate change risks, and discourage growth in projected climatic hotspots.

Cities need local governments with the political will and the capacity to act. There is an increasing need for interaction and cooperation between decision makers and multidisciplinary researchers at the local level.

11.6 Defining and Promoting Health Co-benefits

Co-benefits refer to the multiple and varying benefits derived from a single strategy, policy, or action plan (Castillo 2006). Measures that seek to address more than one issue simultaneously (that is, promoting health and cutting GHG emissions in this case) are likely to have widespread support, especially from city officials. Many of the measures that would reduce emissions such as shifting to cleaner energy sources could bring important health co-benefits to individuals and communities.

Measures bringing co-benefits have drawn increasing attention in recent years given the need for a more cost-effective use of scarce natural, financial, technical, and human resources, and the imminent approach of what scientists call the "tipping point" beyond which climate change can no longer be mitigated. Due to the integration of a range of issues and potential positive impacts, such measures are also known as "co-control measures," "no-regret strategies," or "synergistic measures."

The health co-benefits approach can potentially serve as a policy guide, given a means of determining and quantifying associated costs and benefits.

Although not yet backed by a wide evidence base to vouch for its success as an approach, co-benefits, at least in theory, are regarded as a win-win strategy (Castillo 2006).

The GHG mitigation measures that lead to improved air quality have multiple benefits—better air quality and adaptation to climate change, both in developing and industrialized countries (WHO 2008c). Shifting from less than five kilometers of personal daily car travel to a more active means of transport would deliver significant benefits on obesity and its related negative health effects (for example, heart disease and diabetes) and costs. Nevertheless, negative tradeoffs can emerge in the quest for energy efficiency, for example, some people living in highly insulated new zero-energy pilot buildings have complained of headaches and asthma (WHO 2008b).

Some of the recommendations for a way forward include *(i)* further conceptual clarification and consensus on co-benefits; *(ii)* generation of knowledge through research; and *(iii)* systematic documentation of co-benefits-related research along with capacity building, policy dialogue, and networking.

11.7   Conclusion

Mainstreaming climate change adaptation and mitigation in city development planning is a feasible target requiring intersectoral action and resulting to health co-benefits. Cities have an important role to play in both adaptation and mitigation where they also stand to reap the biggest benefits. The role of a research organization such as the WHO Kobe Centre is to generate and build up the evidence base for informed policymaking.

Appendix 11A

SIXTY-FIRST WORLD HEALTH ASSEMBLY WHA 61.19

Agenda item 11.11
24 May 2008

Climate Change and Health

The Sixty-first World Health Assembly,

Having considered the report on climate change and health;

Recalling resolution WHA51.29 on the protection of human health from risks related to climate change and stratospheric ozone depletion and acknowledging and welcoming the work carried out so far by WHO in pursuit of it;

Recognizing that, in the interim, the scientific evidence of the effect of the increase in atmospheric greenhouse gases, and of the potential consequences for human health, has considerably improved;

Noting with concern the recent findings of the Intergovernmental Panel on Climate Change that the effects of temperature increases on some aspects of human health are already being observed; that the net global effect of projected climate change on human health is expected to be negative, especially in developing countries, small island developing States and vulnerable local communities which have the least capacity to prepare for and adapt to such change, and that exposure to projected climate change could affect the health status of millions of people, through increases in malnutrition, in death, disease and injury due to extreme weather events, in the burden of diarrhoeal disease, in the frequency of cardiorespiratory diseases, and through altered distribution of some infectious disease vectors;

Noting further that climate change could jeopardize achievement of the Millennium Development Goals, including the health-related Goals, and under-mine the efforts of the Secretariat and Member States to improve public health and reduce health inequalities globally;

Recognizing the importance of addressing in a timely fashion the health impacts resulting from climate change due to the cumulative effects of emissions of green-house gases, and further recognizing that solutions to the health impacts of climate change should be seen as a joint responsibility of all States and that developed countries should assist developing countries in this regard;

Recognizing the need to assist Member States in assessing the implications of climate change for health and health systems in their country, in identifying appropriate and comprehensive strategies and measures for addressing these implications, in building capacity in the health sector to do so and in working with government and nongovernmental partners to raise awareness of the health impacts of climate change in their country and take action to address them;

Further recognizing that strengthening health systems to enable them to deal with both gradual changes and sudden shocks is a fundamental priority in terms of addressing the direct and indirect effects of climate change for health,

1. REQUESTS the Director-General:

1. to continue to draw to the attention of the public and policymakers the serious risk of climate change to global health and to the achievement of the health-related Millennium Development Goals, and to work with FAO, WMO, UNDP, UNEP, the United Nations Framework Convention on Climate Change secretariat, and other appropriate organizations of the United Nations, in the context of United Nations reform initiatives, and with national and international agencies, to ensure that these health impacts and their resource implications are understood and can be taken into account in further developing national and international responses to climate change;

2. to engage actively in the UNFCCC Nairobi Work Programme on Impacts, Vulnerability and Adaptation to Climate Change, in order to ensure its relevance to the health sector, and to keep Member States informed about the work program in order to facilitate their participation in it as appropriate and access to the benefits of its outputs;

3. to work on promoting consideration of the health impacts of climate change by the relevant United Nations bodies in order to help developing countries to address the health impacts of climate change;

4. to continue close cooperation with Member States and appropriate United Nations organizations, other agencies and funding bodies in order to develop capacity to assess the risks from climate change for human health and to implement effective response measures, by promoting further research and pilot projects in this area, including work on:

    (a) health vulnerability to climate change and the scale and nature thereof;

    (b) health protection strategies and measures relating to climate change and their effectiveness, including cost-effectiveness;

    (c) the health impacts of potential adaptation and mitigation measures in other sectors such as marine life, water resources, land use, and transport, in particular where these could have positive benefits for health protection;

(d) decision-support and other tools, such as surveillance and monitoring, for assessing vulnerability and health impacts and targeting measures appropriately;

(e) assessment of the likely financial costs and other resources necessary for health protection from climate change;

5. to consult Member States on the preparation of a workplan for scaling up WHO's technical support to Member States for assessing and addressing the implications of climate change for health and health systems, including practical tools and methodologies and mechanisms for facilitating exchange of information and best practice and coordination between Member States, and to present a draft workplan to the Executive Board at its 124th session;

2. *URGES Member States:*

1. to develop health measures and integrate them into plans for adaptation to climate change as appropriate;

2. to build the capacity of public health leaders to be proactive in providing technical guidance on health issues, be competent in developing and implementing strategies for addressing the effects of, and adapting to, climate change, and show leadership in supporting the necessary rapid and comprehensive action;

3. to strengthen the capacity of health systems for monitoring and minimizing the public health impacts of climate change through adequate preventive measures, preparedness, timely response and effective management of natural disasters;

4. to promote effective engagement of the health sector and its collaboration with all related sectors, agencies and key partners at national and global levels in order to reduce the current and projected health risks from climate change;

5. to express commitment to meeting the challenges posed to human health by climate change, and to provide clear directions for planning actions and investments at the national level in order to address the health effects of climate changes.

Eighth plenary meeting,
24 May 2008 A61/VR/8

References

Campbell-Lendrum, D. and C. Corvalan. 2007. Climate Change and Developing-country Cities: Implications for Environmental Health and Equity. *Journal of Urban Health: Bulletin of the New York Academy of Medicine.* 84 (1). pp. i109–i117.

Castillo, C. K. 2006. *A Synthesis of Co-benefits Discussions at the Better Air Quality (BAQ) Conference*. Manila: Clean Air Initiative-Asia.

Confalonieri, U., B. Menne, R. Akhtar, K. L. Ebi, M. Hauengue, R. S. Kovats, B. Revich, and A. Woodward. 2007. Human Health. In M. L. Parry, O. F. Canziani, J. P. Palutikof, P. J. van der Linden, and C. E. Hanson, eds. *Climate Change 2007: Impacts, Adaptation and Vulnerability—Contribution of Working Group II to the Fourth Assessment Report of the Intergovernmental Panel on Climate Change*, Chapter 8, pp. 391–432. Cambridge, UK: Cambridge University Press.

Intergovernmental Panel on Climate Change (IPCC). 2007. *Climate Change 2007: The Physical Science Basis*. New York, US: Cambridge University Press.

McGranahan, G., D. Balk, and B. Anderson. 2007. The Rising Tide: Assessing the Risks of Climate Change and Human Settlements in Low Elevation Coastal Zones. *Environment and Urbanization*. 19 (1). pp. 17–37.

McMichael, A. J., and A. Githeko. 2001. Human Health. In J. J. McCathy, O. F. Canziani, N. A. Leary, D. J. Dokken, and K. S. White, eds. *Climate Change 2001: Impacts, Adaptation and Vulnerability*, pp. 451–485. Cambridge: Cambridge University Press.

McMichael, A. J., D. Campbell-Lendrum, C. F. Corvalan, K. L. Ebi, A. Githeko, J. D. Scheraga, and A. Woodward. 2003. *Climate Change and Human Health: Risks and Responses*. Paris: WHO/WMO/UNEP.

Ono, M. 2009. Heatstroke in Japan: Current Situation and Its Determinants. National Institute of Environmental Sciences, Tsukuba, Japan. Unpublished.

Patz, J. A., D. Campbell-Lendrum, T. Holloway, and J. A. Foley. 2005. Impact of Regional Climate Change on Human Health. *Nature*. 438 (7066). pp. 310–317.

Satterthwaite, D. 2008. Cities' Contribution to Global Warming: Notes on the Allocation of Greenhouse Gas Emissions. *Environment and Urbanization*. 20 (2) October. pp. 539–549.

World Health Organization (WHO). 2008a. *Closing the Gap in a Generation: Health Equity through Action on the Social Determinants of Health*. WHO Commission on Social Determinants of Health Final Report. Geneva: WHO.

———. 2008b. *Draft Meeting Report: First Meeting of the Senior Representatives in Charge of Organizing the Health Systems Response to Climate Change 7–9 April 2008*. Bonn, Germany: WHO Regional Office for Europe.

———. 2008c. *Protecting Health from Climate Change*. World Health Day. Geneva: WHO.

———. 2009a. *Climate Change Exposures, Chronic Diseases and Mental Health in Urban Populations: A Threat to Health Security, Particularly for the Poor and Disadvantaged*. Kobe, Japan: WHO Centre for Health Development.

———. 2009b. *City Health System Preparedness for Changes in Dengue Fever Attributable to Climate Change: An Exploratory Case Study of Bangkok*. Kobe, Japan: WHO Centre for Health Development.

WHO Kobe Centre. 2004. *Health in Development: Healthier People in Healthier Environments—A Proposed Research Framework for the WHO Centre for Health Development*. Kobe, Japan: WHO Centre for Health Development.

World Meteorological Organization (WMO). 1996. *Climate and Urban Development*. No. 844. Geneva: WMO.

PART III

Successful Programs and Measures of Adaptation in Vulnerable Areas

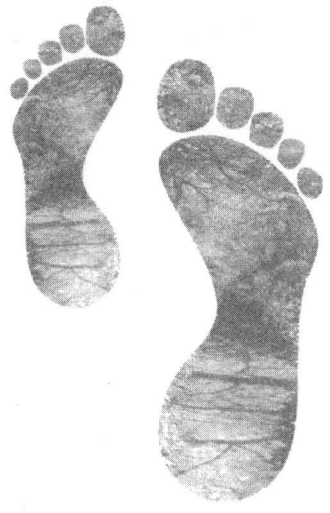

PART III

Successful Programs and Measures
of Adaptation in Vulnerable Areas

Key Messages

The Asia and Pacific region is giving serious consideration to early warnings systems to counter fatalities and flood damage. The regional flood forecasting ability is gradually improving, but local flood forecasting remains a big challenge. The maintenance of ground-based measurement stations is a particular challenge. Recently, advanced methods to introduce early warning systems by remotely sensed precipitation data in combination with a digital elevation model were introduced. Appropriate warnings could be given well ahead in time without the establishment of local climate and flood measurement stations. The social aspect of flood warning systems needs further improvement. For example, formulating emergency action plans involving all concerned groups remain to be established for the most flood prone areas.

Sector-specific insurance models based on the principles of public–private sector partnerships are evolving in the region. Financial relief to climate change damage could be provided by insurance companies in Asia and the Pacific. So far this possibility is underexploited, in particular, in the economically disadvantaged parts of the region, where many poor people cannot pay the premiums for catastrophic damage. In these cases, conditional cash-transfer schemes such as subsidies to the premiums, should be paid by governments until appropriate insurance schemes are established. Weather derivates are agreements where farmers get paid. For example, people are compensated if a previously defined period is too dry, or if an extreme precipitation level or a disastrous wind velocity is exceeded. A current problem in introducing insurance based options is the lack of a dense climate station network in wide areas of the region, which eludes the valuable measurement of climate-induced damage.

Community-based approaches considerably reduce the degree of local vulnerability to climate change by bringing appropriate awareness to local people. Examples of such approaches are the successful lowering of river erosion in Bangladesh, avoiding expected desertification in Mongolia, and introducing renewable energy systems in Indonesia. The projects used a number of innovative pro-poor technologies in agriculture, natural resource use, and disaster management that have since been replicated outside the project implementation areas. The people involved received economic incentives that led to improved household-level consumption, sales of farm products, higher food security, as well as better management of basic health, water, sanitation, and livestock diseases.

Including gender dimensions in adaptation planning improves the equity and efficiency of adaptation measures. Gender issues are of great importance to climate change adaptation at the local level, in disaster prevention, and in disaster relief. Neighborhood associations in Japan provided an example where the role of women in the community saved many lives during earthquakes. Empowering women to key positions in emergency action planning is one efficient measure to improve local readiness, avoid fatalities, and property loss. The UN Manila Declaration for Global Action on Gender, Climate Change and Disaster Risk Reduction of 2008 is a useful document to mobilize action in the Asia and Pacific region.

Chapter 12

Flood Disasters and Warning Systems in Northern Thailand

Thada Sukhapunnaphan and Taikan Oki

12.1 Introduction

Climate induced and natural hazards such as floods, landslides, and debris flow have increased in Thailand in recent years. These disasters have caused both economic and life loss in many areas, especially in the northern provinces which consist of slope and mountainous upstream areas (Teerarungsigul, Chonglakmani, and Kuehn 2007). Humidity brought from the ocean onto the land by the southwest monsoon during the wet season is the main source of flooding, while tropical storms and depression troughs with high intensive rainfall over a widespread area are the triggers of these phenomena. Meanwhile factors such as deforestation, encroachment of the upstream area for settlement, and cropping including the extended settlement into the vulnerable part of the urban area resulted by the population growth. In both the mountains and plains, infrastructure development such as road construction, bridge piers, dams and weirs, could become obstructions for runoff drainage after a storm event.

12.2 Types of Floods in Upper Northern Region

Most of the flash floods, landslides, and debris flow tend to occur in the slopes and foothill areas while overbank flow inundation frequently occurs on the flood plains and lowland. Flooding in the northern region can be classified as two main types described below.

12.2.1 Overbank Flow Inundation

Normally, flooding on the plains or lowland is overbank flow inundation type, causing economic losses but very few of casualties. A noticeable change about this flood type during the last 10 years is its frequency and increasing severity especially in the vulnerable parts of urban areas.

12.2.2 Flash Floods

Flash floods mostly occur in the slope areas such as foothills and valleys where the runoff is rapid. For the last six years, flash floods have increased both in location and severity. Many parts of the region experienced serious losses. The examples below highlight the high frequency of occurrence of flash floods in the northern region (Figure 12.1). On May 3, 2001, a flash flood occurred on the two adjacent tributaries of the Yom River in the Wang Chin district, downstream of Phrae. Some villages were destroyed and 30 villagers were killed. On August 11, 2001 at Nam Koh and Nam Chun villages in Lom Sak district, Phetchabun Province, debris flows destroyed these two villages within two hours (Yumuang 2006). In that disaster 131 people were killed and 5 went missing. Further, 207 houses were totally destroyed while 515 houses were partially damaged. Approximately 191 acres of agricultural land were damaged. The total loss was estimated at about US$4.7 million (Prachansri 2007). On September 2, 2002 at Ban Kongkha village in Mae Sariang district, Mae Hong Son Province, 10 people were killed and 26 went missing. In all, 202 houses were totally damaged and 121 houses partially damaged including a hospital, 2 churches, and 4 schools. On September 16, 2002, a flash flood at Mae Raek village,

| Figure 12.1 | Location of Flash Floods and Overbank Flow Inundation in Recent Years |

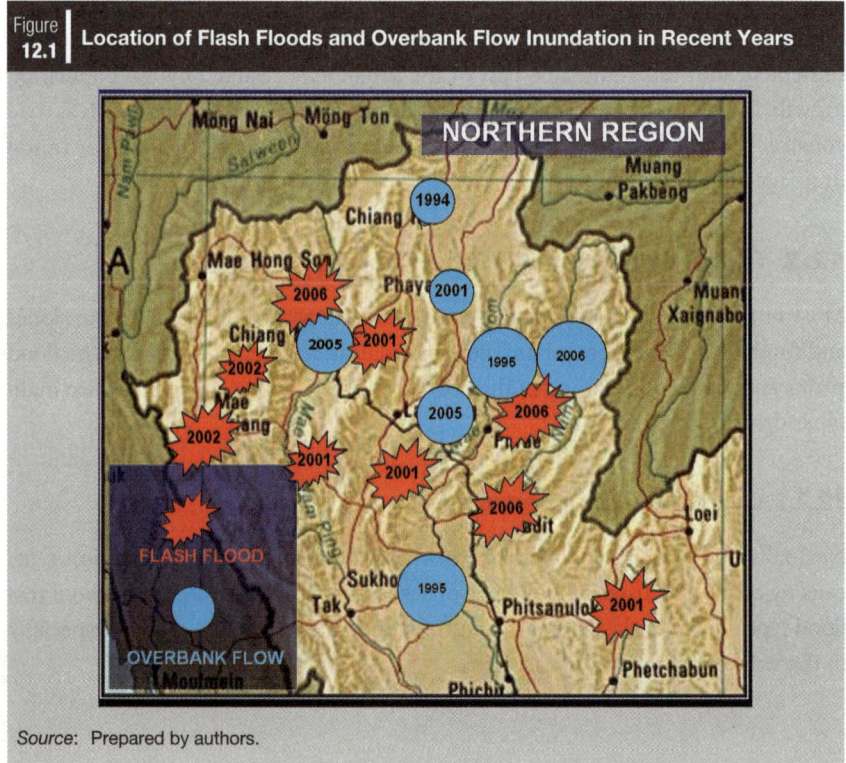

Source: Prepared by authors.

Mae Chaem district, Chiang Mai, totally damaged 12 houses and partially damaged 55 houses (*Health Messenger* 2003). Various organizations collaborated to study the causes and mitigation measures in response to this flood event at Mae Raek. The study concluded that the main causes of the flash flood were related to the geological structure of the catchment, deteriorating forest cover, and land use changes (Thomas, Preechapanya, and Saipothong 2004).

12.3 Factors of Flooding

Land use change is one of the main factors that cause flooding. Growing single crops such as corn, cabbage, or oranges in upstream areas encroaches on water source areas and deforested steep slope areas. These factors favor soil erosion. With deforestation, stumps left in the ground provide immense debris flow that is transported by runoff along the slope and sometimes blocked by obstructions such as roads, bridges, or weirs. The debris accumulates behind the structures until reaching breaking point and then flies out smashing everything in front of them—villages, farms, plants, and cattle. It destroys all property, causes many casualties, and leaves behind a thick mud-covered area. The sediment brought by the flood causes silting and decreases the river capacity, giving way for the next overbank flow to happen. In some cases excavation near the bank makes weak points for the river to cut a new way for it to shortcut its flow.

12.4 Prediction and Warning Based on Rainfall, River Water Level, and River Discharge

12.4.1 Rainfall

Rainfall can be regarded as the main factor that triggers flooding. Related factors are the physical characteristics such as gradient, distance, river meandering, channel friction, surface coverage, and types of land use and structures. However, rain is the first factor to monitor.

12.4.1.1 Periodic Monitoring of Rainfall

- *Monthly*: Six months of wet season are influenced by the southwest monsoon from the Indian Ocean that brings the humidity to the region during May to October. Heavy rainfall normally occurs during the early months of the season but sometimes is reinforced by tropical storms and depression troughs from the South China Sea during July to September causing the possibility of serious floods. These periods of the year are considered for flood monitoring and alert.
- *Weekly*: Weekly weather forecasting, updated weather maps, and animated satellite images by the meteorological department are good sources for monitoring the direction and movement of storms.

- *Daily*: Daily reports from the rain gauge network in the upstream area at 7:00 every morning provide the total volume of rainfall during the past 24 hours. These reports are used for calculating the runoff and discharge of the river. For public understanding, the following criteria can be used if average rainfall in 24 hours is:

  - lower than 35 mm   = safe from flooding
  - 35.1 mm–90 mm   = possibility of flood—stand by and alert
  - higher than 90.1 mm = critical status—warning for preparation (however checking the discharge data is needed for confirmative warning and to classify the scale of flood).

- *Hourly*: The rainfall radar image is updated every hour showing the location, density of rainfall, and its direction of movement. In the case of heavy rain in the upstream area, staff may be stationed at the manual rain gauge sites to report hourly to the head office. The flood and landslide warnings will be given when rainfall is over 100 mm in 24 hours. If rainfall is more than 150 mm, residents must be ready for evacuation, and if over 200 mm evacuation to the safe places and shelters must begin.

Flooding rainfall consists of three main factors:

1. Density: Heavy rain over than 100 mm/24 hours.
2. Time: Continuous and long enough for accumulated runoff.
3. Area coverage: Mostly rain from tropical storms or depression troughs that cover the vast area or the whole basin.

Automatic rain gauge measuring is a convenient way to obtain rainfall data with scheduled interval collection such as every hour or every 15 minutes. The sites are directly connected to the monitoring center so that real time data along with rainfall radar images can be checked. Predictions can be made by analyzing the relationship between both sources to find the volume of current rainfall. Moreover, if the storm path moves along the length of the flood risk area it is possible that the continuous rain will add to the runoff, giving a greater chance to increased size and severity of the flood. This can be monitored by following the rainfall radar image updates and the reports from the rain gauge network.

12.4.2 River Water Levels

The Hydrology and Water Management Center for Thailand's upper northern region has developed correlation equations from the flood levels recorded by the upstream and downstream hydrological stations. The flood travel time or lag time was investigated and the critical upstream water level was defined for announcement of flood warning to the downstream area. The lag time relationship between river water level peaks of upstream and downstream stations can be used for predicting the flood and normally the latter station is located in the flood risk part of the city area. The traveling time of the peak volume from the upstream to the downstream site gives time to alert the public and allow them to prepare themselves for possible flooding (Figure 12.2).

| Figure 12.2 | Time Lag of Peak Water Levels between Upstream and Downstream Stations Used for Flood Early Warning |

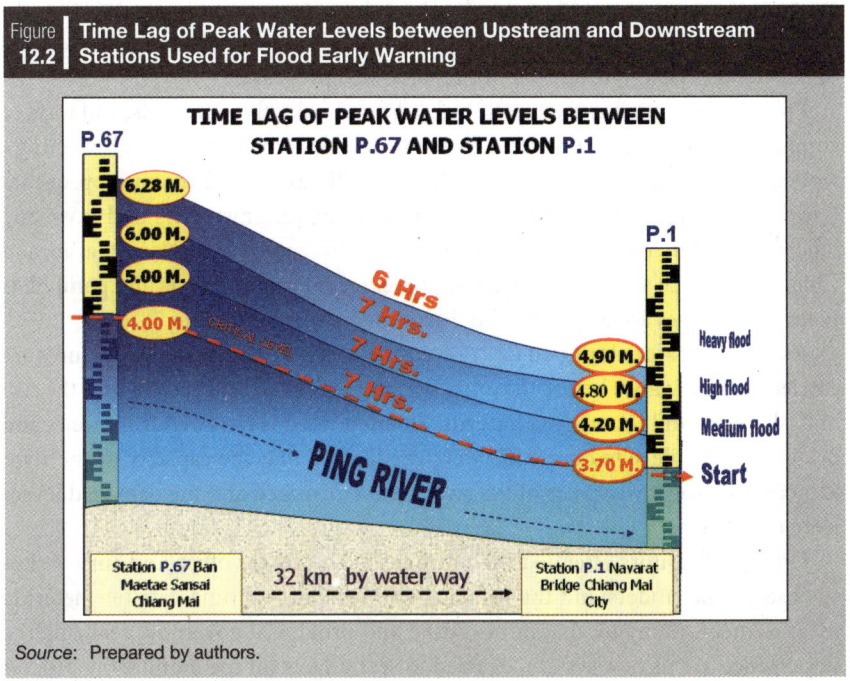

Source: Prepared by authors.

12.4.3 River Discharge

The channel capacity value at the flood risk area by cross section survey can be used for warning. If it is known that the coming volume of discharge from upstream is over the limit of its capacity there will be a probability of overflow bank inundation so the residents must be given a warning.

12.5 Warning Information Dissemination

Residents in flood-risk areas are the main targets for the warnings. For example, in the Chiang Mai city area, Station P.1 at the Nawarat Bridge on the Ping riverside is a good landmark to place an information board and moving signboard for public relations because it is convenient for the people in the flood-risk area to get updated information directly besides through the website (www.hydro-1.net) and other mass media such as radio, television broadcasting, and car announcements by the municipality.

12.5.1 Public Awareness

It is necessary to ensure that all people, not only those with actual experience, are aware of the dangers posed by floods. The public perception of risks and the efficiency of various options that are available will ultimately determine the success of any flood disaster prevention or preparedness program.

Public information and education about flood risks, flood hazards, and ways of coping with flooding are provided for the people by campaigns through meetings, workshops, press, radio and television, especially as a flood season approaches (*Health Messenger* 2003). Meetings with the local administrators and involving offices are held in main cities such as Lamphun, Lampang, Phrae, Nan, Phayao, and Chiang Rai which experienced heavy flood in recent years. Posters and pamphlets in local languages are displayed and distributed at the same time.

As for the strategy for flood loss prevention and mitigation, some measures are carried out in advance before the onset of rainy season each year to reduce the risk of flash floods and inundation hazards. Floating weed is removed from weirs and bridge piers while silted channels were dredged so that the capacity of the river is increased. The lower parts of the river were embanked and the damaged levees were mended.

Groups of visitors attended the meetings at the center office and survey sites to observe and understand the principles and use of real time data telemetering measurement. Many local administrators and provincial governors realizing the importance of this new system, decided to set budgets for the monitoring process by themselves without waiting for assignment from the central government. Flood measurement and monitoring devices such as colored gauges that can be easily built and used by the local people are established.

12.6 Telemetering Research and Pilot Project in Mae Wang Basin

Realizing the importance of increasing flash flood hazards that result in great losses each year, the Hydrology and Water Management Center for Upper Northern Region of the Royal Irrigation Department with support from the University of Tokyo, Japan is undertaking a pilot project of telemetering system research program by installing automatic rain gauge networks at 16 sites (4 super-telemetering sites and 12 rain gauge sites) in the Mae Wang Basin, Mae Wang district of Chiang Mai.

This research aims to process the relationship among factors related to flood such as rainfall, runoff, water level, ground water level, temperature, humidity, evaporation, radiation and wind. The data are real time collected by automatic data logger and directly transferred to the center office by modem connection. Through this project, the hydrology center has developed a measurement and warning system technique from a basic manual to an advanced method with multiple high technology instruments. At the same time, experts from the university have trained the center's staff with new knowledge. With these changes, the hydrology center is coming to a new milestone and hopes that the flood monitoring and early warning system will be more reliable with greater accuracy and efficiency to protect the people from losses by flood disaster. Furthermore, the measurements and process will lead to creating a hydrological model or software that can be generally applied for other areas affected by floods in the future.

12.7 Conclusion

The research and development of the flood monitoring and early warning systems aim to prevent and mitigate the hazards of flood disaster from any losses. People need to be informed with quick, accurate, and reliable information and be able to estimate the scale of flood for preparation and deal with the situation in any stages of pre-flood, during flood, or post-flood without panic. Furthermore the results from the warning system research and development may lead us to find answers about global warming and climate crisis in long-term and other dimensions.

References

Health Messenger. 2003. Issue 20 (June). Burmese edition.

Prachansri, S. 2007. *Analysis of Soil and Land Cover Parameters for Flood Hazard Assessment: A Case Study of the Nam Chun Watershed, Phetchbun, Thailand.* Enschede, The Netherlands: International Institute for Geo-Information Science and Earth Observation.

Teerarungsigul, S., C. Chonglakmani, and F. Kuehn. 2007. *Landslide Prediction Model Using Remote Sensing, GIS and Field Geology: A Case Study of Wang Chin District, Phrae Province, Northern Thailand.* Presented at the GEOTHAI'07 International Conference on Geology of Thailand: Towards Sustainable Development and Sufficiency Economy. November 21–25. Bangkok.

Thomas, D. E., P. Preechapanya, and P. Saipothong. 2004. *Landscape Agroforestry in Northern Thailand: Impacts of Changing Land Use in an Upper Tributary Watershed of Montane Mainland Southeast Asia. Synthesis Report: 1996–2004.* http://www.worldagroforestrycentre.org/water/downloads/Watershed%20Publications/Thomas-et-al-2004-ASB-Thailand%5B1%5D.pdf

Yumuang, S. 2006. 2001 Debris Flow and Debris Flood in Nam Ko Area, Phetchabun Province, Central Thailand. *Environmental Geology.* 51. pp. 545–564.

Chapter 13

Integrated Flood Analysis System: An Efficient Tool to Implement Flood Forecasting and Warning Systems

Kazuhiko Fukami, Tomonobu Sugiara, Seishi Nabesaka, Go Ozawa, Jun Magome, and Takahiro Kawakami [1]

13.1 Introduction

Water-related disasters are one of the challenges that need to be overcome to achieve sustainable development and poverty reduction. Their effect is increasing with population growth and concentration and increasing value of assets in flood plains in recent years. In countries where river improvements are not sufficient, smooth evacuation from flooding is important for limiting the loss of life and property. The timing and magnitude of flooding should be clearly identified in advance. Dissemination of risk by hazard maps, among others, and the direction of evacuation by issuing flood forecasts and alerts are necessary. This is very crucial, in particular, in flash floods that often occur in small river basins with the scale less than several tens of thousands square kilometers. In the case of larger river basins, especially in monsoon regions, floods are regarded as annual events, that is, seasonal floods. In these cases, normal flooding is sometimes regarded as a beneficial regular event for water resources including agriculture, fishery, and aquatic ecosystems. A typical example is the Tonle Sap Lake and its surrounding and downstream areas along the Mekong River. Even for those seasonal flood events, mid-term flood forecasting is very important to schedule not only evacuation from flood risk areas but also water-related activities such as planting. It is important to forecast not only the timing the flood arrives beyond some warning level but also the duration of the flood beyond some water level. Such information should be the minimal requirement so that the "living with floods" concept is tolerable.

As described above, flood forecasting and warning are the indispensable triggers to activate any disaster prevention and water-utilization activities. However, in reality, the development of flood forecasting and warning systems in developing countries has not advanced effectively because of a variety of problems such as:

---

[1] The authors thank the Japanese Ministry of Land, Infrastructure, Transport, and Tourism (MLIT) and the Water Resources Agency of Taipei,China for their provision of hydrological data.

1. poor installation of hydrological observation stations and their poor maintenance;
2. lack of telemetry system and real-time hydrological data;
3. lack of historical hydrological databases for flood modeling;
4. lack of geophysical databases for distributed-parameter hydrologic modeling;
5. lack of financial and/or human resources to build up hydrological monitoring/modeling systems and flood forecasting/warning systems; and
6. lack of local communities for flood warning dissemination, risk communication and/or flood fighting.

In order to tackle the first five problems listed above, the International Centre for Water Hazard and Risk Management (ICHARM) under the auspices of UNESCO and hosted by the Public Works Research Institute (PWRI), developed a concise flood-runoff analysis system—Integrated Flood Analysis System (IFAS)—as a tool kit for effective implementation of flood analysis such as flood forecasting and warning systems. The IFAS implements interfaces to input not only ground-based but also satellite-based rainfall data, geographical information system (GIS) functions to create flood-runoff models, multiple default runoff analysis models, and interfaces to display output results in order to establish flood-forecast systems rapidly and effectively. Satellite-based rainfall data supplement the poor network of real-time hydrological monitoring system in a river basin for emergency cases. The IFAS is a very easy tool to handle and therefore expected to supplement poor financial and/or human resources to build up flood-runoff analysis model and flood-forecasting and warning systems. This chapter introduces major features of IFAS, applicability of satellite-based rainfall data for flood forecasting, and applications of the IFAS to real basins.

13.2 Basic Concept of the IFAS

The major design concepts for developing the IFAS are described below (Fukami et al. 2006; Sugiura et al. 2009).

1. *Utilization of satellite-based rainfall as input data*: Real-time satellite-based rainfall information that covers almost the entire world (60N–60S) is provided from the websites of NASA[2], NOAA[3], JAXA[4], among others. The International Flood Network (IFNet) offers easy to understand and

---

[2] NASA website ftp://trmmopen.gsfc.nasa.gov/pub/merged/mergeIRMicro
[3] NOAA website ftp:// ftp.cpc.ncep.noaa.gov/precip/qmorph/30min_8km
[4] JAXA website http://sharaku.eorc.jaxa.jp/GSMaP/index.htm

Figure
13.1 Integrated Flood Analysis System (IFAS)

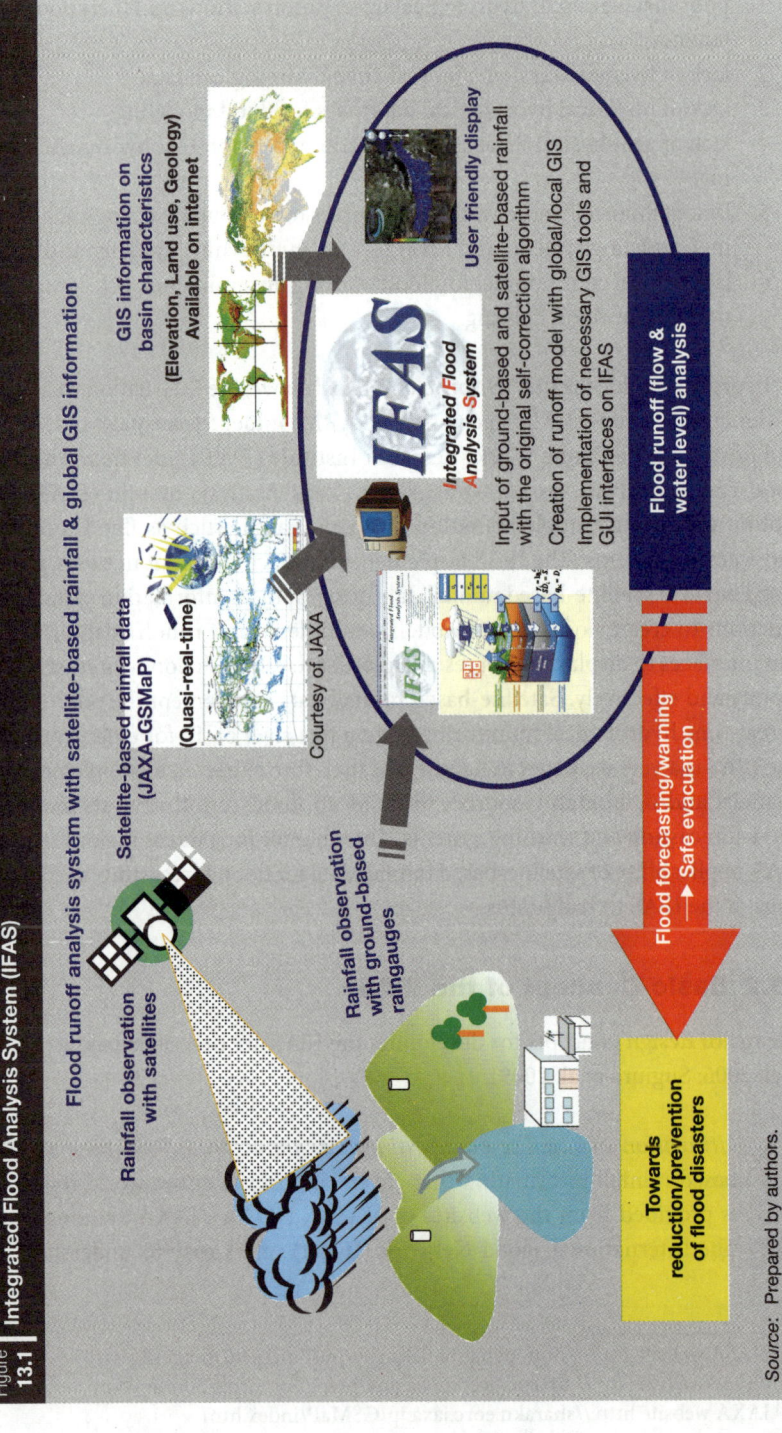

Source: Prepared by authors.

operate interfaces to view and download satellite-based rainfall data (NASA-3B42RT) in text format, that is, Global Flood Alert System (GFAS) for end users.[5] Therefore, satellite-based rainfall data with some time delay are freely available in most areas of the globe. These can be more useful for flood forecasting instead of the very sparse and poorly maintained ground-based rain gauges.

2. *Implementation of multiple precipitation-runoff analysis models*: Two types of runoff analysis models are implemented in the IFAS in advance to cope with a variety of flood characteristics in the world. One model is a conceptual distributed-parameter hydrological model—PWRI-distributed hydrological model (PDHM) Ver.2—and the other one is a physically based distributed-parameter hydrological model (BTOP model). Most of their parameters are related to physical characteristics of the basin such as land use and soil type, which are also globally available in the public domain as GIS data. Guideline parameters are prepared based on past simulation results and therefore, the application can be extended to any poorly gauged river basin as the first approximation.

3. *Implementation of GIS analytical tool for model building and parameter estimation*: The IFAS implements GIS analytical modules in the system to set up the parameters for the flood forecasting/analysis model using GIS databases as described above. Therefore, users do not usually need any help of external GIS analytical software.

4. *Visualization of flood forecasting results*: The IFAS has visual interfaces to display output results to identify a risk of flood easily, including the function to output in the format of Google earth.

5. *Free distribution*: ICHARM distributes the IFAS free of charge, from its website http://www.icharm.pwri.go.jp/research/ifas/index.html

Figure 13.2 shows the main structure and model-building procedure of the IFAS. The IFAS was developed by a cooperative research project (2005–2007) between ICHARM/PWRI, the Infrastructure Development Institute, and major private consultancy companies such as CTI Engineering Co., Ltd, Pacific Consultants Co., Ltd., Nippon Koei Co., Ltd., NEWJEC Inc., CTI Engineering International Co., Ltd., Yachiyo Engineering Co., Ltd., IDEA Consultants, Inc., and Tokyo Kensetsu Consultants Co., Ltd.

13.3 Main Function of the IFAS

The main features and functions of the IFAS are introduced in this section (Fukami et al. 2009).

---

[5] GFAS website http://gfas.internationalfloodnetwork.org/gfas-web/

Figure **13.2** IFAS Main Structure and Functions

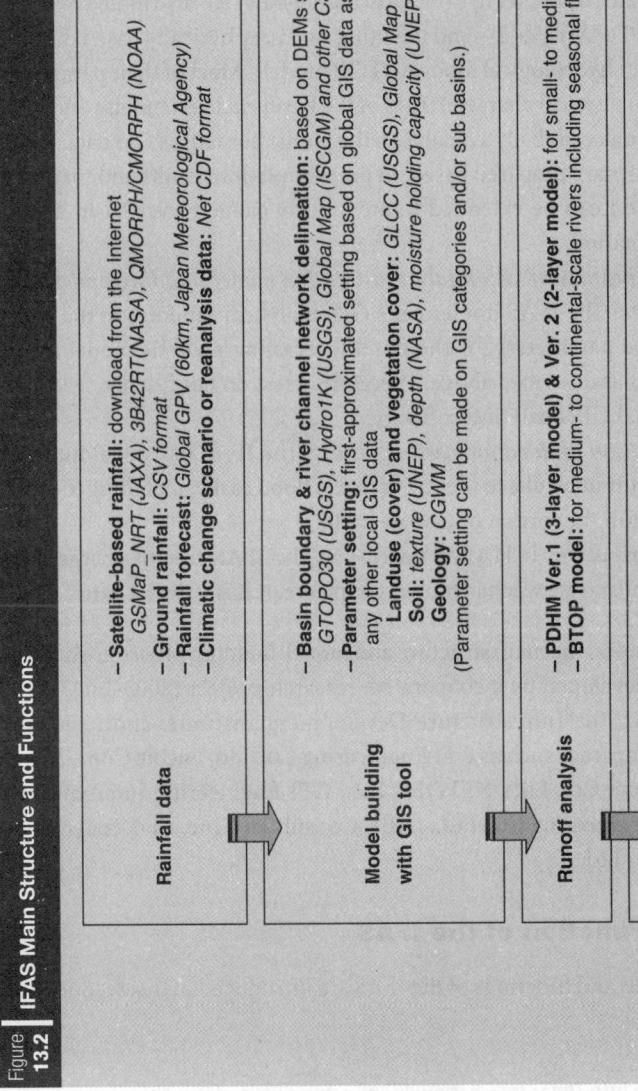

Rainfall data
- **Satellite-based rainfall:** download from the Internet
  GSMaP_NRT (JAXA), 3B42RT(NASA), QMORPH/CMORPH (NOAA)
- **Ground rainfall:** *CSV format*
- **Rainfall forecast:** *Global GPV (60km, Japan Meteorological Agency)*
- **Climatic change scenario or reanalysis data:** *Net CDF format*

Model building with GiS tool
- **Basin boundary & river channel network delineation:** based on DEMs such as
  GTOPO30 (USGS), Hydro1K (USGS), Global Map (ISCGM) and other CSV/shp data
- **Parameter setting:** first-approximated setting based on global GIS data as follows, or
  any other local GIS data
  Landuse (cover) and vegetation cover: *GLCC (USGS), Global Map (IGCGM)*
  Soil: *texture (UNEP), depth (NASA), moisture holding capacity (UNEP)*
  Geology: *CGWM*
  (Parameter setting can be made on GIS categories and/or sub basins.)

Runoff analysis
- **PDHM Ver.1 (3-layer model) & Ver. 2 (2-layer model):** for small- to medium-scale rivers
- **BTOP model:** for medium- to continental-scale rivers including seasonal floods

Visual Presentation
- **Graph outputs:** time series (ex. hydrograph), plan- or layer-view, tables, animation
- **Output to general GiS including Google Earth**

Source: Modified from Sugiura et al. (2009).

13.3.1 Utilization of Satellite-based Rainfall

To do runoff calculation for flood forecasting, it is essential to have real-time rainfall data (or semi–real-time rainfall data). Products of global rainfall information (Table 13.1) observed by earth observation satellites are currently available on the Internet.

Table 13.1	Examples of Satellite-based Rainfall Data Products Available on the Internet			
Product name	3B42RT	CMORPH	QMORPH	GSMaP_NRT
Developer and provider	NASA/GSFC	NOAA/CPC	NOAA/CPC	JAXA/EORC
Coverage		60N~60S		
Spatial resolution	0.25°	0.25°	0.25°	0.1°
Temporal resolution	3 hours	3 hours	30 minutes	1 hour
Delay of delivery	10 hours	15 hours	2.5 hours	4 hours
Coordinate system		WGS		
Data archive	Dec. 1997~	Dec. 2002~	Recent 2 days	Dec. 2007~
Data source	TRMM/TMI	Aqua/AMSR-E		TRMM/TMI
	Aqua/AMSR-E	AMSU-B		Aqua/AMSR-E
	AMSU-B	DMSP/SSM/I		ADEOS-II/AMSR
	DMSP/SSM/I	TRMM/TMI		SSM/I
	IR	IR		IR
				AMSU-B

Source: Modified from Sugiura et al. (2009).

Satellite-based rainfall data have several advantages compared with rainfall information taken from gauges on the ground.

1. As the accuracy of satellite-based rainfall data is never sufficient for flood forecasting purposes on a river-basin scale, ICHARM has developed a "self-correction" method to improve the accuracy of JAXA's GSMaP_nRT. The details are described later. On the other hand, it is important to understand that hydrologists cannot expect highly accurate flood analysis or forecasting without using sufficient ground-based rainfall data. But it is not advisable to wait until implementation of a full network of rain gauges in a river basin. Coupling satellite-based rainfall data with insufficient ground-rainfall data enables hydrologists to start to make flood forecasts with a computer and Internet connection, with a step-by-step gradual improvement of observational network on the ground. Then we will fully understand the importance of in situ hydrological monitoring systems on the ground at the next stage. This understanding is a very important accelerator to install and maintain a good network of hydrologic monitoring systems on the ground.

2. Products of satellite-based rainfall data cover almost the entire world (60N–60S) with relatively uniform accuracy and resolution. The data enable

us to acquire rainfall data of upper stream areas located in other countries. This is an advantage of satellite-based rainfall data for international and transboundary rivers.

3. Since satellite-based rainfall observation started in 1997, past rainfall data can be restored. Such historical data are useful to compare past flood events quantitatively with relatively uniform quality of data and to identify flood risks in a river basin on the basis of past experiences.

13.3.2 Implementation of a Runoff Analysis Engine

The scarcity of hydrological observation facilities means the lack of a historical hydrological database. Parameters of a data-driven rainfall-runoff hydrological model are usually determined by historical hydrological (rainfall and river discharge) data. Therefore, ICHARM considered the utilization of distributed hydrological models as pre-installed precipitation runoff analysis models, the parameters of which can be set on the basis of empirical correlation with geophysical conditions as the first approximation.

One of the runoff models is the PWRI-distributed hydrological model (PDHM) with two versions: Ver.1 and Ver.2. The PDHM Ver.2 (Figure 13.3) (Suzuki and Terakawa 1996) with a two-layer model for flood runoff analysis, in particular,

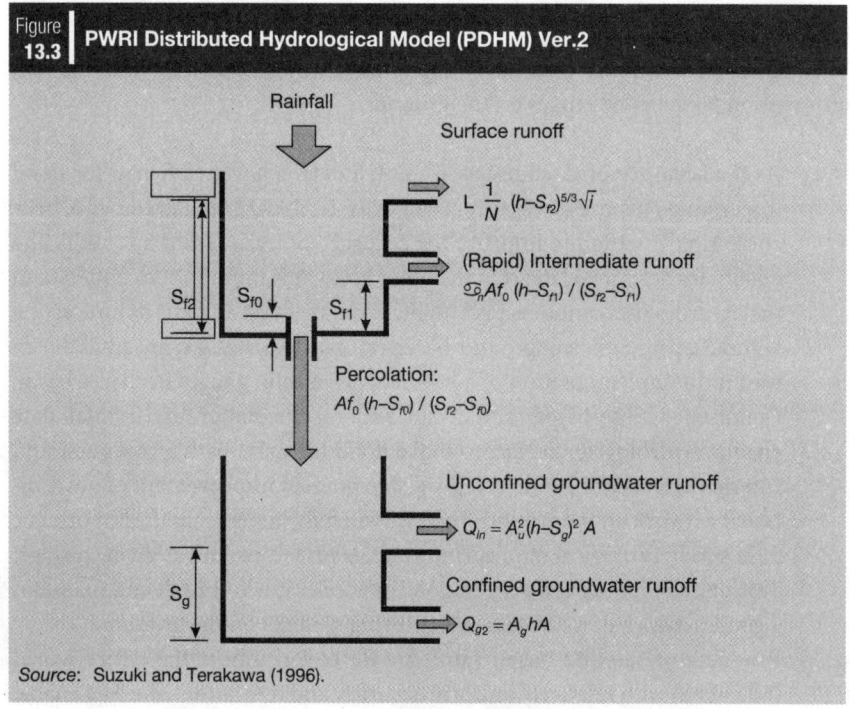

Figure 13.3 PWRI Distributed Hydrological Model (PDHM) Ver.2

Rainfall

Surface runoff

$L \frac{1}{N} (h-S_{f2})^{5/3} \sqrt{i}$

(Rapid) Intermediate runoff

$\sigma_{f1} A f_0 (h-S_{f1}) / (S_{f2}-S_{f1})$

S_{f2} S_{f0} S_{f1}

Percolation:
$A f_0 (h-S_{f0}) / (S_{f2}-S_{f0})$

Unconfined groundwater runoff

$Q_{in} = A_u^2 (h-S_g)^2 A$

Confined groundwater runoff

S_g

$Q_{g2} = A_g h A$

Source: Suzuki and Terakawa (1996).

for flash floods is a simplified version of the PDHM Ver.1 with a three-layer model for both high- and low-flow analysis. These models do not have enough applications worldwide, but are excellent in terms of a variety of applications in Japan (for example, Inomata and Fukami 2007) in attaining high evaluation on stability, easiness to use, handle, and tune, linkage with GIS, and the copyright issue for ICHARM and PWRI.

The PDHM divides a basin into cells and calculates each hydrological process for each cell. The vertical infiltration and percolation are expressed with two (Ver.2) or three (Ver.1) kinds of tanks. Although parameters required for runoff calculation need to be set up for each cell, the IFAS can categorize all cells into several classifications based on globally available GIS data such as land use, geology, and soil, and estimate parameters based on this categorization. Regarding the PDHM Ver.2, the upper tank simulates the surface flow, rapid intermediate flow (subsurface flow), and percolation to the lower tank. The lower tank simulates the unconfined groundwater flow and the confined groundwater flow. Since the PDHM Ver.2 was originally developed for short-period (flash) flood events in mountainous river basins, evapotranspiration was originally ignored. The IFAS, however, has a function to incorporate monthly-averaged evapotranspiration for PDHM Ver.1 and Ver.2 on the basis of a climatological database. River flow is generated after multiple-cell lateral flow of each hydrological process and is routed using kinematic-wave method.

Another type of hydrologic model pre-installed on the IFAS is the Block-Wise TOPMODEL (BTOP model) (Takeuchi et al. 2008). The BTOP model is an extension of the TOPMODEL. The TOPMODEL uses a topographical index derived from a simplified steady state assumption of mass balance and empirical equations of motion over a hill slope. The original TOPMODEL has been applied to many small basins (catchment areas of the order of 2–500 km^2). Therefore, BTOP was modified to apply TOPMODEL in a grid-based framework for distributed hydrological simulation of large river basins (catchment areas in the order of 10,000–100,000 km^2). This extension was made by redefining the topographical index by using an effective contributing area af(a) per unit grid cell area instead of the upstream catchment area per unit contour length and introducing a concept of mean groundwater travel distance. Further the transmissivity parameter T_0 was replaced by a groundwater dischargeability D that can provide a link between hill slope hydrology and macro hydrology. Based on those redefinitions of a few parameters, the BTOP model uses all the original TOPMODEL equations in their basic form. The BTOP model can be used for both short-term (event-based) and long-term (continuous) hydrological simulations, that is, from flood to drought. The BTOP has been widely applied to worldwide river basins with a variety of climatic and hydrologic conditions, including semi-arid and cold regions and large-scale rivers such as the Mekong River.

13.3.3 Function of Model Creation and Parameter Estimation

The IFAS prepares GIS tool kits to analyze geophysical data and to estimate parameters of the hydrological model. In order to secure the worldwide applicability of the IFAS, it was designed to construct flood runoff models based on globally available GIS databases such as digital elevation models (DEM) of USGS (US Geological Survey) GTOPO30[6] or ISCGM (International Steering Committee for Global Mapping) Global Map,[7] land use of USGS–GLCC (Global Land Cover Characterization)[8] or Global Map, and, if required, soil texture of UNEP/DEWA/GRID (United Nations Environment Program/Division of Early Warning and Assessment/Global Resource Information Database)[9] or geology of UNESCO–CGWM (Commission for the Geological Map of the World).[10] These data can be acquired through the Internet and the inputted data are automatically assigned to each cell of runoff models by the IFAS. By using such GIS data on IFAS, it is possible to create a runoff calculation model and estimate parameters at a basin which locations of a river are not identified and also at a basin without the past hydrological information.

The IFAS uses DEM data given as input data, and creates a basin boundary automatically according to altitude data. The target basin is divided into a cell according to the length inputted beforehand, and calculation is performed for each cell. When a user chooses a cell of the outlet, a flow direction of each cell is determined automatically, and creates a river channel network according to altitude data. If a depressed cell is created in the basin and a river channel is divided, modification of the altitude data of the cell is automatically performed so that a flow direction of all the cells may become settled toward the outlet (Figure 13.4).

Parameters are needed to set up for all the cells in the runoff model. In the IFAS, it is possible to set parameters of each cell automatically based on the land-use classification of USGS–GLCC or Global Map given as input data. At present, the classification (Table 13.2) and standard guideline values of parameters, which the authors determined beforehand, are implemented in IFAS to set up parameters without past hydrological information. The user can use those parameter values as the first approximation. However, those values are just based on past experiences of estimation in other river basins, and therefore, are required to be tuned up with in situ hydrological observational data at the next step so as to improve the accuracy of runoff analyzes.

---

[6] http://eros.usgs.gov/#/Find_Data/Products_and_Data_Available/gtopo30_info

[7] http://www.iscgm.org/cgi-bin/fswiki/wiki.cgi

[8] http://edc2.usgs.gov/glcc/glcc.ph

[9] http://edc2.usgs.gov/glcc/glcc.ph

[10] http://www.grid.unep.ch/data/download/

Figure
13.4

Automatic Delineation of Basin Boundary and River Channel Network with DEM

Globally/locally available DEM

Delineated quasi-flow networks

Delineated river channel networks
and watershed boundary

Source: Prepared by authors.

Table 13.2	Parameter Classification (PDHM, Global Map)	
Example of Land Cover Classification (Global Map)	**Example of IFAS Default Classification**
Broadleaf Evergreen Forest	
Braodleaf Decidous Forest	
Needle Evergreen Forest	
Needle Deciduous Forest	1: Forest
Mixed Forest	
Tree Open	
Shrub	
Herbaceous	
Herbaceous with Spare Tree/Shrub	
Sparse vegetation	2: Grassland
Bare area (gravel, rock)	
Bare area (sand)	
Cropland	
Paddy field	
Cropland/Other vegetation mosaic	3: Cultivated land
Mangrove	
Wetland	
Urban	4: Urban area
Snow, ice	
Water bodies	5: Water bodies

Source: Fukami et al. (2009).
Notes: IFAS = integrated flood analysis system; PDHM = PWRI-distributed hydrological model.

13.3.4 Implementation of Graphical User Interface

Calculation results are displayed using a hydrograph, a plan view, a table, and so on. It is also possible to display results of discharges from each mesh and each layer of the mesh. Since calculation is performed with distributed hydrological model to all cells, it is possible to display calculation results of any points in any hill slopes and river channels. Moreover, the display of calculation conditions such as data sources like DEM, land use, rainfall distribution, and parameters and/or internal variables are possible. A time-series display of any variable using animation is also possible. Each chart displayed is linked to each other, and if the location point is moved on a plan view, the content of a chart will also be changed in accordance with the movement of a location point. It is also possible to display calculation results of two or more points simultaneously even in different conditions to compare with each other. Runoff calculation may take about several minutes and it varies according to a number of meshes of a calculation model, a period of flood, and a capability of a computer. It is possible to save calculation results and it is also possible to display them afterwards.

13.4 Applicability of Satellite-based Rainfall Data and an Engineering Approach to Improve Their Accuracy

In this section the authors show the reality and potential of satellite-based rainfall data. In fact, the raw data of satellite-based rainfall are not as accurate as shown in Figure 13.5.

From these figures, it is noticeable that NASA-3B42RT is relatively coincident with ground-based rainfall data as a whole but with larger scatters, and that JAXA-GSMaP underestimates ground-based rainfall as a whole but seems higher correlations. On the other hand, it is very attractive for flood runoff analysts that JAXA-GSMaP is very good in temporal and spatial resolutions of data provision and also in terms of very short delay of data delivery (four hours later than real snapshot from space). Therefore, the authors looked into the details of the underestimation of JAXA-GSMaP using hydrological data in a river basin of Japan with good ground-based observational network (Yoshino River, 3,750 km^2). Then the authors found out the degree of underestimation could be correlated with the spatial correlation characteristics corresponding to the movement of rainfall fields (Shiraishi, Fukami, and Inomata 2008). Based on this empirical relationship, the "self-correction" algorithm for GSMaP was developed successfully as shown in Figure 13.6.

The most important point of this self-correction method is that we do not have to use any external, that is, ground-based observational data to correct GSMaP data. This is a big advantage when conducting flood-runoff analysis and forecasting in poorly gauged rivers.

This methodology was also applied to a disastrous flood event, Typhoon Morakot, in Taipei,China in August of 2009. Raw GSMaP_nRT (hourly) data show much underestimation (Figure 13.7). If the ICHARM's self-correction method (using only just GSMaP_nRT data themselves, without any other external/ancillary/ground-based data) is applied to the raw data, the overall underestimation is improved. However, we can still see large scatters (Figure 13.7). This scatter was caused by the data contamination through very long-time extrapolation of direct microwave radiometric observational data with geostationary meteorological satellite (infrared) data. Direct microwave radiometric observation of rainfall is generally expected to be more accurate than infrared radiometric observation. But the temporal frequency of microwave radiometric observation from the space is not that often due to the limitation of the number of microwave radiometers aboard satellites. Therefore, direct microwave radiometric observation data are extrapolated using the rainfall-field movement information from geostationary meteorological satellite data during the time without any direct microwave observation for any target mesh. If we get rid of very long-time extrapolated data, then the correlation between ground-based rainfall data and self-corrected satellite-based rainfall data becomes dramatically better (Figure 13.8). If the Global Precipitation Measurement (GPM) mission is realized in near future (summer of 2013, in plan), then

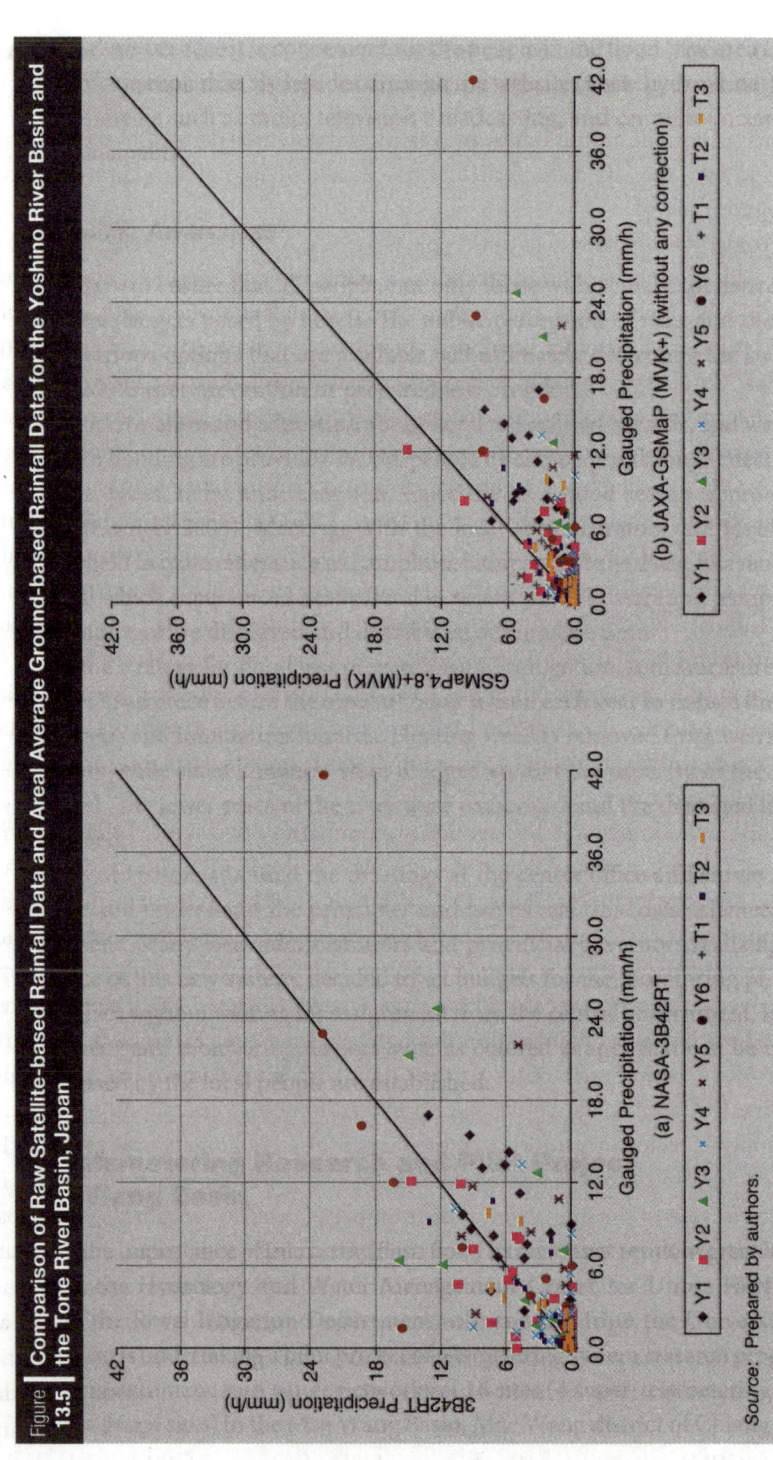

Figure 13.5 | Comparison of Raw Satellite-based Rainfall Data and Areal Average Ground-based Rainfall Data for the Yoshino River Basin and the Tone River Basin, Japan

Source: Prepared by authors.

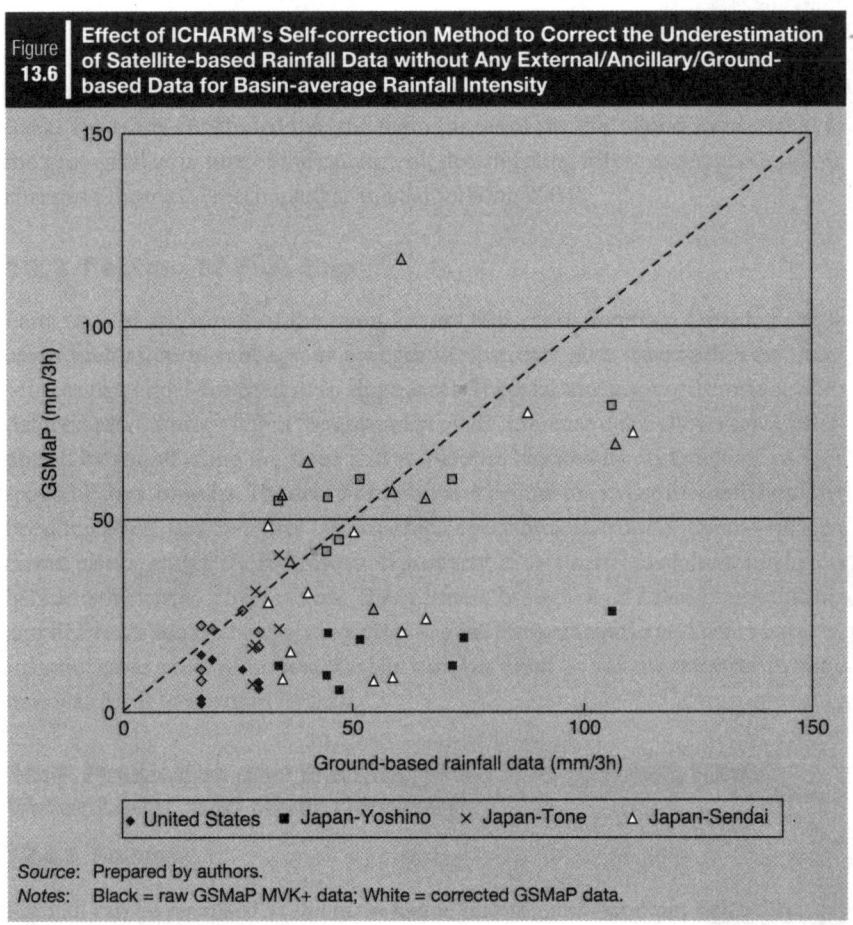

Figure 13.6 Effect of ICHARM's Self-correction Method to Correct the Underestimation of Satellite-based Rainfall Data without Any External/Ancillary/Ground-based Data for Basin-average Rainfall Intensity

Source: Prepared by authors.
Notes: Black = raw GSMaP MVK+ data; White = corrected GSMaP data.

we can expect a much better effect of this self-correction methodology to improve satellite-based hourly rainfall data.

13.4.1 Verification of the IFAS

Preliminary investigations of the applicability of satellite-based rainfall (NASA-3B42RT) data and the IFAS with guideline parameters (2) clarified that a timing of a peak discharge and durations of floods are in qualitatively good agreement and its quantitative accuracy depends on the quality of satellite-based rainfall data and of parameter tuning. Typical examples of runoff simulation using JAXA-GSMaP_nRT with self-correction method are shown in Figure 13.9. The top figure shows a successful example to get a very consistent calculational result with observed river discharge due to the good match between self-corrected satellite-based and ground-based rainfall data. The bottom figure shows an unsuccessful example due

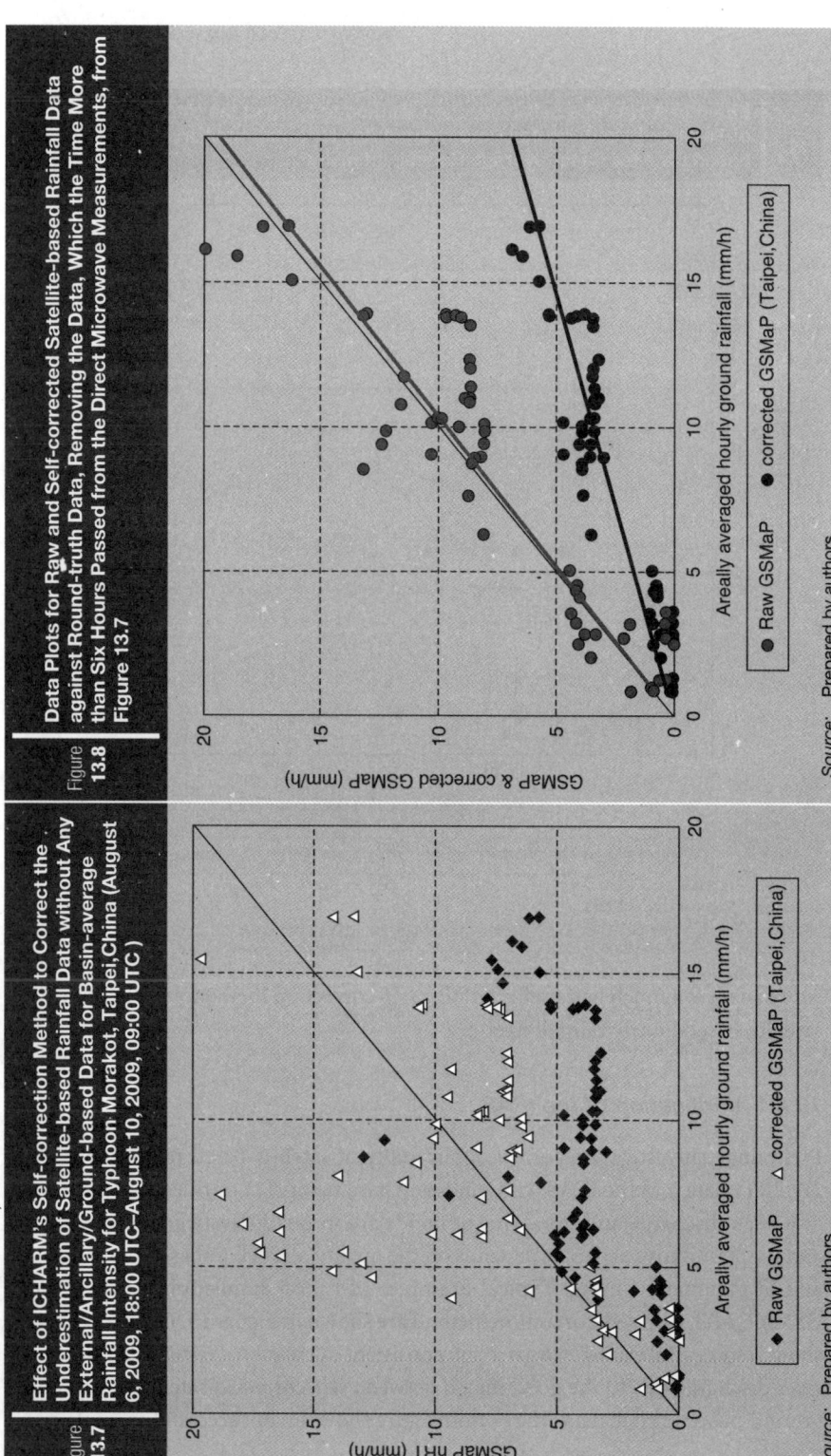

Source: Prepared by authors.

Source: Prepared by authors.

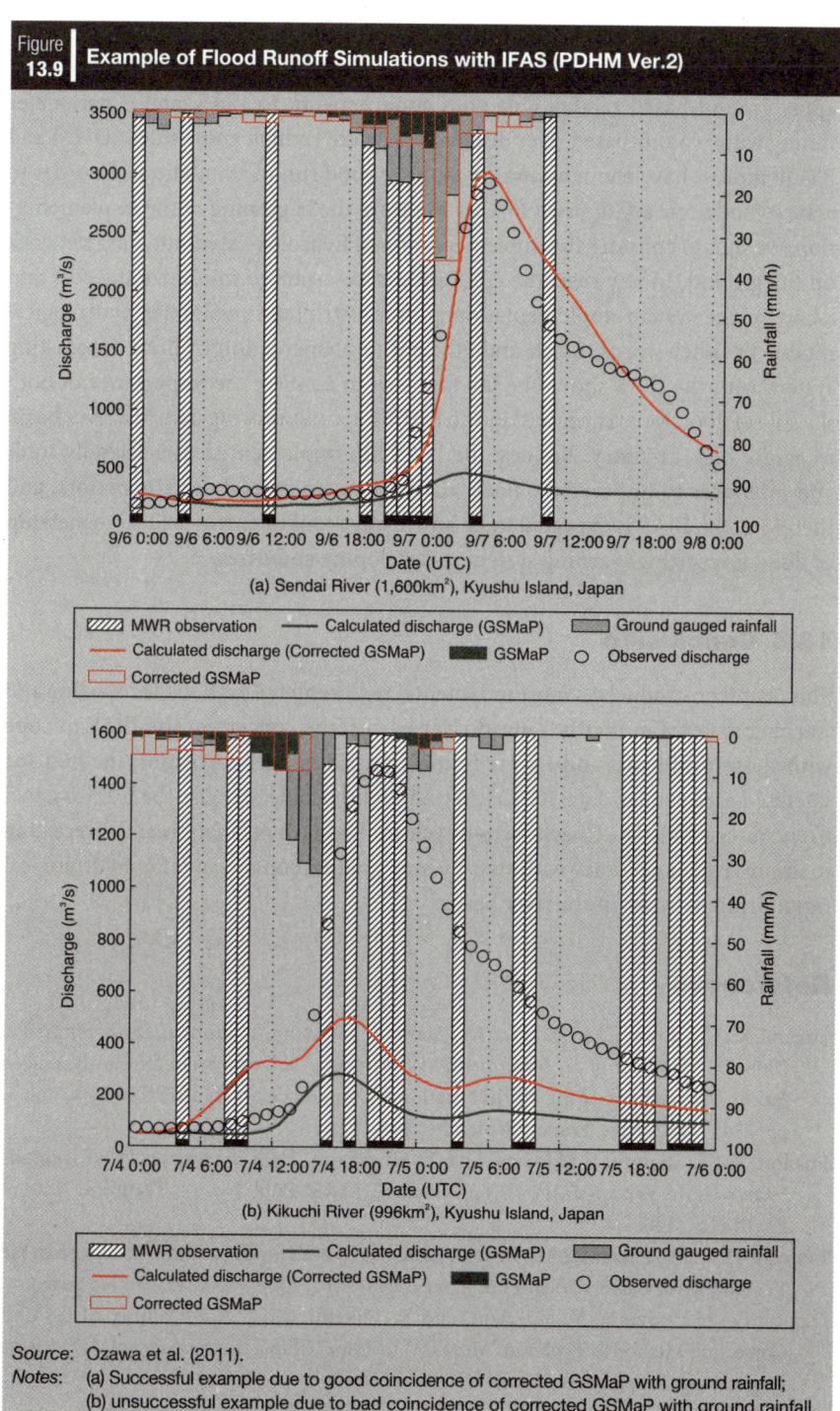

Figure
13.9

Example of Flood Runoff Simulations with IFAS (PDHM Ver.2)

(a) Sendai River (1,600km²), Kyushu Island, Japan

(b) Kikuchi River (996km²), Kyushu Island, Japan

Source: Ozawa et al. (2011).
Notes: (a) Successful example due to good coincidence of corrected GSMaP with ground rainfall;
(b) unsuccessful example due to bad coincidence of corrected GSMaP with ground rainfall.

to very sparse microwave direct measurements from the space during the rainfall increasing period. It depends on the quality of satellite-based rainfall data. If we apply ground-based rainfall data with good network to IFAS, and if parameter tuning with ground-based river discharge data are enough good, then PDHM and BTOP models have enough capacity to make good runoff simulations. In order to assure good accuracy of flood forecasting from the beginning of the implementation, we should consider the implementation of hydrological monitoring network on the ground. In any case, the IFAS should be useful to minimize the cost and labor for the system implementation with a distributed-parameter hydrological model. In other words, quick and efficient implementation of flood forecasting system with the IFAS should be the first step to confirm the importance of both the role of flood forecasting and in situ hydrologic monitoring data in a river basin to assure high accuracy. Besides, the IFAS has implemented user-friendly tools and interfaces to improve the flood analysis system by end users/operators, and therefore, the IFAS is expected to be an effective tool to enhance local ownership of flood forecasting/warning system in developing countries.

13.5 Conclusion

This chapter introduced some fundamental requirements for flood forecasting and warning systems in poorly gauged river basins, the concept of the IFAS to cope with those problems, and its verification including self-correction method for satellite-based rainfall data. ICHARM wants local people to utilize the IFAS as a tool to encourage Plan-Do-Check-Action cycle for flood runoff analysis and forecasting to improve their accuracy with step-by-step gradual enhancement of hydrological monitoring network in the river basin.

References

Fukami, K., N. Fujiwara, M. Ishikawa, M. Kitano, T. Kitamura, T. Shimizu, and S. Hironaka. 2006. *Development of an Integrated Flood Run-off Analysis System for Poorly-gauged Basins*. Proceedings of the 7th International Conference on Hydroinfomatics, Vol. 4, pp. 2845–2852. Nice, France. September 4–8.
Fukami, K., T. Sugiura, J. Magome, and T. Kawakami. 2009. *Integrated Flood Analysis System (IFAS Version 1.2) User's Manual*. ICHARM Publ. No. 14, Technical Note of PWRI No. 4148.
Inomata, H. and K. Fukami. 2007. Development of a System for Estimating Flood Risk in the Yoshino River Basin. *Advances in River Engineering*. 13. pp. 433–438 (in Japanese).
Ozawa, G., H. Inomata, Y. Shiraishi, and K. Fukami. 2011. Applicability of GSMaP Correction Method to Typhoon "Morakot" in Taipei,China. *Annual Journal of Hydraulic Engineering*. 55, pp. 445–450.

Shiraishi, Y., K. Fukami, and H. Inomata. 2008. The Proposal of Correction Method Using the Movement of Rainfall Area on Satellite-based Rainfall Information by Analysis in the Yoshino River Basin. *Annual Journal of Hydraulic Engineering.* 53. pp. 385–390 (in Japanese).

Sugiura, T., K. Fukami, N. Fujiwara, K. Hamaguchi, S. Nakamura, S. Hironaka, K. Nakamura, T. Wada, M. Ishikawa, T. Shimizu, H. Inomata, and K. Itou. 2009. *Development of Integrated Flood Analysis System (IFAS) and Its Applications.* Proceedings of the 8th International Conference on Hydroinfomatics. Concepción, Chile. January, 12–16.

Suzuki T., and A. Terakawa. 1996. Development of Physics-based Distributed Model for Operational Hydrological Forecasting. *Civil Engineering Journal.* 38 (10). pp. 121–126 (in Japanese).

Takeuchi, K., P. Hapuarachchi, M. Zhou, H. Ishidaira, and J. Magome. 2008. A BTOP Model to Extend TOPMODEL for Distributed Hydrological Simulation of Large Basins. *Hydrological Processes.* 22 (17). pp. 3236–3251.

Chapter 14

Effectiveness of Early Warning Systems and Monitoring Tools in the Mekong Basin

Guillaume Lacombe, Chu Thai Hoanh, and Thierry Valéro

14.1 Introduction

The Mekong River is the 12th longest river in the world in terms of length (about 5,000 kilometers[km]) and mean annual flow (475 km³). From the above-5000 meter Tibetan plateau to the Mekong delta in Viet Nam, the Mekong Basin (795,000 km²) includes six countries: Cambodia, the People's Republic of China (PRC), Lao People's Democratic Republic, Myanmar, Thailand, and Viet Nam.

The climate in the basin is mostly controlled by the Southeast Asian monsoon. During the wet season from May to October, southwest wet winds bring heavy rains totaling about 90% of annual rainfall. The dry season from November to April is characterized by northeast dry winds. In August and September, tropical cyclones originating from the North Pacific Ocean and the South China Sea induce long-lasting heavy rainfall interacting with the monsoon. Mean annual rainfall ranges from about 1,000 millimeters (mm) per year in northeast Thailand to 3,000 mm per year along the mountainous border between Lao PDR and Viet Nam (Figure 14.1). The flow regime of the Mekong River and its tributaries is characterized by a flood pulse that generally starts in June and ends in November. This six-month period accumulates more than 90% of annual flow (MRC 2010). For centuries, populations have adapted to this natural annual cycle of floods and droughts.

Each year, between 2.6 and 4.5 million hectares of rice are inundated in the Cambodian lowlands and the Viet Nam delta. These floods are essential for rice production as they provide nutrients through silting and the required amount of water. While lowland areas are cultivated with rain-fed lowland and flooded rice, villages are traditionally installed in the upper areas in order to avoid possible damage from flooding. In the Mekong delta, the flood path is controlled by canals conducting the river flows into the cultivated plains. Such canals also allow the drainage at the downstream end of the cropped area. Although this water infrastructure is sufficient to divert river flow for agricultural use, it does not allow the controlling of extreme floods with large peak discharge and long duration. With the recent acceleration of human development in the lower Mekong Basin, flood-prone areas are now partly inhabited and extreme floods tend to become harmful

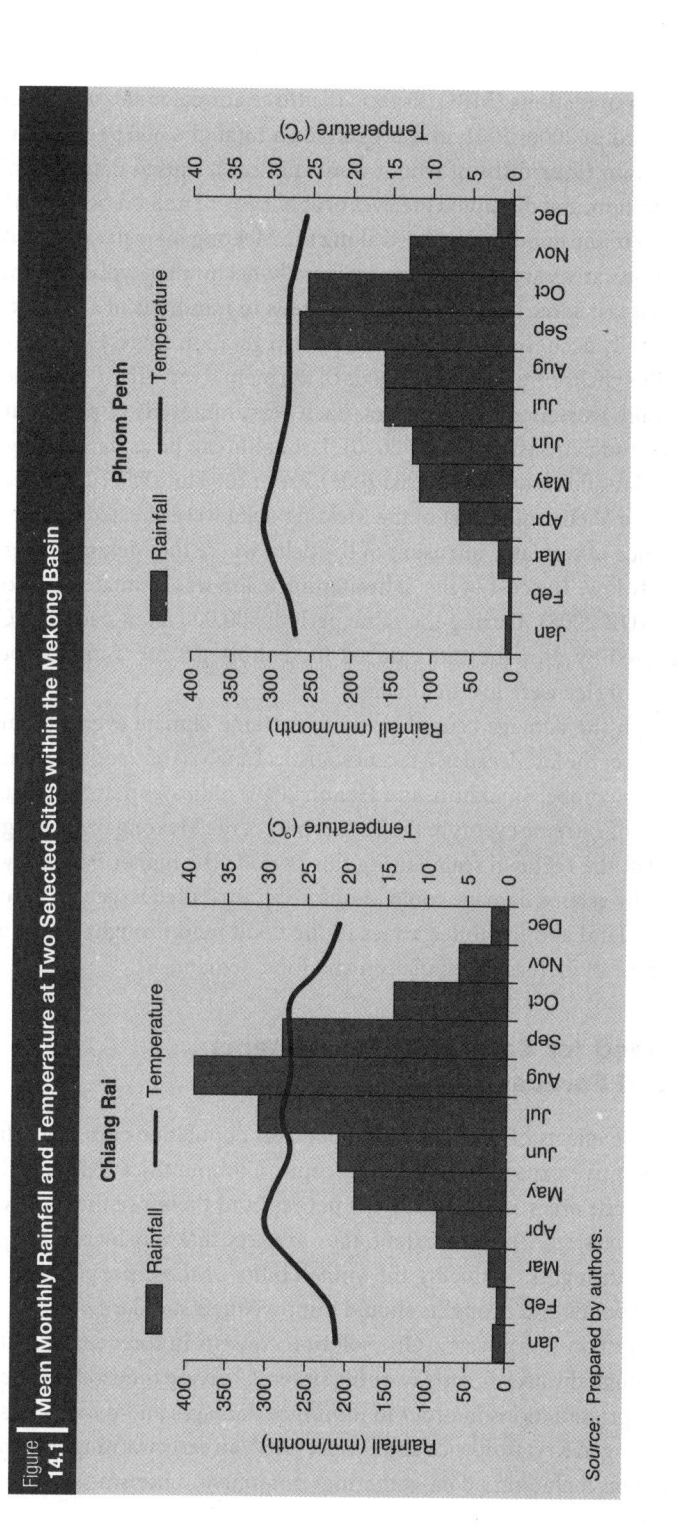

Figure 14.1 | Mean Monthly Rainfall and Temperature at Two Selected Sites within the Mekong Basin

Source: Prepared by authors.

to lowland populations (MRC 2009a). The three annual mainstream flood events that occurred in 2000, 2001, and 2002 killed a total of 1,300 people in Cambodia and Viet Nam (four-fifths of whom were children), caused damage estimated at US$600 million, and disrupted the lives of about 10–12 million people (ADB 2007). In addition to the flooding localized along the Mekong mainstream, flash floods in mountainous areas are sporadic and generally occur after typhoons and tropical storms that last some hours in areas from tens to hundreds of square kilometers. They generally occur in steep areas with poor vegetation cover. Flash floods are not necessarily synchronous with the floods of the main rivers. Their deleterious effects are generally worsened by landslides. Each year, hundreds of people are injured or killed by such disasters (MRC 2010). Droughts can be as devastating as floods (Schiller, Hatsadong, and Doungsila 2006). From 2003 to 2005, 10,000 hectares (ha) of rice in the Vietnamese part of the Mekong delta were affected by saltwater, as a consequence of seawater intrusion in the delta where the Mekong River flow was particularly low. The cost of this saltwater intrusion was estimated at about US$60 million (ADB 2008). During the same period, 500,000 ha of paddy in Cambodia were affected by drought and induced food shortages for 2 million people and 650,000 ha of rice were lost in Thailand.

Although the damage caused by those extreme climate events seems to have increased over the last decades, a recent statistical analysis of trends in several rainfall variables (Lacombe, Smakhtin, and Hoanh 2010) indicates that the frequency and magnitude of extreme events in the central part of the Mekong Basin insignificantly increased at the regional scale during this period. The higher frequency of flood- and drought-related damage could therefore be attributed largely to the expansion of cropped land and inhabited zones in the flood- and drought-prone areas, as a consequence of increased population and food demand.

14.2 Need for Early Warning Systems for Flood Forecasting

The negative effects of droughts and floods on populations are magnified by the limited adaptive capacity of affected people. Communities living in flood- and drought-prone areas are generally the poorest and therefore the most vulnerable to natural disasters. To some extent, they are kept that way by recurring cycles of floods and droughts. Reducing the vulnerability of these people to the negative impacts of floods and droughts should improve their standard of living and assist them to climb out of poverty. One solution consists in forecasting such extreme events through the use of "early warning systems," giving time for the population to take appropriate actions in order to minimize damages and possible casualties.

According to Krysztofowicz and Davis (1983), an early warning system is a chain of subsystems including a data-gathering component, a transmission system of the data to the forecasting center, and a forecast preparation module. The forecasting center converts the data into forecasts that are transmitted to the decision maker. The decision maker uses the forecasts to prepare and release a warning, depending on his evaluation. The warning is transferred to the local authorities who have to take appropriate preventive actions to help flood- and/or drought-prone communities to avoid danger. Unlike "predicting," which is not a real-time computation, forecasting estimates the magnitude of an event that is expected to occur in the near future. Thus, forecasting is a process of decision making under uncertainty. The shorter the forecast time, the more accurate will be the forecast. Flood forecasting models involve the calculation of water stages from rainfall observations and the conversion of simulated flows into water levels. The information obtained in the forecasting procedure is used to design a warning system and to draw up hazard maps, in which the magnitude of a potential extreme event and its frequency of occurrence are indicated. Early warning systems for flood forecasts generally operate at the spatial scale of watersheds that means that targeted surface areas range from tens to thousands of square kilometers. The action units of the early warning systems (alert and evacuation) generally follow the administrative boundaries (country, province, district, sub-district, village), the lowest level being the best one to provide an appropriate answer to a flash-flood alert. Although mountainous and lowland areas are subject to different flood types occurring at contrasted spatial and temporal scales—the damage caused by localized flash floods depends on the flow velocity while the damage occurring in the plains depends on the elevation of the water level and the time length of the flood—emergency action plans are similar and consist in temporarily relocating communities out of the endangered and/or damaged areas. Although early warning systems for drought mitigations are currently being developed in the lower Mekong Basin, this chapter describes in detail the early warning systems for flood forecasting as they are the most advanced in the lower Mekong Basin.

14.3 Development of Early Warning Systems in the Lower Mekong Basin

The first early warning system in the Mekong Basin was set up in the early 1970s, after the large flood event of 1966. The forecasting system, including the major tributaries, was progressively improved after the 1978 and 1981 extreme floods. After the 2000 and 2001 extreme floods that caused US$400 million damage and 800 fatalities, 21 key hydrological stations along the Mekong mainstream and 19 stations for river monitoring were included in the system to improve forecasts during both the wet and the dry seasons. In 2005, the council of the Mekong River Commission (MRC) established a basin-wide flood management and mitigation (FMM) program. Its objectives are grouped into five components:

- The first component includes the establishment of the FMM center in Phnom Penh and its maintenance.
- The second component studies the effects of water infrastructure (reservoirs, embankments, waterways) on floods. The objective of this component is to encourage the adoption of good practices. Flood-proofing measures are developed in flood-prone areas as a cost-effective means of flood mitigation at local level. Poor communities are targeted when considering building design guidelines, financing mechanisms, and raising awareness.
- The third component deals with the mediation and coordination of trans-boundary flood-related issues. The aim is to facilitate dialogue and resolution of issues on land-use planning, infrastructure development, and cross-border emergency management.
- The fourth component focuses on capacity building, knowledge sharing, and public awareness campaigns to improve existing emergency management systems in riparian countries at the provincial, district, and community levels. This component includes strengthening the provincial and district disaster management offices in each pilot province, especially in the Mekong delta in Cambodia and Viet Nam which are the most natural flood-prone areas of the Mekong Basin and where flood control and disaster reduction are put at a high priority by the governments. As a result, the Committee for Disaster Management in Cambodia and the Committee for Flood and Storm Control in Viet Nam are institutionally present at each administrative level (MRC, GTZ, ECHO, and ADPC 2009).
- The fifth component aims at achieving sustainable natural resource management through the development of techniques to assess the probabilities of floods and to improve land-use management policies.

14.4 Theoretical Functioning of the Current Early Warning System in the Lower Mekong Basin

14.4.1 The Stations Network

The early warning system collects hydro-meteorological data from 22 rainfall and 37 hydrological stations on the Mekong River system in the four riparian countries, members of the MRC (Cambodia, Lao PDR, Thailand, and Viet Nam) and two hydrological stations in the PRC. Forecasted rainfall from the National Oceanic and Atmospheric Administration (NOAA) are also sent into the system. Among all hydro-meteorological stations belonging to the network, 17 hydro-meteorological stations installed along the Mekong mainstream are part of the Appropriate Hydro-logical Network Improvement Project (AHNIP) network. Fifteen other stations are located in the four member countries. The stations automatically record river water level and rainfall at hourly interval and generally transmit the recorded data one time a day via global system for mobile (GSM) communication in the morning to the four national data terminals, to the MRC secretariat in Vientiane, Lao PDR, and to the regional FMM Center of the MRC in Phnom Penh, Cambodia. Thirty other hydro-meteorological stations, belonging to the Hydrological Cycle Observing System (HYCOS) (WMO and MRC 2006), have been installed since 2005 along the Mekong tributaries. The HYCOS stations are designed to automatically collect water level and rainfall data at 15-minute interval and transmit recorded data directly to the MRC secretariat and to the regional FMM center servers, at an hourly interval via General Packet Radio Service (GPRS) system, then through the Internet, using the file transfer protocol (ftp). The HYCOS network also includes two additional stations for navigation purpose located in coastal zones of Viet Nam to record seawater level. The MRC secretariat plans to upgrade the equipment of AHNIP stations of the four MRC member countries to align them with the HYCOS technology. With the new HYCOS network, all key tributaries can be monitored, thus enhancing a better near real-time river monitoring and flood forecasting for the mainstream and tributaries of the lower Mekong Basin. In parallel to the automated system for data transfer, observers at the hydrological stations send hydro-meteorological data to their national agencies by short message service (SMS). A quality control process in the line agencies is implemented to cross check the consistency between both data sets (either collected automatically or manually). In case of discrepancies, the observers are recalled in order to identify the origin of the error and should correct it. This double-check is the first and mandatory step of the data quality process, under the responsibility of the line agency.

14.4.2 Forecasting Model

Hydro-meteorological data collected from the AHNIP and HYCOS networks are used as input to run the MRC forecasting models. Rainfall data are adjusted for topographic effects by MRC flood forecasters according to past experience and hydrological observations that are updated annually. Flood forecasting for the middle reaches of the lower Mekong Basin is based on the Streamflow Synthesis and Reservoir Regulation (SSARR) model developed by the United States Army Corps of Engineers. Outflow derived for each sub-basin from rainfall is based on soil moisture accounting models in the upper and middle parts of the basin. In the lower reach (including the Mekong delta), multiple regression models are used. Artificial neural networks are used in both upper and lower reaches of the Mekong River for the flood forecasts. Results are cross validated using the output from the SSARR and regression models. Rating curves are used to convert estimated discharges into estimated water levels. The MIKE 11 hydrological model is used to map floods in the Mekong delta from flood water levels. Parallel to the SSARR model, a new Mekong Flood forecasting system based on the Unified River Basin Simulator (URBS) model developed by the Queensland Department of Natural Resources and Mines, Australia, and the Deltares Flood Early Warning System developed by the Deltares Research Institute, the Netherlands, has been established at the regional FMM center. First tests indicate that the system provides satisfying results, especially between Luang Prabang and Mukdahan. However, further improvements are required at some locations, especially in the area downstream of Pakse (Tospornsampan et al. 2009).

14.4.3 Flood Forecast Dissemination

Five-day forecasts are disseminated through the Internet and transmitted to the member countries. Flow- and water-level forecast are disseminated daily (or four times daily in August and September) to the national Mekong committees, national line agencies, national disaster management committees, mass media, and UN organizations. These organizations disseminate information to other concerned parties including local communities via the website, email, telephone, facsimile, and radio. The regional FMM center website (http://ffw.mrcmekong.org/) is updated every day. During the dry season, low-flow forecasts are conducted and updated weekly. The MRC secretariat forwards operational data to countries prior operation of regional forecast.

14.5 Effectiveness of the Early Warning System in the Lower Mekong Basin

Although the early warning system has operated for many years to provide useful forecasts during the flood season to people in the lower Mekong basin, in particular in downstream countries as Cambodia and Viet Nam, the 2008 flood of the Mekong River revealed some flaws in the early warning system. The poor data coverage associated with inaccurate hydrological models for flash-flood simulations introduced large errors in the conversion of rainfall into simulated flows, estimated water level and flooded areas. In addition, the global scale US geological survey/NOAA satellite images were found to be too coarse for the lower Mekong Basin. Finally, the information transferred to local flood-prone communities was found to be inaccurate while the coordination between the MRC country members appeared to be weak and inefficient. This is partly due to the lack of capacity of local authorities and communities to react efficiently in the case of emergency situations. Although these inaccuracies were still acceptable in the past as population densities in flood-prone areas were low, rapid population growth in the region, intensification of agriculture, changes in land use and river morphology, and rapid technology development makes it imperative that the system be upgraded, and a forecasting system, based on modern technology combined with a more effective warning system, be installed.

14.6 Improvements for Future Early Warning Systems in the Lower Mekong Basin

14.6.1 Improving Forecast Accuracy

The 7th Annual Mekong Flood Forum held in Vientiane in 2009 (MRC 2009b) concluded that new accurate and effective models should be coupled with geographical information systems, digital elevation models in order to provide flood maps and risk and/or vulnerability maps. Such maps would be helpful to assess the damage risks and thus reduce the vulnerability of people to flooding. It also concluded that the quality of the early warning systems depends on the database and density of the hydro-meteorological station network (Airaksinen et al. 2009). One priority for improvement consists of installing more stations along the mainstream and major tributaries. Including additional data from stations in the PRC is also important as the PRC section of the Mekong River Basin contributes to nearly two thirds of the Mekong discharge during the dry season. As tropical storms and typhoons arriving from the South China Sea cause the most severe flooding situations in the Mekong River Basin, enhanced observations, tracking, and modeling of these storms is recommended for accurate medium-term flood forecasts. The improvement of flash-flood forecasts should rely on past observations in order to identify flash-flood–prone areas and install additional rain gauges in these areas. Meteorological forecasts will have to rely on a combination of satellite images, weather forecasts, and measurements from ground-based radar and rain gauges (Schultz 1996). Such a network should help to validate information from satellite images.

14.6.2 Improving Forecast Dissemination

Improvement of forecast dissemination is also required as the improved technology of flood forecasting and warning systems is not sufficient to provide a good warning. The August 2008 flood event was the first regional flood episode for which active forecasting was investigated by the Regional FMM Center (MRC 2009a). Therefore real examples of flood events that were actually monitored and forecasted are still very limited. Consequently, improving forecast dissemination relies on theory and past failures rather than referenced successful cases when early warning systems operated well. This chapter presents some general guidelines that should help improving forecast dissemination. Flood forecasts must be transferred to people at risk through a chain of actions: conversion of forecasts into warning by a decision-making process involving different stages, and finally, warnings converted into appropriate actions by local authorities and the people at risk themselves. This requires to precisely identify the path of warning from forecast to persons responsible for actions, while avoiding duplication of tasks. A way to reach this goal is to encourage sharing of data and information, especially in border areas where international cooperation is required to establish evacuation plans. Clear roles and responsibilities have to be defined for each line department of the provincial and district disaster management offices. The capacity analysis in the national meteorological centers will have to be improved as well as the institutional memory. The implementation of quality controls of the early warning systems should help improving the emergency management of local authorities and communities. The coherence of approaches between different countries is very important to address transboundary issues. This could be reached through mutual respect and cooperation between neighboring countries. People-centered approaches have to be prioritized as they are low-cost, effective, and relevant to local conditions in flood-prone areas.

14.7 Conclusion

The management and mitigation of flood risks and damage are of increasing concern for the authorities and communities of the flood-prone areas in the lower Mekong Basin. Flood vulnerability of the population has been increasing rapidly over the recent decades, in response to demographic growth and to the acceleration of urbanization and rural development, especially in flood-prone areas. While improvements to the structural components of the early warning systems—which mainly involve the accuracy of the flood forecasts—are required, nonstructural components (forecast dissemination and evacuation plans of flood-prone communities) are seen as the weakest link in the chain of actions of the actual early warning systems in the lower Mekong Basin. Thus, their improvement should be prioritized. This task is a long-term and challenging process requiring the coordination of stakeholders that may not be used to collaborating together and often belong to different socio-cultural backgrounds.

References

Asian Development Bank (ADB). 2007. *Kingdom of Cambodia, Lao People's Democratic Republic, Kingdom of Thailand, and Socialist Republic of Viet Nam—Preparing the Greater Mekong Subregion: Flood and Drought Risk Management and Mitigation Project.* Manila: ADB.
———. 2008. *Preparing the Greater Mekong Subregion: Flood and Drought Risk Management and Mitigation Project.* Regional Technical Assistance Report, Project Number 40190. Manila, April.

Airaksinen, P., J. Ikonen, N. W. S. Demetriades, and H. Pohjola. 2009. *Real-time Hydro-meteorological Observation Networks: Development Possibilities for the Early Warning System of the Mekong River Basin*. In 7th Annual Mekong Flood Forum. Integrated Flood Risk Management in the Mekong River Basin. May, 13–14. Bangkok: Mekong River Commission.

Krzysztofowicz, R., and D. R. Davis. 1983. A Methodology for Evaluation of Flood Forecast Response Systems: 1. Analyses and Concepts. *Water Resources Research*. 19 (6). pp. 1431–1440.

Lacombe, G., V. Smakhtin, and C. T. Hoanh. Forthcoming. Rainfall Trends in Central Mekong Basin: 1953 to 2004. *Theoretical and Applied Climatology* (Manuscript under review).

Mekong River Commission (MRC). 2009a. *Annual Mekong Flood Report 2008*. Vientiane: Mekong River Commission.

————. 2009b. *Integrated Flood Risk Management in the Mekong River Basin*. Proceedings of the 7th Annual Mekong Flood Forum, May, 13–14. Bangkok: MRC.

————. 2010. *State of the Basin Report*. Vientiane: Mekong River Commission.

MRC, GTZ, European Commission Humanitarian Aid department (ECHO), and Asian Disaster Preparedness Center (ADPC). 2009. *Manual on Flood Preparedness Program for Provincial and District Level Authorities in the Lower Mekong Basin Countries*. Manual produced under Component 4 of the Flood Management and Mitigation Program. Vientiane.

Schiller, J., M. Hatsadong, and K. Doungsila. 2006. A History of Rice in Laos. In J. M. Schiller, M. B. Chanphengxay, B. Linquist, and S. Appa Rao, eds. *Rice in Laos*. Los Banos, Philippines: International Rice Research Institute.

Schultz, G. A. 1996. Remote Sensing Applications to Hydrology: Runoff. *Hydrological Sciences Journal*. 41 (4). pp. 453–475.

Tospornsampan, J., T. Malone, P. Katry, B. Pengel, and H. Pichan. 2009. *Short- and Medium-term Flood Forecasting at the Regional Flood Management and Mitigation Centre*. In Proceedings of the 7th Annual Mekong Flood Forum. Integrated Flood Risk Management in the Mekong River Basin. May, 13–14. Bangkok: MRC.

World Meteorological Organization (WMO) and MRC. 2006. *Project Document: Establishment of a Hydrological Information System in the Mekong River Basin*. Vientiane, Lao PDR: World Meteorological Organization, MRC, and Agence Française de Développement.

Chapter 15

Insurance Solutions to Climate Change in Asia and the Pacific

Yuri Murayama, Harumi Yashiro, and Hideki Kimura

15.1 Introduction

Insurance companies have comparatively less negative impact on the environment than other industries such as manufacturing, and even less impact than other service industries. Furthermore through their insurance and investment business, insurance companies are contributing to lead society to become less vulnerable to climate change. According to the Intergovernmental Panel on Climate Change (IPCC) (Parry et al. 2007) and the Association of British Insurers (2005), climate change affects are more severe in developing countries with worsening water shortages, lower crop productivity, and increasing floods and storms than developed countries. At a side event of COP15, Munich Re stressed the current climate situation: storm and flood damage has increased three times and the average temperature has risen 10% since the 1980s (Munich Re 2009a). According to UNEP-FI, financial damage caused by climate change will double every 10 years: it will be US$150 billion in the next decade as this trend continues (UNEP-FI 2002).

The Stern Review (2007) stated that environmental protection is less costly than the damage caused by global warming. The Kyoto Protocol and later conferences focused more on climate change mitigation such as the Clean Development Mechanism (CDM), joint implementation (JI), and emissions trading (ET). The Bali Roadmap, also known as the Bali Action Plan (UNFCCC 2008), however, included adaptation to climate change. Insurance companies are expected to play a key role in adaptation—especially in developing countries—by pooling and transferring the risks through insurance schemes. Climate change brings insurers not only risks (such as excess insurance payments against loss, bankruptcy, and loss of the appraised value for the investment) but also business opportunities (such as investment on renewable energy, new insurance products, and new services).

In developed countries, damage from natural disasters can be large since economic activities are concentrated in the big cities, most of which are located in coastal areas. Although property values are less in Asian developing economies (and most property is uninsured), financial damage from natural disasters still cost 50% of world financial loss (ADRC 2008). In addition, the loss from catastrophes in developing countries could be bigger than their GDP. In this vulnerable situation, natural disasters could cause further poverty and the vicious cycle of it. Therefore, insurance has enormous potential for economic sustainability especially in these areas.

15.2 Damage Estimation

Insurance companies face tremendous risks of increased insurance payments against the increased damage caused by natural disasters such as typhoons, floods, and others. Figure 15.1 shows the number of natural disasters increased four times between 1975 and 2000. Though the increment of events has slowed down, Figure 15.2 shows that financial damage has maintained an upward trend. According to the Swiss Re focus report (Heck, Bresch, and Trober 2006), the winter storm in Europe is expected to cause 16% to 68% additional economic damage from 1961 to 2100. The loss caused by the abnormal climate would be US$1 trillion in 2040 (UNFCCC 2007). The Association of British Insurers (2005) reported that temperatures would increase 2°C–3°C this century. If they increase 5°C–6°C next

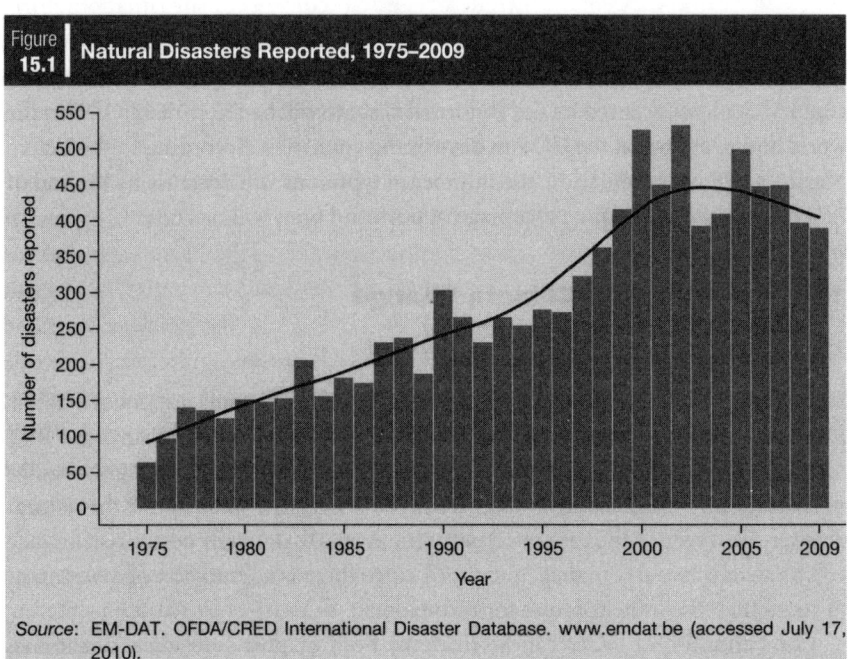

Figure 15.1 Natural Disasters Reported, 1975–2009

Source: EM-DAT. OFDA/CRED International Disaster Database. www.emdat.be (accessed July 17, 2010).

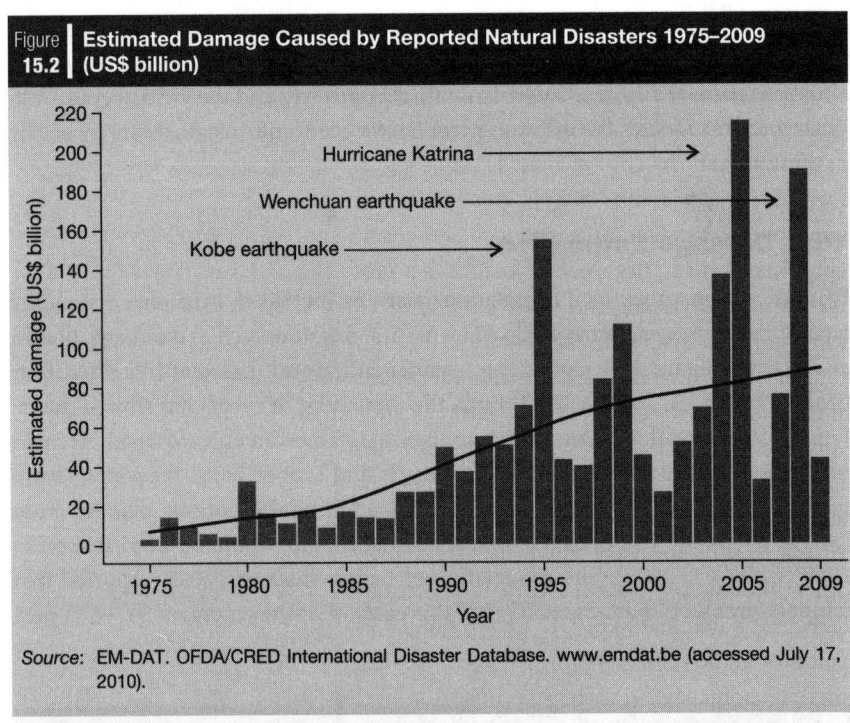

Figure 15.2 | Estimated Damage Caused by Reported Natural Disasters 1975–2009 (US$ billion)

Source: EM-DAT. OFDA/CRED International Disaster Database. www.emdat.be (accessed July 17, 2010).

century, the losses caused by the abnormal climate will be 5%–10% of GDP in the world and over 10% of the GDP in developing countries. According to the Tokyo Marine's typhoon simulation, the number of typhoons will decrease by the end of 21st century, however, the percentage of powerful ones will increase.

15.3 Insurance on Climate Change

15.3.1 Insurance as a Risk Management

Some risks are preventable, some can be minimized, and some are not avoidable. Risk management is divided into two groups: "risk control" and "risk finance." Risk control is fundamental for risk management. If infrastructure is built and regulation sufficiently established, damage would be less for the same size of the natural disaster. However, in the case of catastrophic events that merely occur, risk finance will be an effective risk management tool, since there is a limitation of investment to strengthen the infrastructure for rare events.

Compensation of losses can be made by both ex-post (subsidies, donations, and funds) and ex-ante (insurance). Ex-post financing, however, cannot guarantee that each victim will receive the support to recover from the loss, even though there are sufficient donations and/or assets. In contrast, the insured is guaranteed to be financially compensated to some extent, if the event satisfies the insurance conditions (Brauner 2005). Currently, most governments' financing measures for disasters are taken after the occurrence of disasters. As Arnold (2008) stated, "Ex-post funding approaches are inefficient, often poorly targeted, and insufficient. Moreover, they provide no incentives for proactive risk reduction measures such as improved urban planning, higher construction standards." To take proactive risk management, insurance can be a major scheme of risk finance against catastrophe. Risk-averse individuals and institutions prefer to pay an appropriate premium to hedge risks. Insurance companies estimate the damage, set the appropriate premium, pool the risks, and disperse the risks among the portfolio. If insurance companies can estimate the damage, they can prepare for the payout. However, insurance companies have tremendous risk to leave the market because the claims caused by the catastrophes can be an extreme amount. In the case of disasters that merely occur, the law of large numbers will not be applied to estimate the risk and damage. In addition, catastrophe insurance faces the capacity limitation to be dealt with in the market. Recent frequent catastrophes (e.g., Hurricane Ivan [2004], Hurricane Katrina [2005], UK Severn flood [2007]) have shrunk the traditional insurance market underwriting capacity. However the development of securitization of insurance will transfer the risks from the insurance market to the financial market. Measures of the risk transfer are called securitization (or catastrophe [CAT] bond) derivative, or alternative risk transfer (ART). Since financial markets have a bigger capacity than insurance markets and no credit risk, they can hedge the risk more stably. Therefore, it is expected that the catastrophe insurance market will expand further.

15.3.2 Insurance and Risk Transfer: Insurance, Reinsurance, and CAT Bonds

15.3.2.1 Insurance

To date, major climate-related insurance has been indemnity type insurance. Indemnity type insurance pays off the insurance according to the actual loss. Therefore it consumes time and cost for the evaluation of the actual loss. In addition there is a risk of moral hazard and no incentive to reduce the risks. However, an index type of the insurance, similar to the weather derivatives in financial markets, pays off the insurance according to the observed index such as rainfall, wind velocity, and temperature. This index type insurance makes prompt payouts and also motivates the insured to minimize the risks. However it remains a basis risk that is the difference between the payout and the actual loss. In the case of catastrophes, prompt payment is necessary for early reconstruction.

There is capacity concern for both types of insurance mentioned above. If the number of policyholders is large and the risks are regionally concentrated, insurance companies will try to transfer the risks to reinsurance companies or to the financial market, though ART has higher premium or payout costs.

15.3.2.2 Reinsurance

Traditionally insurance companies transfer the risks to the reinsurance market and buy the others' risks to have better portfolio. Reinsurance companies also transfer their risks among the reinsurance market or capital market. When disasters occur, the insurance will be paid out. Since the higher risks are more likely to be transferred to the reinsurance market, the premiums tend to be higher. Recent successive disasters (e.g., Hurricane Katrina and Hurricane Ivan) have shot up reinsurance premiums. Cummins, Doherty, and Lo (1999) mentioned that although the demand–supply gap in the reinsurance market is not huge, raises in premiums seem to be unavoidable, if the risks are transferred only to the reinsurance companies. Transferring the risks to the reinsurance markets stirs the concerns of credit risk, moral hazard, adverse selection, risk selection, as well as high reinsurance premiums.

15.3.2.3 Catastrophe Bonds

As mentioned above, insurers try to disperse the risk by transferring their risk to, and purchasing other risks from reinsurers. However recent successive disasters have shrunk the capacity of the reinsurance market; thus insurers and reinsurers are not able to disperse their risk satisfactorily.

In order to transfer the risk to bigger "capital markets" where there is no credit risk, CAT bonds have recently attracted investors' attention. The sale of CAT bonds grew rapidly until 2007, while from 2007 to 2010 the growth of sales has decreased or stagnated. The size of the traded CAT bond in total has been growing steadily and is expected to increase with the expansion of the insurance market. It is because the correlation between climate insurance and other derivatives is less, thus investors can have a more diversified portfolio. In addition, higher payouts for CAT bonds are expected. Investors buy the bonds from enterprises, insurance companies, or reinsurance companies. If the triggered event occurs, the partial or full principal will be paid to enterprises, insurance companies, and/or reinsurance companies and investors have risk for a loss of principal. If the triggered event does not occur, comparatively higher payouts will be paid to investors. Figuring out the correlation between the risks is essential for risk finance. If there is an inverse correlation or decorrelation between the risks, thus most probably losses will not occur in a serial manner. Watanabe, Suzuki, and Yashiro (2007) examined the correlation between the number of typhoons and losses from typhoons in Asian countries. They found that not all the countries have similar correlation patterns between the number of the events and the losses. They concluded that the risk finance to the countries where there is a higher correlation between the number of attacks and losses needs to be carefully chosen.

So far most of the CAT bonds dealt in the market are covering catastrophes in the United States and Europe. Even though the correlation between risks in western countries is weak, bringing CAT bonds in Asia to the portfolio has a possibility to disperse the regional risks.

High payouts cause higher insurance premiums. In developing countries higher premiums will eliminate potential policyholders. In order to control the appreciation of the premiums, subsidies, or other financial mechanisms from the community level, governments, international organizations, and enterprises should be considered. In addition, CAT bonds also have basis risks (trigger-base CAT bond), information asymmetry, moral hazard, and adverse selection.

15.3.2.4 How to Assess the Risks

For insurers, getting a clear grasp of the worst case scenario is necessary to prepare for losses and to maintain their business. In order to estimate the risk cost, insurers assess whether the risk is adequate, not excessive, or fairly discriminatory, and its availability, affordability, and flexibility. Although insurance companies run risk assessments statistically with past data, in the case of catastrophes and in a changing climate, statistics alone are not enough to assess the risk. Catastrophes occur rarely (longer return period), and frequency and damageability of the catastrophe will be different in a new climate. Recently insurers combined three risk-assessment tools for natural disasters: traditional statistical risk assessment, engineering risk assessment, and computer simulation.

Described below is a risk assessment on typhoons to examine the wind velocity in the future climate. An increment of maximum wind velocity in the future climate was simulated with existing data on the number and size of typhoons and the wind velocity in both the current and future climate. In the future climate, the central pressure of typhoons was set as 14% less than the average of its probability distribution from 1932 to 1996, and the number of occurrences is set as 30% less than the average of its probability distribution from 1932 to 1996 (Tokyo Marine unpublished data). Standard deviation is not modified. (Generally it is said that the number of the typhoons will decrease and the frequency of strong typhoons will increase.)

Actual simulation will combine all impact factors, each of which are modeled to reflect the past events and future climate, in order to improve the accuracy.

Although climate insurance demand is higher in the rural areas, where farming is concentrated, less information on weather and damage is available. Thus observation systems need to be installed in these areas.

15.4 Insurance in Asia and the Pacific

15.4.1 Vulnerability of the Asia and Pacific Region

The damage from natural disasters in Asia in the last 32 years has been the largest in the world (ADRC 2008). In all, 90% of the victims, 50% of the injured, and 50% of the economic loss in the world are from Asia (Table 15.1). Major natural disasters in Asia in 2007 were earthquakes (43.8%), floods (25.5%), and wind storms (30.7%) (CRED–EMDAT 2010). For countries that have a smaller economic scale, even one small disaster may severely impact a country's economy. "Although many initiatives have been launched and investments made in developing countries in regions vulnerable to disasters, the increasing frequency and magnitude of natural catastrophes that result in economic loss and human casualties have hindered those initiatives" (ADRC 2008).

Table 15.1	Natural Disasters, 1975–2007			
	Number of disasters	Number killed	Total affected	Total damage US$ ('000s)
Asia	3,439	1,281,189	5,047,632,951	594,334,344
	37.34%	57.24%	88.94%	44.57%
World	9,207	2,238,319	5,675,595,783	1,333,357,184
Source: CRED-EMDAT (2007).				

15.4.2. Insurance in Asia and the Pacific

15.4.2.1 Insurance Coverage

Although developing countries are the most vulnerable to climate change, in these countries insurance schemes have not yet extended to every corner of the nation nor have the schemes themselves matured.

The Nat Cat Service database (Munich Re 2009b) shows that 34% of the natural disasters in 2008 occurred in Asia, causing 60% of the loss in the world. Insurance, however, covered only 5% of the loss in Asia. Munich Re stressed that only 3% of the poor in the 100 poorest countries could afford insurance in 2009. According to Arnold (2008), the insurance market in most developing countries is undeveloped. The coverage for natural disasters is extremely limited, and it is limited to industry's property. Due to the weak financial sector in developing countries, insurance has not been able to take the role. Basic risk assessment and loss prevention are the priority for public sector funding followed by training and education for local insurers, intermediaries, and at-risk parties (Dlugolecki 2009).

In addition, although there is a need for climate insurance, people in developing countries prefer not to take out insurance even if the premium is affordable. Xavier Gine, a World Bank economist in Malawi, has seen "microinsurance sputter time and again, even in areas where microloans thrive" (Kiviat 2009). As was seen in Bangladesh, the insured may not be paid out because of the credit risk or risk of unilateral tearing up of the contract by the insurer. This trust issue deters would-be policy holders. Gine found that, in reality, those people who were more willing to buy insurance policies were risk takers.

15.4.2.2 Possibility of Climate Insurance in Asia and the Pacific: Challenges and Solutions

Although mutual aid in communities has been taking the major role of safety nets up to now, the function will be lost in the wake of a disaster. Thus a well-designed insurance product against disaster should be developed.

Table 15.2	Challenges to Propagate Climate Insurance in Asia and the Pacific	
Challenges	**What is needed**	**Solution**
Development of the insurance	Weather data and observation system for R&D of the insurance product and payoff	Development of the weather observation system
	Research on correlation of the weather index and damage	Pooling the data Development of the R&D system
	Infrastructure and budget for the research, collaboration, and coordination among ministries and agencies concerned	Coordinating the stakeholders
Penetration of the insurance	Affordable premium for local farmers Penetration of the insurance	Control of the premium with subsidy and with risk control
	Optimal size of the policyholders to pool the premium and disperse the risk	Selling the insurance combined with micro credit Promotion to the farmers
Risk transfer	Capacity development and risk transfer to re-insurance market or to capital market	Risk transfer to the capital market or to the international organization

Source: Adapted by authors from UNEP-FI (2006).
Note: R&D = research and development.

As mentioned above, poor infrastructure for observing the weather and poor data hinder insurance development in the area. In order to estimate the impact more accurately, weather data, damage data, and crop-growth data in the area should be obtained. Since data collection requires time, budget, and great effort, a database should be published. Though it is a part of the core competency of the institutions, Munich Re (Nat Cat) and Swiss Re (Sigma) open some data to the public and also publish their research.

In addition, the insurance system in developing countries has not matured. Developing countries need to open their markets to foreign capital, to learn the know-how from senior insurance companies in the world, to build and pool knowledge on climate change, to develop the product together with insurance companies in developed countries, and to provide access to international risk transfer markets. Since the premiums are not affordable for a higher risk segment, subsidies should be applied to make coverage wider.

The Munich Climate Insurance Initiative (MCII) was established in 2005 with forefront financing institutions, governments, and enterprises (Munich Re, World Bank, UNFCCC, TERI, and others) to promote financials solution for the poor against damage from climate change in developing countries by utilizing the pool through emissions trading. Past and current carbon polluters in developed countries try to compensate for damage caused by their emissions. Thus it will hold down the premiums and make the policies affordable for more people. Microinsurance is one form of insurance to sell with small premiums, which MCII also promotes. Although bad harvests cause farmers to fall behind in their debt payments, the insured will hedge the loss of drought with prompt payouts. Microinsurance can be sold to farmers in combination with microcredit. Although index-type insurance has the basic risk of a gap between the actual loss and payout, it realizes the prompt payment and also gives incentives to farmers to engage in loss reduction measures and to switch to more robust crops.

15.5 Public–Private Partnerships

[S]afety nets for high risk poor communities will not work without public–private partnerships, as no one partner can operate without the assistance of the others: highly exposed and fiscally unstable developing country governments cannot fully absorb the risks, informal community solidarity and family systems are overtaxed by large covariant losses, and private insurers cannot offer low cost policies, given the need for expensive reinsurance and large uncertainties in the projected loss estimates. (Arnold 2008)

In order to develop insurance products against the risk caused by climate change in Asia and in order to let the insurance work as financial protection, coordination between the public and private sectors need to establish databases, develop insurance products, transfer the risks, set affordable premiums, and reduce the risk through infrastructure (Table 15.3).

Table 15.3	Public–Private Partnership (PPP) for Climate-related Insurance	
Issues	Role of governments and international organizations	Role of insurance companies
Database	Basic data and research Awareness raising	Risk modeling Sharing the research outcome
Resilience enhancing measures	Regulation and enforcement	Incentives in product design Knowhow transfer from developed country to developing country
Vulnerable sectors and communities	Infrastructure Pilot adaptation scheme funding Diminishing livelihood support	Microcredit and microinsurance Risk transferred to reinsurance Pooled development funds
Risk transfer	Guarantee fund Volatility smoothing	Insurance if conditions of insurability are met otherwise services for public schemes
Disaster relief	Restricted Using hazard reduction and pre-funding	Flexibility of business during emergency Services for public schemes Claims under climate-impact insurance

Source: Adapted by authors from UNEP-FI (2006).

15.6 Conclusion

Insurance companies in Asia and the Pacific have a variety of financial solutions to climate change. In the conventional insurance business, they offer insurance products that cover damage caused by natural disasters. In addition, as mitigation measures they also invest in the clean technology sector to promote a low-carbon society. These activities in turn contribute to risk management to a certain extent.

The risk management offered by insurance companies, however, is not sufficient in Asia and the Pacific, since the infrastructure to collect precise weather data is not fully constructed. In addition, the catastrophe risk in Asia and the Pacific is too high to set affordable premiums for the low-income segment without subsidy. As UNEP-FI raised concerns, by 2025 insurers may withdraw from some markets, if the risks become too high (UNEP-FI 2006).

The establishment of weather observation systems, deregulation of the insurance industry as well as appropriate financial support by governments and international organizations will accelerate the growth of the insurance market in Asia and the Pacific. The control of the premium price will be the key challenge in microinsurance.

References

Arnold, M. 2008. *The Role of Risk Transfer and Insurance in Disaster Risk Reduction and Climate Change Adaptation.* Policy Brief. Stockholm: Commission on Climate Change and Development. http://www.ccdcommission.org/Filer/pdf/pb_risk_transfer.pdf

Asian Disaster Reduction Center (ADRC). 2008. *Natural Disasters Data Book 2008: An Analytical Overview.* Kobe, Japan: ADRC.

Association of British Insurers. 2005. *Financial Risks of Climate Change.* London: Association of British Insurers.

Brauner, C. 2005. *Tsunami in South Asia: Building Financial Protection.* Swiss Re Focus Report. Zurich, Switzerland: Swiss Reinsurance Company.

CRED-EMDAT. OFDA/CRED International Disaster Database. www.emdat.be (accessed 17 July 2010).

Cummins, J. D., N. Doherty, and A. Lo. 1999. *Can Insurers Pay for the "Big One"? Measuring the Capacity of the Insurance Market to Respond to Catastrophic Losses.* Philadelphia, US: The Wharton School, University of Pennsylvania.

Dlugolecki, A. 2009. The Climate Change Challenge. *Risk Management.* SCI (45, May), pp. 1–1245 (May).

Heck, P., D. Bresch, and S. Trober. 2006. *The Effects of Climate Change: Storm Damage in Europe on the Rise.* Zurich, Switzerland: Swiss Reinsurance Company.

Kiviat, B. 2009. Why the World's Poor Refuse Insurance. *Time.* September 21.

Munich Re. 2009a. *Climate Change and Impacts. Fact Sheet.* http://www.munichre.co.jp/public/PDF/2009_11_26_app1_en.pdf

———. 2009b. *Natural Catastrophes 2008.* Nat Cat Service. http://www.munichre.com/app_pages/www/@res/pdf/natcatservice/annual_statistics/2008_Natural_Disasters_perc_distrib_continent_touch_en.pdf

Parry, M. L., O. F. Canziani, J. P. Palutikof, P. J. van der Linden, and C. E. Hanson, eds. 2007. *Contribution of Working Group II to the Fourth Assessment Report of the Intergovernmental Panel on Climate Change.* Cambridge, UK: Cambridge University Press, UK, and New York, US.

Stern, N. 2007. *The Economics of Climate Change: The Stern Review.* Cambridge, UK: Cambridge University Press.

United Nations Environment Programme Finance Initiatives (UNEP-FI) Climate Change Working Group. 2002. *Key Findings of UNEP's Finance Initiatives Study.* CEO Briefing, Paris, France. http://www.unepfi.org/fileadmin/documents/CEO_briefing_climate_change_2002_en.pdf

UNEP-FI Climate Change Working Group. 2006. *Adaptation and Vulnerability to Climate Change: The Role of the Finance Sector.* UNEP-FI CEO Briefing, Geneva, Switzerland. http://www.unepfi.org/fileadmin/documents/CEO_briefing_adaptation_vulnerability_2006.pdf

United Nations Framework Convention on Climate Change (UNFCCC). 2007. *Climate Change: Impacts, Vulnerabilities and Adaptation in Developing Countries.* Bonn, Germany: UNFCCC. http://unfccc.int/resource/docs/publications/impacts.pdf

———. 2008. Report of the Conference of the Parties on its 13th session, Bali, Indonesia. December, 3–15. http://unfccc.int/resource/docs/2007/cop13/eng/06a01.pdf#page=3

Watanabe, H., K. Suzuki, and H. Yashiro. 2007. *Risk Management with Natural Disaster Information in Asia.* http://www.tokiorisk.co.jp/consulting/natural_disaster/pdf/ 070122_1.pdf

Chapter 16

Community-based Approaches to Climate Change Adaptation: Lessons and Findings

Masanori Kobayashi and Ikuyo Kikusawa

16.1 Introduction

Climate change has been widely recognized as a threat to ecosystems and human well-being. While measures for reducing GHG emissions need to be further insti-gated more intensively, it has also become a reality to adopt measures to adapt to climate change impacts, namely, increased climatic variability particularly flood and drought.

This chapter presents three case studies on community-based approaches to cli-mate change adaptation and/or co-benefit for mitigation and adaptation supported under the framework of the Asia-Pacific Forum for Environment and Development (APFED).[1] These case studies demonstrate the key elements to undertaking effective activities to tackle climate change particularly adaptation in the overall context of promoting community empowerment and sustainable development, and to provide perspectives for recommended policies and measures.

16.2 Community-based Approach to Adaptation: Case Studies

16.2.1 Case Study 1

Disappearing Lands: Supporting Communities Affected by River Erosion, Gaibandha District, Bangladesh.

16.2.1.1 Background

"Disappearing Lands: Supporting Communities Affected by River Erosion" is a five-year project implemented by Practical Action–Bangladesh, which was selected for the APFED Knowledge Initiative Awards for Good Practices in 2007 (IGES, 2009a). The project implementation region (Gaibandha district) is located in

---

[1] APFED was established as an Asia-Pacific regional group of eminent experts in 2001 with the support of the Government of Japan. The Institute for Global Environmental Strategies (IGES) serves as a secretariat for APFED. For further details, refer to www.apfed.net

north-eastern Bangladesh at the confluence of two major rivers: Tista and Brahmaputra. This geographical location makes the area vulnerable to floods and river erosion. Frequent disasters make life difficult in this district where stable land, employment, and infrastructure are lacking. People often take shelter in very marginal areas, where basic services such as safe water, sanitation, health, and education are minimal or non-existent. The remoteness of the district and the complexity of problems result in a high degree of social marginalization, human migration, child labor, exploitation, high mortality rates, early pregnancies, and violations of human rights.

16.2.1.2 Main Objectives and Activities

The project aimed to address the needs of the community based on four main objectives: (*i*) preparation of inhabitants to withstand the impacts of natural disasters; (*ii*) provision of basic services to people displaced by river erosion; (*iii*) provision of alternative livelihood options to people at risk; and (*iv*) improvement of social and political rights of disadvantaged people living in disaster-prone areas.

Key project activities included training communities in the technical skills of agriculture, fisheries, livestock keeping, agro-processing, light engineering, small enterprise development, volunteer group organization, community-based extension development, training of trainers, as well as the establishment of community services such as health clinics, refuge shelters, schools, and vocational centers.

16.2.1.3 Linkage to Biodiversity Conservation

Access to the common property resources particularly on water bodies and sand bar islands by resource-poor households has been created through negotiation with government administrations. The major achievements of the project are some innovations and adoption of agricultural practices, such as floating vegetable gardening and crop production in barren and unfertile sandbars utilizing pit cultivation techniques. These practices are now being replicated in other parts of the country, and in some cases, internationally.

16.2.1.4 Outcome

The immediate impacts of this project included improved household-level consumption, sales, and higher food security resulting from improvements in production techniques, as well as better management of basic health, water, sanitation, and livestock diseases. The training of professional groups led to a significant and diversified income source for farmers, fishermen, and livestock keepers, as well as local agro-processing groups. Adding to the generation of employment, the project helped to create volunteer groups equipped with rapid evacuation materials and early warning systems for action against sudden flooding. Households were resettled in cluster villages, which reduced migration rates as a result of the acquisition of stable livelihood options. The project facilitated the innovation of a number of pro-poor technologies in agriculture and disaster management, which have been replicated by interested nongovernment organizations (NGOs) outside the project implementation area. Additionally, some of the practices have been recommended by policymakers for integration in national policies, and dissemination to other areas of the country. The project has attracted interest from international NGOs in sharing the lessons learnt during its implementation, and has resulted in a number of publications in leading newspapers and journals both at home and abroad. This has also led to the establishment of a database and systematic monitoring and evaluation procedure. The project has successfully influenced national and international policymakers to act against global warming, climate change and poverty issues.

16.2.1.5 Lessons

This project has shown that a holistic risk-reduction approach complemented with alternative livelihood options, infrastructure support, and access to basic services has the potential for long-term impacts in reducing poverty. Participatory designing, planning, and development of the project ensured leading results, and high satisfaction of beneficiaries and other stakeholders in sustainable development. Technology-focused programs were able to bring about long-term change by building skills, knowledge and capacity, and promoting self-reliant development rather than being externally driven. Implementing projects in partnership with other stakeholders helped to intensify operations within a limited timeframe in a cost-effective way. The full-time presence of technical personnel in the field and direct communication with stakeholders added to the quality of support provided to partners and project beneficiaries. Networking with local governments and service providers proved to reinforce project implementation and the flexibility and cordial cooperation of donors was essential in producing positive results. Providing learning opportunities for project members has been proven to foster creativity and innovation for further achievements.

16.2.2 Case Study 2

Rehabilitating Desert Zone Ecosystems and Promoting Sustainable Alternative Livelihoods in Gobi Protected Areas, Buffer Zones and Peripheral Communities in Mongolia, South Gobi Province (Umnogobi Aimag), Mongolia.

16.2.2.1 Background

Mongolia occupies a critical ecological transition zone where the Siberian taiga forest, the Central Asian steppe, the high Altai Mountains, and the Gobi Desert converge. Due to the severe climate and low rainfall, soils are, however, thin and grass and pasture lands are becoming increasingly fragile. Land degradation and desertification have been identified as a top priority in the 2002 Mongolia State of Environment Report (UNEP 2001). One of the obstacles identified in the context of promoting the implementation of the protected area in the Gobi dry zone is the lack of incentives for local stakeholders to support natural and ecosystem conservation and sustainable management (IGES 2009a, 2009b). The protected areas have not provided sufficient incentives to local stakeholders and have failed to capture the nexus of the government's interests in promoting ecosystem protection and the local people's interest in improving their livelihood. Climate change is a threat to the agrarian society of Mongolia. There has been a gradual increase in the average temperature. More importantly, the level of precipitation that remains stable in the national average is on a declining trend in the South Gobi region (Figure 16.1). Increasing aridity further exacerbates desertification and drought, coupled with other anthropogenic factors such as overgrazing and over-use of underground water.

16.2.2.2 APFED Showcase Project

Under the APFED Showcase Programme of 2006, the project intended to promote sustainable ecosystem management and sustainable alternative livelihoods in

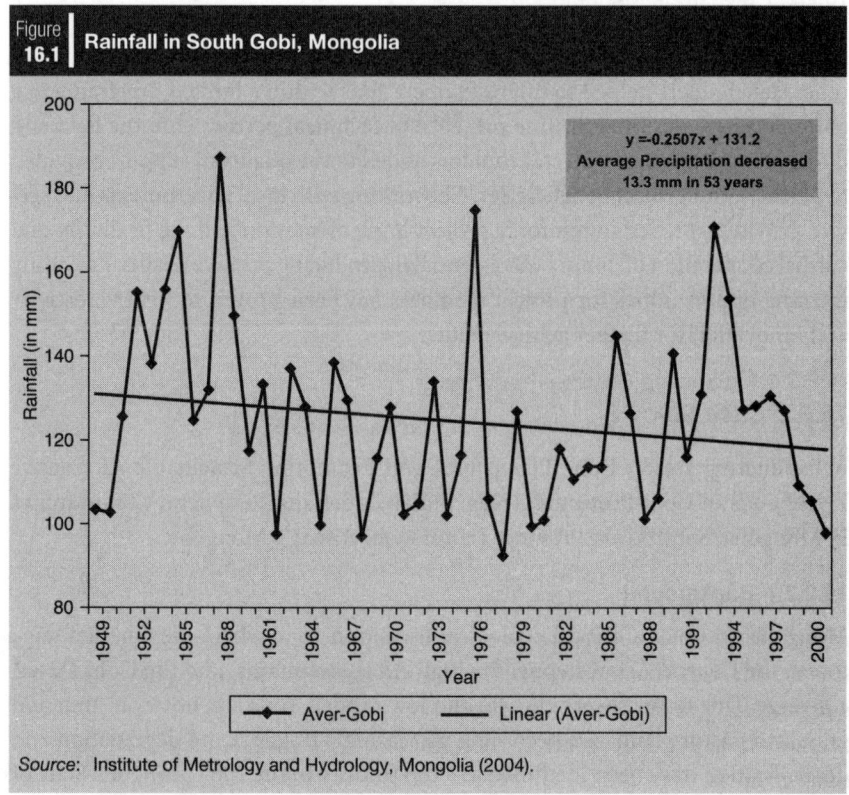

Figure 16.1 Rainfall in South Gobi, Mongolia

y =-0.2507x + 131.2
Average Precipitation decreased
13.3 mm in 53 years

Aver-Gobi   Linear (Aver-Gobi)

Source: Institute of Metrology and Hydrology, Mongolia (2004).

Gobi-protected areas, buffer zones, and peripheral areas. Project components were endorsed by the stakeholders in Bulgan Soum, Umnogobi Aimag in July 2007. It was agreed to pursue:

1. afforestation, reforestation, and agro-forestry on degraded land using selected drought resistant species;
2. protection and improvement of steppe and rangeland through construction of forage farms;
3. water supply system improvement (for example, restoration of abandoned wells and malfunctioning irrigation schemes);
4. sand dune stabilization by utilizing mechanical, botanical and engineering approaches;
5. establishment of sand breaks and windbreaks and natural forest protection;
6. control of carrying capacity of grassland and prevention of overgrazing; and
7. capacity building throughout the project activities.

16.2.2.3 Livestock Management

The increase in livestock numbers and the proportion of exploitative goat species are believed to be one of the major causes for desertification and pasture degradation. The total livestock number fluctuates in a certain cycle. However, it is notable that in Bulgan Soum, the proportion of goat species is higher than the national average, and is steadily increasing due to rising cashmere prices. The activities intended to promote alternative income generation provide local herders and villagers with alternative ways of life other than increasing goat species for selling cashmere. Such alternatives will reduce the grazing pressure in these areas, and help mitigate pasture degradation and desertification.

16.2.2.4 Optimizing Grazing Practices

Local concentrations of livestock and herders are also considered as causes of pasture degradation. Changing grazing patterns, namely a conversion from frequently relocating nomadic grazing patterns to settled grazing, is an underlying change in grazing practices over the decades. In order to restore traditional nomadic grazing practices to optimize grazing pressures across pasture land in proportion with pasture-carrying capacities, it is proposed to encourage local herders to prepare a pasture-use map highlighting pasture availability and a plan for pasture use in each season with maps that indicate the locations of water points and wells as important information. The challenge is that such a map is alterable over time, and herders need to observe it collectively in coordination with other groups that operate in adjacent areas.

16.2.2.5 Protecting Saxaul Woodland

Saxaul is a unique, but endangered shrub that was once densely growing in the Gobi region, but became scarce in the 1990s when poaching of saxaul trees was rampant due to excessive extraction for fuel wood. In the early 1990s, Bulgan Soum introduced a regulation to declare saxaul woodlands as protected areas, and the poaching of saxaul was curbed substantially. However, there emerged a group of poachers for a plant called Goyo. Goyo is a bulbous grass in the sandy soils used to lower blood pressure, relieve liver and spleen ailments and control internal bleeding, particularly in the PRC. It is collected in April and September. When mixed with goose-foot and Mongolian badana (bergenia crassifolia), Goyo is an important ingredient in a number of medicines. As Goyo grows in the root systems of saxaul, poachers damage saxaul root systems, and it is alleged that such poaching results the further destruction of the saxaul woodlands. It is thus important to control Goyo collection, and to provide poachers with alternative income-generating opportunities.

16.2.2.6 Farming as a Source of Alternative Income to Grazing

Local farmers try to boost income by increasing livestock numbers. One of the ways to restrain the increase of livestock is to provide incentives to engage in income-generating activities other than livestock grazing. One family rents a farmland of 0.07 ha in Bulgan Soum, growing tomatoes, watermelon, onions, potatoes, carrots, paprika, radishes, and melons. Elderly parents stay home to look after vegetable gardens, while young sons look after livestock. By having multiple sources of income, the demand for increasing livestock can be mitigated.

However, that would not necessarily eliminate the desire of livestock farmers to increase numbers. It serves only a mitigating factor. In addition, arable land space and irrigation water are limited. Water springs sporadically available in the areas provide a precious source of irrigation water. Open-air irrigation canals have a substantive loss of water through evaporation and penetration into the ground. Drip irrigation is another technique, but materials and technology still need to be further replicated. Droughts and unpredictable front in late summer also threaten farms. Remoteness from the market, limited seed supply, and the need for spices/breeding improvement are some of the other challenges that local farmers face.

16.2.2.7 Key Points

In light of various observations and discussions, it was agreed to pursue the following additional components:

1. sustainable pasture management: awareness raising of herders, dialogues on response options;
2. fenced wind-breaking woodland and pasture restoration zones;

3. water management: restoring obsolete wells, dispersing concentrated grazing pressures;
4. farmland creation;
5. saxaul protection: fenced zones, patrolling, awareness raising on illegal Goyo collectors, providing them with alternative livelihood;
6. handicraft;
7. solar cooking system demonstrations; and
8. educational panel on Gobi ecosystems.

As the project addresses multi-faceted issues of ecosystem management and community empowerment, various environmental and socioeconomic factors will be monitored and evaluated.

16.2.3 Case Study 3

Community-based Educational and Partnership Actions–Carbon Neutral Initiative for Community Empowerment and Climate Change Mitigation in Lombok and Bogor, Indonesia.

The 2007 APFED showcase project, "Community-based Educational and Partnership Actions—Carbon Neutral Initiative for Community Empowerment and Climate Change Mitigation in Indonesia," was launched with the Indonesian Institute of Science (LIPI), Bogor Institute of Agriculture, and local communities (IGES 2009a).

16.2.3.1 Background

Diversifying energy sources by promoting micro-hydropower energy demand has increased while the supply of electricity remains stagnant. Energy is supplied by oil (54.4%), gas (26.5%), coal (14.1%), PLTA (hydropower, 3.4%), geothermal (1.4%), and others (0.2%). The potential of renewable energy is estimated at 845 million barrel oil equivalent (BOE) by water, 219 BOE by geothermal and 500 MW by micro-hydro followed by biomass (49,81 GW), solar (4,80 kWh), and wind (2–6 m/s). Such renewable potential has not yet been utilized substantively. Seventy micro-hydro sites exist in Indonesia. A community-driven process for bidding to select a constructor also helps community empowerment. The government adopted a national energy policy with the Presidential Decree No.5 of 2006 to reduce the use of oil from 54.4% to 20%, increase coal from 14.1% to 33%, increase natural gas from 26.5% to 30%, and increase renewable energy from 0.2% to 17% 2025. Electrification growth rates have also been on the decline since 2000.

16.2.3.2 Cultivating Micro-hydropower

Only 15% of micro-hydropower potential is currently used. There are 6,200 villages that are not connected to the electric grid; the country is expected to continue to face difficulties in access to electricity. The Ministry of Energy Resources and Minerals provides training on energy generation techniques, including micro-hydropower. Gajamada University also offers a masters degree on micro-hydropower. In rural areas, micro-hydropower will help meet energy demand in agriculture, livelihood and social activities (school, health, cultural, and religious activities [IGES and LIPI 2008]).

1. *Energy and rural agricultural communities*: *Energy input in Indonesia's agricultural* sector is marginal. In rural households, firewood and charcoal are major sources of energy. The ratio of agricultural area to households has been decreasing. Micro-hydropower provides a useful solution to increase agricultural productivity with enhanced input of energy and to improve sustainability.

2. *Integrated energy supply*: Commercial, agro-industry (fertilizer, pesticides, and medicinal), bioenergy, and biomass all linked to agriculture must be developed in an integrated manner.

3. *Enabling macro-policy*: While the central government is required to advance enabling macro-policies, local governments also need to play a role. A target of 90% access to electricity by 2025 is being vigorously pursued, and micro-hydropower is considered important to achieve this objective. To do so, the central government has been intensifying collaboration with local governments. The central government encourages local governments to apply micro-hydropower at the community level. The central government provides assistance to install micro-hydropower generation systems, and operations can be carried out by communities. Local communities can gain economic benefits and use revenue for their own purposes. Such successful schemes can be widely seen in many countries.

4. *Cost-benefit of micro-hydro power*: Initial investment costs are a major constraint in developing micro-hydropower generation systems. However, as energy prices increase, there can be incentives for investors to cultivate renewable energy potential. However, as the government continues to pro-vide subsidies and stabilize the energy price, it does not stimulate action by investors. In remote areas which have difficulty in securing connectivity with a grid, there can be substantive interest in cultivating micro-hydropower potential.

16.2.3.3 Challenges

Geographic distance between micro-hydropower potential areas and villages in need of electric supply poses a major constraint in applying micro-hydropower. The volume of electricity generated is also limited and does not provide an optimal level of electric supply. The limited involvement of local villagers causes difficulty in supporting the proper management of micro-hydropower. Limited availability of technical experts is another constraint.

1. *Integrated natural resource management*: Forest management and reforestation are important for sustaining micro-hydropower. Degradation of surrounding forests lowers water availability and micro-hydropower generation, and reduces the income of local communities.

2. *Tenure*: Land tenure and land ownership inheritance are also important factors. When children inherit land, it is often divided, which poses problems in the use of river water for micro-hydropower. The division of land also reduces agricultural productivity.

3. *Land use change*: Sales of land to convert land use to industrial purposes also pose a constraint in developing micro-hydropower generation systems. Government regulations have been introduced to restrain the conversion of land. Illegal logging control, awareness raising and community-based law enforcement would be useful tools to contain illegal logging.

4. *Conflict of interests*: The methods of mediating conflicts of interest between the national electric company and micro-hydropower proponents remain an issue. The electric company buys electricity produced with micro-hydropower at a price slightly higher than the average sales price of 400 rupees as a voluntary action. With this system, micro-hydropower generation has a potential to bring profits to the community.

5. *Water-use competition*: Riparian villages tend to have conflicts over water use.

6. *Methodology of micro-hydropower*: Micro-hydropower follows two major approaches, mainly off-grid (a village that is not connected with a grid) and on-grid (a village that is already connected with a grid).

7. *Technological and engineering challenges and their socioeconomic implications*: Due to the change of water availability, micro-hydropower is still unstable in terms of electricity supply. This technical constraint still prompts local residents to access the grid of national electric companies. There are two major types of micro-hydropower turbines. One is called "crossflow" that has valves and a turbine situated in a top-down vertical direction. Water flows vertically and the turbine turns horizontally. It requires a height difference of eight meters. The other type is an "open flow." Valves are placed around the turbine in such a way that the turbine runs vertically in the same direction as waterfalls.

8. *Traditional knowledge*: In promoting micro-hydropower, traditional knowledge must be also utilized.

9. *Ownership of micro-hydropower and cooperatives*: Ownership of micro-hydropower generation systems is important to ensure the continuous commitment of local people. A cooperative scheme is useful. When there

is conflict between the central government and local government, this can make a cooperative relationship difficult. In national laws, for example, the government is supposed to provide 20% of the national budget for education, while local government gives priority to infrastructure development. There are conflicts over resource allocations of local governments; for example, local governments give priority to health rather than cooperative operation in micro-hydropower generation.

16.2.3.4 Case Studies

1. *Cintamekar*: The National Forestry Corporation (Perhutani) owns the area surrounding a locally operated micro-hydropower generation system, where a community is trying to promote reforestation. However, the community does not have sufficient financial and technical resources to carry out reforestation. On the other hand, Perhutani does not necessarily manage forests and promote reforestation. Forest maintenance being vital to harness water sources for micro-hydropower generation, the government therefore launched in 2005 a program for communities to generate micro-hydroenergy with Kelompok Usaha Bersama (KUBE), a business group supporting society on renewable energy.

2. *Nusa Dua, Bali*: Wind-power generation is promoted and management handled by local communities and rural cooperatives. Through the power generation from these activities, communities can gain benefits to support other income-generating activities.

3. *South Sulawesi En Rekang*: A community-based energy generation program has been undertaken to develop a specific product for a local community in South Sulawesi. However, local communities showed resistance due to distrust. The fact that the program followed a government-driven top-down approach also hindered positive project impacts. Although electricity provision has the benefit of diversifying lifestyles, it has been shown that simultaneous education is important in order to promote energy efficiency and avoid wasteful use of electricity.

4. *Tana Toraja, Sulawesi*: In this JICA project that generates 390 kW from micro-hydropower, awareness raising was key to ensuring effective implementation and gaining support from the community.

5. *People Centred Business and Economic Institute (IBEKA)*: IBEKA, an NGO originating from students' activities at the Bandung Institute of Technology, and promoting appropriate technology and social collaboration, has supported micro-hydropower generation with Swiss Contact (a Swiss NGO) and GTZ. Today a power company supplies electricity to city inhabitants, and a micro-hydropower plant supplies electricity to small villages.

6. *Cintamekar, Subang*: A joint venture run by local communities and a private company operates a micro-hydropower generation system. Initial investment costs were US$225,000 with cost-sharing between UN-ESCAP (US$75,000), Yayasan IBEKA, an NGO (US$75,000) and PT. Hidropiranti, a private company (US$75,000). Sixty percent of the revenue was reserved to repair defects in the turbine. The site was originally used for agricultural irrigation and covers 400 m length from the water-intake point to release point. Illegal logging in the upstream causes water shortages and impairs water availability. The government pays little attention to the repercussions of illegal logging on water harvesting. During the dry season, water becomes scarce. During the rainy season, water mixes with trash and impedes the operation of micro-hydro power generation systems. Sorting and removing garbage from water to be taken in the turbine is one challenge facing stakeholders.

7. *Village Sukaharja*: Village Sukaharja is located northeast of Bogor. There are about 1,000 households in Sukaharja and a total resident population of about 6,000. In Sukaharja, one of the typical agrarian rural communities, 25% of the total population is engaged in agriculture. Other than agriculture, villagers also engage in fabric and textile activities. A micro-hydropower system is under preparation within the house compound of a shoemaker who lives with his wife, a three-year-old daughter, and his parents-in-law. He employs six workers to make shoes for infants at home. He and his staff produce 600 pairs of shoes per week for domestic markets, and his earnings are 100,000 rupiah per week and 400,000 rupiah per month. His father-in-law also earns an equal income from agricultural work. From the aggregated household income of 800,000 rupiah per month from him and his father-in-law, he pays 60,000 rupiah per month for 405 kW required for shoe fabrication and 900 kW for household use, accounting for 7.5% of his household income. In the house compound, the Bogor Institute of Agriculture is in the process of setting up a micro-hydropower generation system. It has a 600 W per second capacity that only offsets a part of the shoemaker's electricity requirement for business and the Cintamekar have water volume of 1,500 liters per second (l/sec). To operate a micro-hydropower system, it is estimated that at least 200–300 l/sec is required.

16.3 Lessons and Recommendations

Through the analysis on the case studies presented in the preceding part of this chapter, a number of points were observed in reference to field-level activities to promote adaptation and mitigation in the context of tackling climate change. To highlight a few, the following points are deemed as worth noting.

16.3.1 Observations

16.3.1.1 Sensed Climatic Variability and Technical Observation

Local people share the perception that climate patterns have changed indicating, for instance, temperature increase, shorter winter, less snow, fewer rainy days, increase in cyclones and hurricanes. However, they often lack the data collected by meteorological monitoring stations. Sensed climatic variability is not sufficiently related to the meteorological data. Lack of such data makes it difficult to plan policy measures and activities for dealing with climatic variability.

16.3.1.2 Fewer Options for Marginalized Communities

Economically better off communities have more options for tackling climate change while disadvantaged and marginalized communities have little to choose due to the limited financial and technical capacities. Disadvantaged communities are in unfavorable climatic and topographical conditions such as dryland, areas affected seasonal aridity, flood prone river basins, and coastal zones with seawater intrusion.

16.3.1.3 Absence of Priority to Support Marginalized Communities

Neither in policy documents nor in financing operations is priority given to marginalized communities. Thus, in the implementation of climate change related policies, fiscal support and private investment are not necessarily directed toward marginalized communities that remain the most vulnerable to climate change. Such communities are not equipped with proper knowledge to materialize supporting schemes.

16.3.2 Recommended Actions

Based on the aforementioned observations, the following are recommended for promoting actions in the future.

16.3.2.1 People-centered Approach

Local stakeholders must be the center of the process for assessing options of actions, planning, and implementing activities. The measurement of impacts must be done by local people with the benchmarks and equipment suitable to be handled by local people. Incentive provisions need to be incorporated in the activities as a strategic tool to ensure the stakeholder involvement and support from the long-term perspectives.

16.3.2.2 Relating Field Actions to Macroeconomic Policy Evolution

It is important to find a way to channel the voices of local stakeholders and communities to national policy and decision-making processes with a view to facilitating macro-policy and institutional evolution.

16.3.2.3 Linking People's Perception with Scientific Data

It is important to provide local stakeholders and communities with climatic information and data and assist them in understanding the long-term trend of climatic variations and planning measures for tackling climate change. The provision of training and monitoring equipment needs to be facilitated as capacity-building activities.

16.3.2.4 Facilitating Stakeholder Dialogue on Options of Actions

To mobilize local stakeholders and communities and catalyze their support to collective activities, it is essential to provide support at the initial stage to stakeholder dialogues. In such dialogues, it is deemed as useful to involve external facilitates who can extract the views commonly shared by stakeholders and provide options of actions with objective information from holistic viewpoints.

16.3.2.5 Co-benefit and Lifecycle Approach

A co-benefit approach must be promoted to create multiple impacts from community and field activities addressing, for instance, mitigation, adaptation, and poverty reduction in an integrated manner. A lifecycle approach is also important, as the cost and benefit for the community must be assessed in the long term.

References

Institute for Global Environmental Strategies (IGES). 2009a. *APFED Message on Climate Change*. Kanagawa, Japan: IGES.

———. 2009b. *APFED Lessons and Findings on Biodiversity and Sustainable Development*. Kanagawa, Japan: IGES.

Institute for Global Environmental Strategies (IGES) and Indonesian Academy of Science (LIPI). 2008. Summary of the Workshop on Progress of Micro-hydro and Its Impacts on Sustainable Development, Bogor, Indonesia. September 28, 2008.

United Nations Environment Programme (UNEP). 2001. *Mongolia. 2001. Mongolia: State of the Environment 2002*. UNEP.

Chapter 17

Learning to Adapt: Case of Gender Alliance in Japan

Midori Aoyagi

17.1 Climate Change Projections by the Intergovernmental Panel on Climate Change

The Intergovernmental Panel on Climate Change (IPCC) Fourth Assessment Report (Solomon et al. 2007) reports on the changing pattern of rainfall and temperature rise, and their consequences. Annual temperatures will increase in the whole Asian region. The condition of this projection is the business as usual case, with the target year of around (or between 2080 and) 2100. Agricultural production will be affected in varying degrees.

In June, July, and August, precipitation in central Asia will very likely decrease, precipitation in East Asia will extremely likely increase, and precipitation in Southeast Asia will likely increase. In December, January, and February, precipitation in eastern Asia will increase extremely, and precipitation in Southeast Asia will decrease significantly. These precipitation changes will affect the agricultural production system and lead to an increase in natural disasters, such as heavy rain, floods, landslides, more powerful typhoons, and extreme heat. An adaptation strategy will aim at reducing damages from these natural disasters, changing the way of the agricultural production system.

In Central Asia and the northern part of Asia, degradation of fresh water resources, as well as climate induces a decline in food production. In South Asia and the People's Republic of China, some people might move their home because of degradation of the environment, and climate induced storm and flood disasters will be increased. In rural Asia, water, food, and natural disasters will increase the suffering of people's everyday lives. In these areas, where people's main income sources are agricultural, forestry or fishery production, suffering causes a degradation of people's living standards.

In the Japanese context, reducing damages from natural disasters will be the most important, as the country has very steep mountains and narrow riverbanks. In the Asian context, adapting agricultural production systems and preventing infectious disease will be the most important. In other words, food security is the most important, in everyday food supply and earning income. From gender perspectives, there are many issues to be discussed in this field. This chapter focuses on the disaster prevention process or disaster recovery process and gender issues in Japan. These processes might be of common interest in Asian countries.

17.2 Japan's Community-level Disaster Prevention (Recovery) System

The Japanese disaster prevention system is based on the Disaster Countermeasure Basic Act and the disaster prevention plan. The plan emphasizes the "participation by all the stakeholders" for decision making and implementation of the regional disaster prevention plan in each region.

For residents' participation, the plan encourages each community to organize "jishu bosai sosiki" (literally translated into English as "self-organized disaster prevention organization") for disaster preparedness and rescue activities at the community level in Japan (Bajek et al. 2008). This organization is usually based on the "chonai-kai" (community council) or "jichi-kai" (neighborhood community association). Kurata (2000) summarized the activities of these community-based organizations as follows: *(i)* a unit of household and compulsory participation; *(ii)* a traditional and uniform system over the country; *(iii)* a cooperative relationship between public sectors; and *(iv)* a base unit for daily and basic community events. As membership is compulsory for residents, these neighborhood-based associations play an important role as an informal "subsidiary organization" of local government.

The need for these neighborhood-based organizations was recognized at the Hanshin-Awaji-Earthquake in January 1995, which was one of the largest earthquake disasters (before the Tohoku-Earthquake in March 2011) in Japan. The experience of this earthquake and recovery from this disaster resulted in a revision of the related laws. In the disaster law, "to foster jishu-bosai soshiki" item was added. The 2007 Whitepaper on Disaster Prevention, Chapter 2, section 2, item 4 stated, "neighborhood organizations rescued about 80% of afflicted people." The reasons why these neighborhood-based communities played such an important role were:

- People knew who lived where: Searching efficiently.
- People knew faces of others: Easy identification of rescued person.
- These organizations work not only in earthquakes, but flood, fire, and other disasters.

According to the White Paper, the average participation rate of jishu-bosai-soshiki was more than 60%, and in several regions, it was more than 90% in 2004 (Japan Cabinet Office 2007).

17.3. Neighborhood Associations and Gender

Table 17.1 shows the membership distribution by sex and age in one of the example organizations in Kishiwada city, Osaka (Bajek et al. 2008).

Table 17.1	Sex versus Age Cross-tabulation					
			Age (years)			
			20–30	**40–50**	**60–70**	**Total**
Sex	Male	Number	0	4	17	21
		% of total	0.0	10.8	45.9	56.8
	Female	Number	2	11	3	16
		% of total	5.4	29.7	8.1	43.2
Total		Number	2	15	20	37
		% of total	5.4	40.5	54.1	100.0

Source: Bajek, Matsuda, and Okada (2008).

The example identifies two groups in the membership of this association. One is the over 60s male group, the other is the 40s–50s female group. This distribution reflects the structure of the neighborhood association itself. As membership is based on the household unit and compulsory participation, the over 60s male group was a group of "representatives" of each member household, and the 40s–50s female group is the women's branch of the community's disaster prevention unit.

17.4 Women's Alliance for Disaster Prevention

Women's alliance for disaster prevention is based on the networks of the women's branches of neighborhood-community associations, or community councils. This alliance is not in charge of disaster preparation or recovery, but handles various community activities. Through these activities, individual members communicate with the local people, and also keep a network of member organizations.

Since its member organizations are based on the community's women's branches, it has the advantage for domestic (home) disaster prevention activities among everyday life, and recovery activities. Women are expected to join activities such as caring for others (younger people, older people, people who need help), doing domestic work (such as preparing meals, for people at evacuation camps) in the recovery process, while men are expected to join the "physical" recovery process, such as repairing roads and bridges, for emergency operation.

Those women's activities are extremely important in an emergency situation. But as these expected roles are based on gender, there are also problems. For example, these roles are regarded as a supporting role, so that women do not usually take part in the decision-making process even during evacuation. Another issue is about women who have jobs outside the community. An example is during the Niigata-Chuetsu earthquake recovery process, one women who was at management-level in a company, could not return to her office until the evacuation camp had closed, and it took about two weeks, while her male colleagues in the same community could return their office after in several days. As a result, she was fired because of her "long days off," there are gaps between the traditional gender role and the contemporary "female social role."

The National Basic Plan for Disaster Prevention emphasizes "the role of women," especially, taking part in the "decision-making" process. But, the current neighborhood-community–based association system is not appropriate for women to join the decision-making process, as women are not regarded as "formal members" or "representatives of the household." Many Japanese government systems are based on this "household" concept, and also a strong "gender role" culture. It is very difficult for women to take part in the decision-making process in every level of the society, thus underutilizing women's ability.

The Anamizu Sengen (The Anamizu Declaration by Anamizu-Town in Ishikawa Prefecture after the Noto-earthquake) ("A Conference for Disaster Prevention women" Anamizu-Town, Ishikawa) is the result of the conference for the disaster prevention by women in 2007. The declaration emphasizes "the need for caring for everyone who are disadvantaged, not only children and older people, but also working mothers, and people who are disadvantaged in various aspects."

17.5 Policy Recommendations

Current strategies have to continuously be revised to take account of the role of women. Government disaster prevention activities tend to use conventional social networks, (for example, a residents' council). But often these networks are not gender sensitive, and underutilize women's abilities.

There are ongoing movements in Asia, one being the Manila Declaration for Global Action on Gender, Climate Change, and Disaster Risk Reduction (Manila, Philippines, October 22, 2008) by UN/ISDR (International Strategy for Disaster Reduction). Its contents are similar to the Anamizu declaration.

References

Bajek, R., Y. Matsuda, and N. Okada. 2008. Japan's Jishu-Bosai-soshiki Community Activities: Analysis of Its Role in Participatory Community Disaster Risk Management. *Natural Hazards*. 44 (2). pp. 281–292.

Japan Cabinet Office. 2007. *The 2007 Whitepaper on Disaster Prevention*. Publishing Office of Ministry of Finance. http://www.bousai.go.jp/hakusho/h19/index.htm (in Japanese).

Kurata, W. 2000. Community Katsudo to Jichikai no Yakuwari (Community activities and the role of residents association) (In Japanese). *Kwansei Gakuin University Bulletin*. (86). pp. 63–76.

Schubert, R., H. J. Schellnhuber, N. Buchmann, A. Epiney, B. Griebhammer, M. Kulessa, D. Messner, S. Rahmstorf, and J. Schmid. 2007. *World in Transition: Climate Change as a Security Risk, Flagship Report 2007—German Advisory Council on Global Change*. London: Earthscan. http://www.crid.or.cr/digitalizacion/pdf/eng/doc17839/seccion-a.pdf

Solomon, S., D. Qin, M. Manning, Z. Chen, M. Marquis, K. B. Averyt, M. Tignor and H. L. Miller, eds. 2007. *Contribution of Working Group I to the Fourth Assessment Report of the Intergovernmental Panel on Climate Change*. Cambridge, UK and New York, US: Cambridge University Press.

United Nations International Strategy for Disaster Reduction (UNISDR). 2008. *Manila Declaration for Global Action on Gender in Climate Change and Disaster Risk Reduction*. http://www.unisdr.org/preventionweb/files/8024_ManilaDeclarationforGlobalActiononGenderinClimate.txt

Chapter 18

Structural and Nonstructural Adaptation Measures of Climate Change in India

Agastin Baulraj

18.1 Introduction

The objective of this chapter is to show how substitution between structural and nonstructural adaptation measures by a community or a donor agency would lead to higher utility or a specified utility at low cost. Natural resource management can be improved by using nonstructural measures along with structural measures.

The magnitude of climate-induced disasters has increased. Myers (1997) believes that the issue of environmental refugees "promises to rank as one of the foremost human crises of our times." According to him environmental change and the natural and man-made disasters associated with it are forcing millions of people to flee their homes and seek refuge in neighboring countries or in other regions of the same country. Myers claims that there were at least 25 million environmental refugees in the mid-1990s, and this unrecognized category exceeded the then 22 million refugees as officially defined (Myers 1997). Further, he argues that when global warming takes hold there could be as many as 200 million people displaced by disruptions to the monsoon system and other rainfall regimes, by droughts of unprecedented severity and duration, and by sea-level rise and coastal flooding.

With the ever-increasing demands of expanding human population, horrifying effects of climate change and irresponsible use of water, we are heading towards a crisis. The picture is bleaker for countries like Pakistan and India, where the economy is based on agriculture. Climate change is a challenge to societies and economies around the globe.

Adaptations occur relative to the stimulus—anticipatory and reactive. Adaptive strategies are based on the time frame of the stimulus: long range, tactical, and contingency. The review of literature provides a list of adaptive management measures: structural or infrastructural, legal and legislative, institutional, administrative, organizational, regulatory, educational, financial, research and development, and technological change.

Policymakers have two options—structural and nonstructural measures of adaptation. Both have some benefits and costs associated with them. The scarcity of resources necessitates allocation of resources between structural and nonstructural measures.

18.1.1 Structural Measures

Structural measures refer to any physical construction to reduce or avoid possible impacts of hazards, or application of engineering techniques to achieve hazard resistance and resilience in structures or systems. For example, structural measures are dams, flood levies, ocean wave barriers, earthquake-resistant construction, and evacuation shelters.

18.1.2 Nonstructural Measures

Nonstructural measures are those that use knowledge, practice, or agreement to reduce risks and impacts, in particular through policies and laws, awareness creation, disaster-preparedness training and education, compilation of data and research, land use planning, and insurance.

It is the duty of governments and other stakeholders to optimize the adaptation measures. All economic agents are optimizers. Any economic agent will choose values for decision variables so as to optimize the value taken by the objective function be it profit or utility function. The aim of the optimization process is to find out what investment should be made in each type—structural and nonstructural—and what output (net benefit) should be set so as to maximize to the greatest degree possible.

When international agencies and governments give humanitarian aid and other developmental loans the need to optimize arises.

18.2 Structural and Nonstructural Measures: Substitute or Complementary?

Are structural and nonstructural adaptation measures substitutes or complementary? They are neither perfect substitutes nor perfect complementarities. At the aggregate level and at the disaggregated level it is different. It is different at global, regional, national, and subnational levels. Social and economic conditions give rise to different configurations of actors in different countries at different times; these societal and economic actors exercise influence not only on the making of policy but also on the making of institutions.

Policy making in any country in the globe cannot be understood without reference to the complex relations between the national government and provinces—which are in turn affected not only by the formal institutions but also by underlying economic social structures throughout the country and so on.

Similarly in Asian countries the choice between the structural and nonstructural adaptation measures can be decided based on the socio-political conditions of the micro, meso, macro, and mega levels.[1]

Table 18.1 shows that enormous funds flow from countries, individuals, and other donor agencies for humanitarian purposes relating to floods, earthquakes, and tsunamis. It includes structural and nonstructural measures in each project. It becomes very important to use the funds efficiently. For maximizing utility the mix of structural and nonstructural adaptation should be identified depending on the local conditions. The limitation is that data is in aggregate form at the macro level. The climate change-related disasters have increased the appeals and consequently the commitments of financial resources to countries.

Figure 18.1 shows the comparative position of countries among the fifteen selected countries that receive aid. Bangladesh, Sri Lanka, Pakistan, and Indonesia obtained more aid in 2010. A few countries like Bhutan, Singapore, Mauritius, and Malaysia have not obtained any up to August 2010. The funds received are used for adaptation measures and relief measures. For example in the tsunami relief fund granted to India, structural measures and nonstructural measures were adopted.

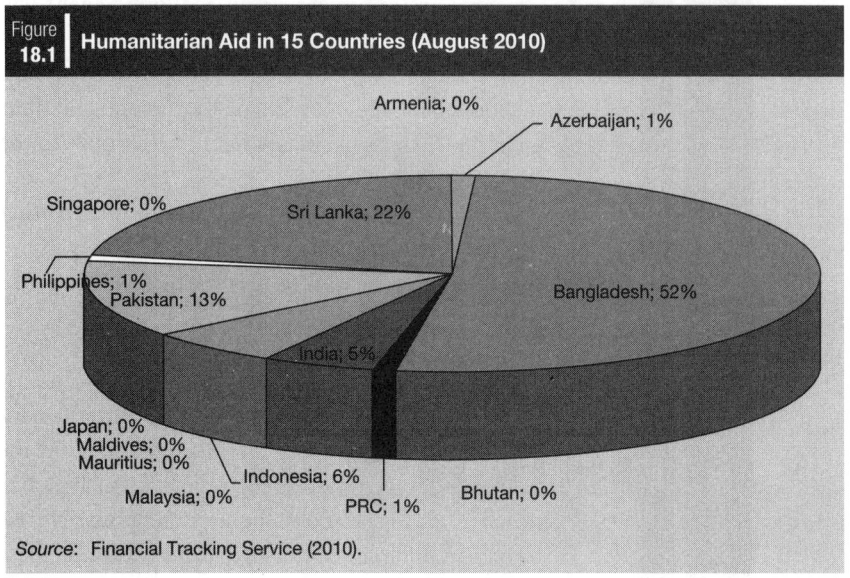

Figure 18.1 | Humanitarian Aid in 15 Countries (August 2010)

Source: Financial Tracking Service (2010).

---

[1] MEGA: Global level (UN, UNFCCC, WTO, WB, international civil society)
Regional level (ADB, ESCAP, SAARC, regional civil society)
MACRO: National level (national government, National civil society)
MESO: Subnational level (district, local river basin, eco-specific)
MICRO: Local level (community, family, individual)

Table 18.1 Humanitarian Aid to Selected Asian Countries from 2001 (August 2010) (US$)

Country/year	2010	2009	2008	2007	2006	2005	2004	2003	2002	2001
Armenia	703,002	328,024	110,063	703,002	3,587,284	2,335,235	933,711	361,784	2,567,270	2,527,378
Azerbaijan	3,224,632	1,125,176	1,828,155	3,224,632	2,095,222	5,065,881	789,589	3,917,060	3,894,646	3,357,484
Bangladesh	306,589,903	29,137,237	42,954,439	306,589,903	12,217,624	6,348,527	108,909,403	3,601,881	6,258,923	1,906,163
Bhutan	43,228	1,928,996	0	43,228	190,000	0	0	0	0	0
People's Republic of China	3,110,613	5,790,346	310,087,160	3,110,613	6,788,123	4,990,572	6,685,809	11,207,128	5,496,143	1,323,665
India	29,434,815	106,888,673	24,525,340	29,434,815	7,667,192	95,827,711	7,866,544	5,396,483	96,500,351	125,526,952
Indonesia	35,272,092	73,123,625	13,014,198	35,272,092	133,473,696	1,426,605,943	24,190,152	45,197,543	84,625,252	13,184,336
Japan	66,667	0	0	66,667	0	0	199,261	0	0	0
Malaysia	0	28,346	171,857	0	538,679	520,000	225,571	0	0	0
Maldives	150,000	0	1,000,000	150,000	654,135	113,015,871	0	0	0	0
Mauritius	0	0	0	0	0	29,000	0	21,529	670,533	0
Pakistan	78,648,535	711,643,181	43,349,159	78,648,535	110,644,834	1,171,548,220	6,957,128	10,191,240	10,705,721	48,246,126
Philippines	4,660,212	117,645,314	22,681,001	4,660,212	32,553,190	3,486,468	19,666,842	1,666,582	587,876	699,406
Singapore	0	0	0	0	4,839	0	0	0	7,517	0
Sri Lanka	132,715,533	263,426,126	172,719,601	132,715,533	70,288,846	704,639,795	25,499,891	20,839,109	8,665,624	7,299,484
Total	594,621,242	1,311,067,053	632,442,981	594,621,239	380,705,670	3,534,415,228	201,925,905	102,402,342	219,981,858	204,072,995

Source: Financial Tracking Service (2010).

18.3 Optimizing the Choice of Structural and Nonstructural Measures

When an individual has to maximize his utility (may be in terms of nutrition) by consuming apples, he has no choice to substitute and it is a case of single optimization. In adaptation measures relating to climate change it could be merely a construction of a check dam and the donor grants funds for that specific project and there is no choice to make. But when an individual is left with a choice of apples and oranges and a budgetary allocation then he has a choice and he can substitute the two, that is, he can either maximize his utility or minimize the cost of obtaining a particular level of utility. The optimization methods could be further classified as constrained and unconstrained. If there are no constraints of any kind then it is a case of unconstrained optimization. When there are constraints like limited budgetary allocation, employment policy, or technology policy, then it is a case of constrained optimization.

In the case of climate change adaptation measures countries—especially developing and least developed countries—are faced with budgetary allocation constraints. Though international agencies and governments give humanitarian aid and other development grants or loans, policymakers have to act within the budgetary constraints and other policies agreed to by them. In the discussion of structural and nonstructural adaptation measures, there could be constraints like "at least one major structural work to be completed" or "a particular percent of the funds allotted should have been spent in housing or other structural measures," or similar constraints. Policymakers at the national level, regional level, district level, or community level will need to optimize their utility subject to constraints. Generally there are constrained optimization situations. One such optimization technique is known as Lagrangian method. It is a method of optimization. Another tool that helps explain maximization is the indifference curve that is used to explain the choice between structural and nonstructural measures of adaptation.

Indifference curve: A basic tool of Hicks–Allen ordinal analysis of demand. Indifference curve represents all those combinations of goods that give the same satisfaction to the consumer. The donor agency or implementing agency will act like a consumer in maximizing the level of satisfaction or utility. Indifference curve (IC) represents consumer's (the donor's) scale of preference between two goods.

Price line or budget line: Price line shows all those combination of two goods which the consumer (donor in this case) can buy by spending his money on the two goods at given prices.

Let us take a hypothetical example that the fund allotted for adaptation measures is US$100 million and a type of structural adaptation measure (such as a check dam, a small dam that is usually two meters high built across small rivers or canals) costs

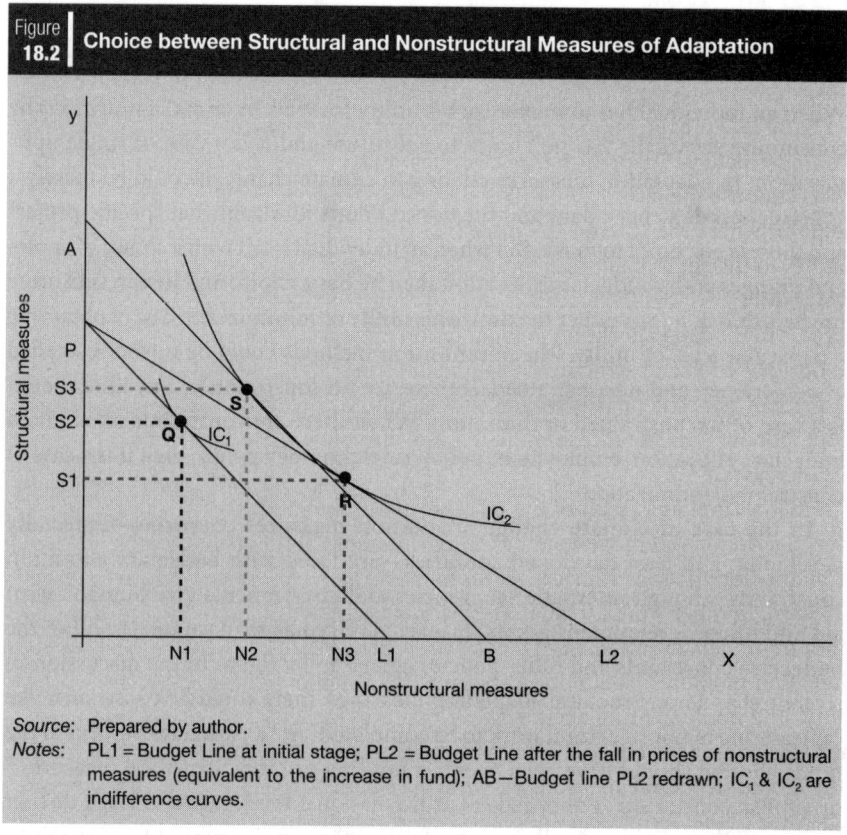

Source: Prepared by author.
Notes: PL1 = Budget Line at initial stage; PL2 = Budget Line after the fall in prices of nonstructural measures (equivalent to the increase in fund); AB—Budget line PL2 redrawn; IC_1 & IC_2 are indifference curves.

US$10 million and a package of nonstructural measures (education, awareness, and training component for different stakeholders) costs US$4 million. If all US$100 million is spent on structural measures, 10 check dams (represented by OP in Figure 18.2) can be constructed. If all the money is spent on, for example, the package of nonstructural measures, then 25 packages of nonstructural measures (OL1 in Figure 18.2) can be adopted. Price line (PL1) is drawn by connecting the points that represent the maximum structural measure (OP) and maximum nonstructural measure (OL1) that can be adopted with the budget allotted.

Similarly when the price of nonstructural measures is reduced by 50% due to the increase in the supply of service providers (in effect this could be equivalent to a higher allocation of budget for adaptation measures), the situation would be that 50 packages of nonstructural measures could be adopted. If the price of the other product (structural, check dam in this case) remains the same, the price line will become PL2 which means that 50 packages of nonstructural adaptation measures could be adopted, if money is spent in full on nonstructural measures. PL2 is drawn by connecting the two points that represent the maximum structural measure (OP) and the maximum nonstructural measure (OL2) that can be adopted with the budget. AB is the line that is a different possibility with the same budget as of PL2 but with different combinations of structural and nonstructural measures.

Figure 18.2 uses an indifference curve to explain how different mix of structural and nonstructural adaptation measures help maximize the satisfaction or utility. Let us suppose the farmers in a river basin enjoyed a specified utility at point Q, the point where the IC_1 and the price line (PL1) are tangent. Indifference curve that lies to the right of another indifference curve shows that it represents higher level of utility. So after the price reduction of nonstructural measures (equivalent to increase in the fund allocation for adaptation measures), the points that represent the optimal choice of structural and nonstructural measures would be R. When the price of nonstructural measures falls the donor or implementing agency can adopt more of both the measures, that is, the purchasing power of the funds available increases.

It follows that as a result of the increase in purchasing power due to the fall in price, the donor will move to a higher indifference curve and will become better off than before. It is as if prices had remained the same, but his money income increased. In other words, a fall in the price of nonstructural measures does to the donor or implementing agency what an equivalent rise in budgetary allocation for adaptation measures would do. At point S, OS3 and ON2 combination gives the utility. The implementing agency will substitute nonstructural measures (which has become cheaper relatively) for structural measures. At R the (OS1 and ON3 is the mix of structural and nonstructural measures) utility is higher than at point Q, because higher indifference curve represents higher utility or satisfaction. S and R could be preferred to the point Q. (Remember the budget for AB and PL2 is the same. The size of squares in the diagram does not represent utility.)

18.4 Indian Experience of Adaptation

In India the experience from the 2004 tsunami and the adaptation measures that are followed help to understand the optimization of structural and nonstructural measures of adaptation. Climate change results in natural disasters that cause problems in natural resource management. The tsunami had its impact on agriculture, fisheries, and livelihoods. The non-fishing community, which includes small and marginal farmers and agrarian workers and other laborers living on the coast, suffered a double blow because their losses were not recognized and counted for a long time. The fishing community was more visibly devastated since structural measures like shelters, fishing boats, and other equipment were destroyed.

The Nagapattinam and Cuddalore districts of India affected by the 2004 tsunami experienced sea water flooding, destroying hectares of agricultural land. The NGO Coordination and Rehabilitation Center (NCRC), Nagapattinam, urged the

Table 18.2	Sources of Long-term Funding and Assistance to Tamil Nadu State		
Name of agency	Fund flow	Term	Activities both structural and nonstructural
World Bank	US$423 million	3 years	Emergency Tsunami Reconstruction Project, housing, infrastructure, shelter belts, repair of public building, mangroves, fisheries
Asian Development Bank	US$143 million	3 years	Tsunami Emergency Assistance Project, livelihood promotion, transportation, restoration including bridges, ports and harbors rural and municipal infrastructure, water supply, sanitation
International Fund for Agricultural Development	US$30 million	8 years	Livelihood program, micro and rural finance and microenterprise development, community resource management
Prime Minister's National Relief Fund	US$5.2 million (233.5 million Indian rupees)		Ex gratia payment
Government of Rajasthan	US$3.9 million (178.4 million Indian rupees)		Infrastructure, education, livelihood assistance for fishermen community

Source: Compiled from Govt. of Tamil Nadu, Tsunami Rehabilitation Programme Office Reports, Chennai.

Tamil Nadu Organic Farmers' Movement with 30,000 members led by a scientist, to take up the issue. Once their problems were brought to the fore, the district authority gave the landowners 1,250 Indian rupees each for a hectare of land (Table 18.2).

The NCRC along with the Tamil Nadu Organic Farmers' Movement was actively engaged in desalination work. The salt was leached, and then deep plowing was done to improve the quality of the soil. Green manure seeds like *daincha* were cultivated which are salt solvent. Likewise, two to five times of organic manure per hectare of land was applied. Then the people started cultivation.

The agricultural sector in India has introduced a number of adaptation measures (structural and nonstructural) such as the introduction of new varieties, biodynamic farming, land reclamation, irrigation facilities, canal desilting, pond desilting, check dams, and water management practice. These measures include the following initiatives.

1. The new crop varieties released to counter climate change in Tamil Nadu, India include Tamil Nadu Agricultural University Rice–CO 50, suitable for cultivation as transplanted rice throughout the rice growing areas of Tamil Nadu; and Rice TRY 3 for adoption. The vegetable crops released were brinjal VRM 1, with a yield potential of 49.88 tonnes per hectare and suitable for adoption in Vellore and Tiruvannamalai districts; tomato hybrid CO 3, a

150-day crop with yield potential of 96.2 tonnes per hectare; chili hybrid CO 1, and celery OTY 1. Farm implements released were the needle type tray seeder for vegetable nurseries and a trailer mounted steering for power tiller–trailer system.

2. Bioremediation of selected water bodies has been carried out in the Tamil Nadu region—Chennai, Cuddalore, Nagapattinam, and Kanyakumari. In Chennai alone, bioremediation has been undertaken in five lakes.
3. Raising mangroves and shelterbelt plantations.
4. Provision of basic amenities and infrastructure like storm water drains to new habitations in 16 tsunami-affected town panchayats (a unit of local administration in India).
5. Desilting, widening, and strengthening of canals and banks in Nagapattinam district, badly affected during the 2004 tsunami.
6. Climate change and other factors affect the livelihood of farmers. An extent of 8460.34 hectares of agricultural land was saline-affected after the 2004 during tsunami. Farmers have been trained in nonstructural adaptation measures through Krishi Vikas Kendras. Soil samples were taken periodically to monitor the progress of reclamation. This is a perfect case of optimization of structural and nonstructural measures.

The structural and nonstructural adaptation measures have to be optimized. Just as an individual maximizes his utility function, the government, agencies, or donors have to maximize their utility function. The right combination of structural and nonstructural measures of climate change adaptation becomes a necessity to maximize the utility function.

To cite a case as an example, the sanitation facilities provided in a community had to be reinforced with training on sanitation for school children and community workers to increase the net benefit. Here the benefits from structural adaptation were further increased by the nonstructural adaptation measures. Under the Total Sanitation Project in India, sanitary complexes (public toilets) were constructed as a common good. Toilets were also constructed in schools and individual houses. The experience in India shows that the structural measures alone did not maximize the utility function, that is, they could not give the desired results. People got subsidies and bank loans and constructed toilets at home. However they did not use them, and continued with their old practices. In some places people used the constructed toilets as small cattle sheds. In other places people constructed toilets with inferior materials and they could not be used for long.

But the nonstructural methods improved the utility function. The screening of films to school children and adults in villages explaining rheumatic heart diseases and other health hazards, the need for washing hands after the use of toilets have rendered good results. Sometimes the school authorities had to be taught the importance of keeping the toilets open during school hours. A scheme of rating schools was used to motivate the administrators of schools. The lessons learnt were that the structural measures and nonstructural measures have to be carried out side by side.

What is the optimum level of structural and nonstructural measures depends on the education level of the community, willingness of the stakeholders to help attain the objectives, among others. Similarly in the tsunami-affected areas displaced people in temporary shelters did not use the toilets. A survey taken showed that toilets were damaged and the people did not know the importance of using toilets. In many places, the maintenance of public good was the problem, though all structural measures were sufficiently provided. The training of field-level workers and then conducting competitions and other events for women and school children separately (nonstructural component) was used to cultivate the habit of using toilets (the structural component).

In the tsunami-affected areas, a donor agency, Church's Auxiliary Social Action (CASA), had given common work-shed facilities and financial assistance before giving villagers the needed skills like business plan preparation, and team building in case of group activities. The activities failed. Later CASA identified that nonstructural measures like training in marketing, inventory management, and team building needed to be provided, to complement the structural measures. There was a turnaround in the functioning of the groups and individuals. Though the common work-shed (structural measures) facilities are important in some livelihood activities, the nonstructural component proved to be of more use in sustaining the activities.

18.5 Constraints in Integration of Structural and Nonstructural Measures

Much of the necessary weather and climate data are not easily accessible to the scientific community either because of security concerns or institutional constraints. The various data are not only being collected by government agencies but the private sector also collects the weather information for their specific research purposes.

To make climate information available to common people, it is necessary to build a comprehensive, robust, and freely accessible weather and climate data portal along with current information on weather and climate. Very few meteorological observatories are in hilly or geographically remote areas to collect atmospheric and surface meteorological variables.

The observational networks should undergo changes according to evolving needs, and should be modernized to a major extent by national agencies. The importance of sea surface temperature (SST) in addition to the other meteorological variables in monsoon and climate simulations is already well documented in scientific literatures. There is an urgent requirement not only to construct other observed climate related data of high resolution, but also, with high accuracy, fine resolution grid data of SST over the Indian Ocean. The scientific community depends very much on international agencies to get reliable high resolution weather and climate data. The Himalayan and Tibetan snow region has a well-established link with monsoon rainfall and water resources on northern India. As the cryosphere is an important component of the climate system, there is a need to organize data on glaciers, glacier lakes, snow cover, and snow depth. The glacial melt and snow impact on water resources can be better studied, if there is remote sensing and satellite data of high spatial resolution over snow covered regions.

Like global reanalysis atmospheric and surface data, the regional reanalysis project for South Asia can provide in-depth understanding of past and present climate variability at a fine spatial scale. In order to understand the atmospheric and oceanic boundary layer processes, the Department of Science and Technology, Government of India, has conducted the monsoon trough boundary layer experiment (MONBLEX), INDOEX. More sophisticated meteorological towers are in need to major fluxes at selected sites and stations to understand the processes of atmospheric boundary layer and meso-scale phenomena.

Providing climate information to the end-user community can be best achieved by collecting and improving fine resolution satellite data for forest sectors, water resources, agriculture, and others. These data sets may provide a unique opportunity to conduct impact and vulnerability studies.

Climate change adaptation should be done at a community level to succeed. The local communities are divided. Local bodies are not resourced with trained personnel in the field of climate change. There is a lack of awareness among local leaders. Climate change and natural resource management is taught in schools using unimpressive traditional ways focusing only on examinations and not on transferring knowledge to society. Using nonstructural adaptation measures of training the other stakeholders would be more effective. Higher educational institutions also do not emphasize the application of knowledge to the field. Though environmental studies are introduced in undergraduate courses, natural resources management at the villages around the institution has not improved. But students are sensitized to the needs of the society.

18.6 Effectiveness of Weather Insurance as Adaptation Strategy in India

Insurance is a nonstructural mechanism that helps private individuals gain resources to recover from disasters such as coastal flooding and is a potential response mechanism. The following are key elements of weather insurance:

1. Provides protection against adverse deviations in a range of weather parameters like frost, heat, relative humidity, and rainfall between December and April.
2. Generic insurance product insuring crops like wheat, potato, barley, mustard, and grain.
3. Maximum liability is linked to the cost of cultivation and varies from crop to crop.
4. Allows for speedy settlement of claims, say within 4–6 weeks after the insurance period.

Weather insurance is a mechanism for providing effective risk management aid to those individuals and institutions likely to be impacted by adverse weather incidences. Though there are some improvements to be made in the functioning of the scheme, it is worth emulating. Table 18.3 shows the claims settled and farmers benefited from the insurance scheme specially designed for the agricultural sector by a separate agency.

The crop insurance in practice is based on an area approach and not on the individual farm approach. This is very useful for farmers who are affected by natural disasters induced by climate change. The participation of farmers is not uniform. In some areas they are willing to join the scheme, while in other areas the agricultural extension officers have to adopt persuasive methods, depending on the personal judgment of risks.

18.6.1 Best Practices by Deccan Development Society

The Deccan Development Society, an NGO, has been working in 75 villages in India on sustainable agriculture and food security issues with marginal farmers from *dalit* (marginalized) groups for the last 15 years. The farmers of the Deccan Development Society emphasize two fundamental aspects that are very critical to pest management.

1. Enhancing soil fertility for building stronger and richer soil that can be the first antidote to the pest attack.
2. Enhancing the biodiversity in the farms as the first defense against insects and pests.

In addition, farmers are following non-pesticide management options, most of which are based on farmers' traditional knowledge systems.

In India, the National Disaster Management Authority has been formed. There has been a shift in emphasis from response and recovery to strategic risk management and reduction and from a government-centered approach to decentralized community participation.

Table 18.3 Claims Settled under Weather Insurance Scheme 1999–2000

Serial number	Name of federal state of India	Number of farmers covered	Number of farmers benefited	Area (in hectares)	Sum insured (in Indian rupee)	Claims paid (in Indian rupee)	Farmers benefited (as percentage of farmers covered)	Claims paid as percentage of sum insured
1	Andhra Pradesh	18,920,261	3,955,805	29,738,526.0	2,929,399.30	252,361.03	20.9	08.61
2	Tamil Nadu	2,388,504	1,143,426	3,445,272.2	469,574.98	103,654.73	47.9	22.07
3	Kerala	317,772	64,252	270,168.57	43,631.53	2,192.38	20.2	05.02
4	Gujarat	9,198,166	3,133,566	21,742,767.0	1,850,185.80	298,205.44	34.1	16.12
5	Maharashtra	22,555,073	7,253,387	21,119,954.0	1,247,794.90	147,623.52	32.2	11.83
6	West Bengal	7,088,296	1,788,894	3,633,598.4	601,479.46	78,021.10	25.2	12.97
7	All India	134,675,464	36,392,486	210,866,855.0	14,829,431.00	1,494,933.00	27.0	10.08

Source: Department of Agriculture and Cooperation, Ministry of Agriculture, Government of India.

The Emergency Management and Research Institute, a public–private initiative with the help of Satyam Computer Services, has been formed. Some of its early efforts involve training, creation of a single emergency telephone number, and the establishment of standards for staff, equipment, and training.

18.7 Recommendations

Structural measures are tangible and eye-catching. The general public will be more impressed with structural measures when compared to nonstructural measures like training, education, community development, and information dissemination. Low cost measures that are appropriate to local, regional, or national conditions could be substituted for high cost structural measures. Nonstructural measures make a project sustainable and increase efficiency. Substitution of nonstructural with structural measures or substitution of structural with nonstructural measures can happen depending on the quality of the institutions, systems and procedures, types of governance, and the attitude of stakeholders.

Costs and benefits of adaptation have to be quantified in a scientific manner for proper data collection or analysis involving the local community.

A common reporting standard has to be adopted. Disaggregated data on structural and nonstructural measures should be compiled. The focus should be on getting estimates at a disaggregated level so that more detailed studies can be undertaken.

The costs and benefits of climate change adaptation have to be widely disseminated and discussed in policy circles. Hence there is a need to design short-term training programs for different stakeholders. A research priority with regard to climate change adaptation would be the assessment of costs of adaptation for countries as well as for regions. This would be important for climate change adaptation negotiations as well as for allocation of resources.

References

Department of Science and Technology, India. www.dst.gov.in (accessed August 18, 2010).

Financial Tracking Service. 2010. www.fts.unocha.org (accessed August 18, 2010).

Government of India, Ministry of Agriculture, Department of Agriculture and Cooperation, Directorate of Economics and Statistics Report (Unpublished).

Government of Tamil Nadu, Tsunami Rehabilitation Programme Office Reports (Unpublished).

Myers, N. 1997. Environmental Refugees. *Population and Environment.* 19 (2). pp. 167–182.

PART IV

Action Plan for Policymakers and
Planners to Reduce Risk Impact

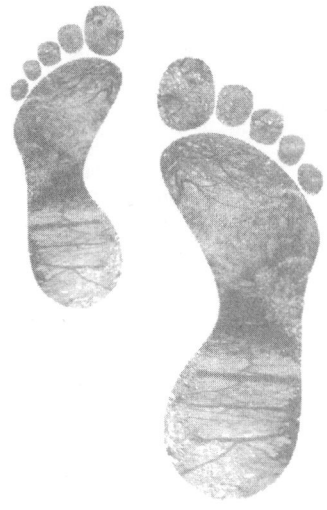

Action Plan for Policymakers and
Planners to Reduce Risk Impact

Key Messages

Analyzing the resilience and vulnerability, understanding the links between disaster risk reduction and climate change adaptation, and downscaling climate change and impact assessment data are essential capacities required to deal with the effects of climate change. Policymakers must have access to adequate operational national systematic observation networks, and access to the data available from other global and regional networks.

Global warming will continue to increase due to existing greenhouse gas (GHG) content in the atmosphere, even if emissions are reduced to acceptable levels. Climate change affects countries and social groups differently. Therefore, local capacity should be developed at the local scale to forecast climate change, to assess the impacts in various sectors, to design appropriate adaptation measures, and to build alternate strategies. Higher education institutions should be mandated to play a major role in this respect.

Consequences of natural disasters are severe, visible, but unpredictable. However, resources available for natural disaster prevention and mitigation are limited. Many financial schemes exist but their effectiveness is low, making it difficult for households or local government authorities to take ex-ante activities against natural disasters. Ex-post support in the form of government assistance and emergency donor relief is inefficient. Therefore, there is a need for financial schemes to insure households that are highly self-reliant, comprehensive, sustainable, and appropriate. Measures in the long term have to be closer to the "market rules" meaning that capital spent can be refunded or profitable. A community managed insurance scheme based on a self reliance fund (SRF) is proposed. Accessing the funds that are available at present is a complex and lengthy process. Even if this process is streamlined, a lot more funding will still be required for adaptation. New international financing mechanisms and sufficient responses to adaptation are needed.

Since agriculture will continue to be the main sector of the economy in many countries, it is important to study the economic impact of climate change on agriculture. Three approaches have been widely used in the literature to measure the sensitivity of agricultural production to climate change: agronomic-economic models, cross-sectional models, and agro-ecological zone (AEZ) models. Even though several models are used to capture the impacts at the macroeconomic level, the Ricardian type model with cross-section data could be applied to assess the impacts of climate change at farm level.

Chapter 19

Adapting to Climate Change:
Developing Local Capacity

Srikantha Herath

19.1 Introduction

The need to adapt to climate change has been recognized as a key development requirement as current predictions show global warming will continue to grow due to existing greenhouse gas (GHG) content in the atmosphere, even if the global community succeeds in bringing down GHG emissions to acceptable levels. Assessing impacts at the local scale and designing appropriate adaptation measures is a key challenge for affected countries. It is important to develop local capacity to support achieving this objective, especially in the relevant institutions in developing countries.

The United Nations University (UNU) has developed programs so that local capacity can be developed in a sustainable manner. This chapter first describes a study of climate change impacts on rice production in Sri Lanka due to global dimming to highlight the complex issues associated with adaptation. Next, it discusses a capacity development program that introduces tools and methodologies for rainfall downscaling and flood risk assessment, and the requirements to make these programs successful.

19.2 Rainfall Changes and Impact on Rice Production in Central Sri Lanka

Observations in Sri Lanka have shown a consistent decrease in rainfall during the inter-monsoonal rain period from February to April since the early 1970s. Herath and Ratnayaka (2004) analyzed the seasonal trends of rainfall using data from 62 rain gauges from the central region covering 1963 to 1993. The data show a decreasing rainfall trend for the first inter-monsoon season from March to April. The same characteristics were observed when rainfall under Atmospheric Brown Cloud (ABC) conditions were simulated using the Weather Research Forecast (WRF) local area model using a modified radiation algorithm to account for absorption and scattering (Pathirana and Herath 2004). Simulations were conducted for cases with and without aerosol impacts in Central Province, Sri Lanka. Only one aerosol level was considered and the parameterization of the limited atmospheric model was adjusted so that the short-wave radiation received at ground had a reduction of about 25%. The resulting data were accumulated at daily scale and the relationships between changes to short wave radiation at ground level, rainfall reduction due to ABC, changes to maximum daily temperature as a function of solar radiation were determined.

In order to identify the main linkages and prioritize the relationship between climate change (CC) and national sustainable development (SD) goals and policies, in vulnerability, impacts, and adaptation (VIA), and to explore possible remedial measures, action impact matrix (AIM) tools were used through interactive sessions and workshops. The AIM-based process facilitates early screening and problem identification by preparing a preliminary matrix that identifies broad relationships, and providing qualitative data of the magnitudes of impacts, as displayed in Figure 19.1. Thus, the preliminary AIM helps to prioritize key relations between policies and their impacts on sustainability. Yet, the AIM provides an integrated viewpoint, meshing development decisions with priority economic, environmental, and social impacts. The organization of the matrix facilitates the tracing of impacts, as well as articulating the links among various development actions, that is, policies and projects. The AIM-building process helps to harmonize views among economists, environmentalists, and other stakeholders thereby improving prospects for successful implementation of climate change responses.

The AIM approach was applied to water using sectors at the country level to better understand interactions among three key elements:

- national development policies and goals
- key SD issues and indicators
- climate change and water resources

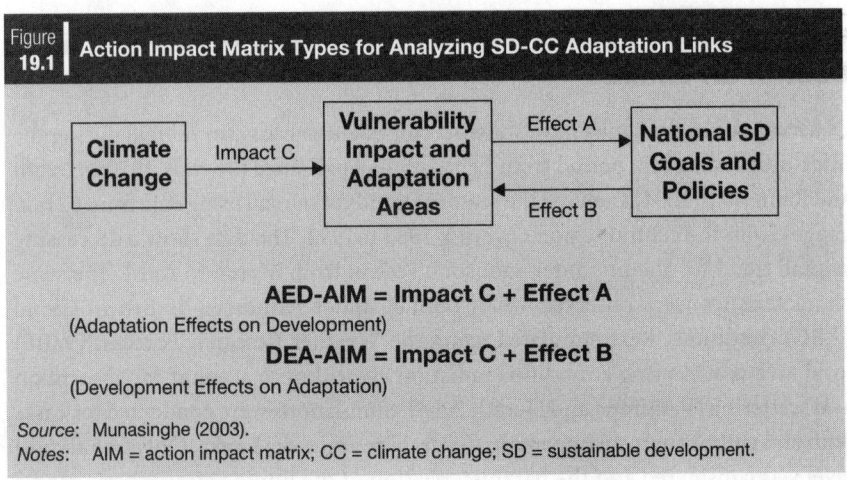

Figure 19.1   Action Impact Matrix Types for Analyzing SD-CC Adaptation Links

AED-AIM = Impact C + Effect A
(Adaptation Effects on Development)

DEA-AIM = Impact C + Effect B
(Development Effects on Adaptation)

Source: Munasinghe (2003).
Notes: AIM = action impact matrix; CC = climate change; SD = sustainable development.

First, the two-way interaction between national development policies and goals with key sustainable development issues and indicators are explored. Then, additional impacts of elements on climate change and water resources on the above interactions are applied. These are summarized in two matrices:

- water using sectors vulnerability, impacts, and adaptation and effects on development (WED)
- development effects on water using sectors vulnerability, impacts, and adaptation (DEW)

Table 19.1 shows the WED matrix for four water using sectors (1–4) in relation to seven development goals (A-G) presented by the government. The value of each cell of the table is dependent on the local physical and social conditions and the importance of the development policies. They were decided by summarizing existing information and studies on each of the items and ranked according to expert opinions. This process demonstrated the localized nature of the impacts and the complex relations among different sectors. Once the agriculture sector emerged as the most vulnerable sector, a detailed modeling study was carried out

Table 19.1	Effects of Water-using Sectors on Development (WED-AIM) in Sri Lanka with Climate Change Impacts				
		Vulnerability, impacts, and adaptation (VIA) in water using sectors				
		(1)	(2)	(3)	(4)	
		Agriculture	Hydro power	Water for humans	Water for bio- & ecological res.	Row totals (with climate change)
(S0)	Status (No climate change impacts)	−1	0	−1	−1	
(S1)	Status (+climate change impacts =>)	−2	−1	−3	−2	
Development Goals/Policies (+climate change impacts)						
(A)	Growth	−3	−1	−2	−2	−8
(B)	Poverty Alleviation	−2	−1	−3	−1	−7
(C)	Food Security	−3	−1	0	−1	−5
(D)	Employment	−2	0	−1	−1	−4
(E)	Trade and Globalization	−1	−1	0	−1	−3
(F)	Budget Deficit Reduction	−1	−1	−1	−1	−4
(G)	Privatization	0	0	0	−1	−1
Column Totals (with climate change)		−12	−5	−5	−7	

Source: UNU-ISP (2008).

to assess impact on rice production using the CERES rice model with two rainfall and temperature time series generated using the modified WRF model with and without atmospheric brown cloud conditions (Swain and Herath 2004). The study has shown that considerable reduction in rice yield can be expected. This reduction also would be dependent on the degree of fertilizer use and management practices. In addition to rice, the study area has tea plantations that would be affected differently by changes in weather, especially around the tea-processing factories. They would be more sensitive to fluctuations in temperature, moisture gradients, and rainfall intensity. Furthermore, rice farming is traditionally carried out by the Sinhalese farmers whereas tea estate workers are mainly Tamils. Adaptation measures should take into consideration not only the impacts and measures to remedy them, but also the social harmony and equity among different affected social groups. Thus, appropriate measures must be developed locally through a bottom-up approach in consultation with the affected communities and stakeholders guided by knowledge of external constraints and local needs.

19.3 Challenges in Mainstreaming Adaptation Strategies

Some capacities required for developing appropriate adaptation strategies that emerged from the above study are the ability to:

- downscale climate change forecasts to local scale;
- assess impacts in various sectors; and
- develop alternate response strategies and assess their appropriateness by analyzing consequences beyond implementation of such adaptation practices.

Developing countries face many challenges in mainstreaming adaptation strategies in to development planning, such as:

- limited number of qualified professionals in the required fields;
- lack of dialog between research and implementing communities to establish a feedback loop to address emerging issues; and
- problems and inadequacy of the higher education sector to take on research programs that enable customization of global knowledge and future projections to suit local conditions.

Thus, capacity-development programs should address the following needs:

- The programs should be based on the needs of the target countries.
- They should aim to increase the number of technically competent persons who can use advanced space based observations and global climate/weather forecasts.

- They should facilitate customizing knowledge to meet local conditions and constraints.

The target communities to achieve these objectives are:

- researchers and scientists who assist in customizing existing knowledge to suit local conditions supported by global experiences;
- professionals and practitioners to introduce new methods, tools, and standards; and
- administrative and local government officials who need to have an overview of technology and science.

19.4 Capacity Development Program

Designing and carrying out capacity development programs that meet the above requirements as well as build a critical mass large enough to continue these activities in an effective manner is a challenging task. The UNU has conducted capacity development programs to achieve this in an effective and sustainable manner. This chapter describes one of the recent capacity development programs that attempts to address these issues.

The UNU convened a workshop in 2003 with experts from 15 countries to discuss extreme flood risk in major Asian cities. The countries represented were: Bangladesh, Cambodia, People's Republic of China, Fiji Islands, India, Indonesia, Lao People's Democratic Republic, Malaysia, Nepal, Pakistan, Philippines, Singapore, Sri Lanka, Thailand, and Viet Nam. The experts recommended setting up a joint action program, focusing on assessing extreme flood risks and developing response plans. As a follow up, two case studies were conducted, in Hanoi, Viet Nam and Bangkok, Thailand. Building on these case studies, UNU developed a training program comprising three modules: *(i)* rainfall downscaling, *(ii)* GIS, and *(iii)* inundation modeling.

To address the above challenges, UNU developed a comprehensive program with the following components.

1. Training workshop for training of trainers to bring senior representatives of leading universities and agencies responsible for implementing flood control measures to go through the three modules, using data from each country. The workshop provides the ability to use methods and tools, and to understand the data and application problems.
2. Conduct national workshops with related stakeholders to discuss the training and application requirements considering gaps identified in the previous program. These entail a roving training program in each country and setting

up a long-term demonstration pilot project to apply methods and evaluate prediction uncertainties.

3. Conduct local training workshops to provide necessary tools and methodologies that address identified needs.

4. Apply these methods and tools to target catchments supported by long-term monitoring and research to resolve the application anomalies and reliability requirements.

19.4.1 Implementation: Phase 1

The UNU, in partnership with UNESCO-IHE, Monash University, Australia, Nippon Koei Co., Ltd., Japan, and the Asian Institute of Technology (AIT), Thailand, conducted the first workshop in two sessions each lasting 3 weeks. The workshop was held at AIT, with the support of the GIS Applications Center of AIT.[1]

In this phase, participants got a basic understanding of the theories and tools used in downscaling using local area weather forecast models in nested schemes, interfacing the weather information to distributed hydrological model for flood inundation estimation and the use of GIS for input data preparation, output verification and risk analysis. At the same time, the sessions highlighted the three important issues related to the use of such approaches. The first is the setting up of the models using local measurements and observations, which required adequate data sets with acceptable resolutions and available in usable digital form. The second issue relates to model uncertainties. To assess this, past observations to calibrate the models and separate data sets for verification are needed. Thus, the first requirement is the availability of field observations of required resolutions and accuracy for calibration and validation. The second aspect is the skill in calibrating models, as well as modifying the algorithms, assumptions, and modeling approaches to suit the conditions at hand. This highlights the need for ensuring data quality as well as availability of research capability to customize models to suit local conditions. The third issue is the need for close cooperation between the researchers who can carry out this customization and the implementing agencies responsible for managing environmental processes monitoring and data use.

---

[1] The following institutions participated in the program, which generally consisted of a faculty from a university and professionals from the organization responsible for flood control: Tsinhua University, Beijing Municipality (PRC), Institute of Engineering, Department of Hydrology and Meteorology (Nepal), University of Philippines, PAGASA (Hydro Meteorological Agency) (Philippines), University of Peradeniya, Irrigation Department (Sri Lanka), Institute of Hydrology and Meteorology, Department of Storm Control and Dyke Management (Viet Nam).

19.4.2 Implementation: Phase 2

The second component of multi-country phase 1 training was completed in March 2009. A follow up national workshop was conducted in Sri Lanka in May 2009. The workshop participants requested the following:

- Intensive training program for professionals to rapidly develop capacity to use the tools and methods introduced in the international program.
- Set up a committee for implementation of long-term application to demonstration catchments supported by university research.

19.4.3 Implementation: Phase 3

In response to the request to develop training programs for local professionals, a third national workshop was organized jointly by UNU, the Civil Engineering Department of the University of Peradeniya, and the Sri Lanka Irrigation Department.

The following activities were carried out at the workshop:

- WRF local area model set up and rainfall forecast for 3 days: 3 layer nested scheme
- Current global weather forecast downscaling
- Past extreme rain event modeling
- Re-analysis data downscaling
- GIS system study
- Set up and simulate demo-data set flood inundation in Yom river basin, Thailand
- Analysis of flood, risk assessment, and loss estimation for demo data sets
- Case study and discussion of Colombo city inundation simulation

19.4.4 Outcomes and Expectations

It is hoped that bringing academicians and practitioners together will produce sustainable relations in sharing data and information and promoting applied research. Hydrologists and meteorologists found the cross understanding and using local area models (LAMs) used in meteorology and distributed hydrological models (DHMs) used in hydrology useful. All participants benefited from the opportunity to use GIS, especially the present system embedded with the capability to handle both time and space domain data with facilities to directly couple to hydrological and atmospheric models.

The UNU's aim is to organize with partner institutions such activities to stimulate and support specialists in order to produce a critical mass of researchers and practitioners in the region who actively engage in using earth observations and global models for analyzing global change and resulting local impact assessment, followed by developing appropriate response strategies to meet response challenges.

19.5 Conclusion

Climate change poses new and unprecedented challenges that need to be carefully analyzed, anticipated, and prepared for. To meet these challenges governments need to invest in developing the capacities of planning sectors to incorporate expected impacts and adaptation measures in development planning. In many developing countries this will require a significant commitment to introduce new technologies including acquiring and use of earth environmental information, use of model based forecasts, downscaling to local conditions and carrying out impact assessments as a prerequisite to planning response strategies and mainstreaming to development planning. The higher education sector in developing countries needs to provide leadership in this process and help to customize the available global knowledge and tools for local conditions. Unfortunately, the higher education sector remains a largely ignored area in development assistance as well as in government spending. This gap will need to be rectified with the recognition that a well-trained and informed professional community is essential to assess environmental changes and their impacts on livelihoods, daily lives, and potential hazards as we move to a future where past experiences cannot be taken for granted, and where continuous research and accumulation of knowledge is essential for survival and development.

The international community—including development assistance organizations and educational institutions in developed countries—has an important role to play in meeting the capacity development needs of developing countries. A coordinated approach should be taken so that programs can complement each other to facilitate sustainable technology transfer. For example, ongoing development projects need to incorporate components to support the higher education sector (such as research and development) in the implementation phase, and to promote the development of training and research programs by other organizations that can become valuable field experimental stations. Platforms for sharing experiences among research and training networks should be established involving both practitioners of implementing agencies and researchers in the higher education sector. Governments and donor agencies need a long-term commitment to develop the required local capacity to meet climate change challenges.

References

Munasinghe, M. 2003. Linking Poverty and Environment in a Sustainable Development Framework. In Hearth, S. and Z. Adeel, eds. *Environmental Dimensions of Poverty.* Proceedings of UNU-RIVM Workshop, pp. 110. Katmandu, Nepal.

Herath, S., and U. Ratnayaka. 2004. Monitoring Rainfall Trends to Predict Adverse Impacts: A Case Study from Sri Lanka (1964–1993). *Global Environmental Change.* 14 (Supplement). pp. 71–79.

Pathirana, A., and S. Herath. 2004. Assessment of Atmospheric Brown Cloud Impacts on Local Climate with a Modified Mesoscale Atmospheric Model. In S. Herath, A. Pathirana, and S. B. Weerakoon, eds. *Proceedings of the International Conference on Sustainable Water Resources Management in the Changing Environment of the Monsoon Region, Vol. 1,* pp. 34–42 . Colombo, Sri Lanka: United Nations University.

Swain, D. K., and S. Herath. 2004. Solar Radiation Stress Assessment on Rice Production Using CERES Model. In S. Herath, A. Pathirana, and S. B. Weerakoon, eds. *Proceedings of the International Conference on Sustainable Water Resources Management in the Changing Environment of the Monsoon Region, Vol. 1.* Colombo, Sri Lanka: United Nations University.

United Nations University Institute for Sustainability and Peace (UNU-ISP). 2008. *Global Warming and Atmospheric Brown Cloud Effect on Local Climate and Rice Production.* Policy Brief.

Chapter 20

Financing Adaptation Responses: Disaster Mitigation in Viet Nam

Bui Duong Nghieu

20.1 Introduction

Consequences of natural disasters are severe, visible, but unpredictable. Recent studies suggest that climate change will create more natural disasters and they will be more severe. Natural disasters lead to diminished economic growth rates and erase achievements of hunger eradication and poverty reduction. As to enterprises, natural disasters delay their production and business processes, thus increasing expenses and losses. For households, natural disasters are catastrophes. After natural disasters, poor people become poorer. Moreover, following a storm or a natural disaster event, many formerly poor households (who have just escaped from poverty) are quickly pushed back into poverty (re-poverty).

Risks and losses following natural disasters are huge, but financial resources to mitigate them are very limited. In the meantime, mechanisms to mobilize, allocate, manage, distribute, and use those scarce financial resources have not been improved in Viet Nam or in other countries.

Every year, natural calamities, mostly water-related types such as floods, storms, and droughts, end hundreds of human lives and cause losses of millions of US dollars. On average, Viet Nam annually suffers about six to ten storms, tropical depressions, and floods that occur suddenly in large areas with different frequencies. In Viet Nam, 70% of the population is prone to natural disasters and more so now with climate change (Government of Viet Nam 2001). On average, over one million persons need emergency aid following a natural disaster. Nevertheless, resources available for natural disaster prevention and mitigation are limited. In fact, many financial schemes exist but their effectiveness is low, making it difficult for households or local governmental authorities to take ex-ante activities against natural disasters.

There is a need for a financial scheme that is highly self-reliant, comprehensive, sustainable, and appropriate to insure households. Also, it is needed to finance ex-ante measures to reduce vulnerability to natural disasters.[1]

---

[1] The idea to set up a self sufficient fund for natural disaster mitigation in Viet Nam was initiated by a research team from the Institute of Financial Science, assisted by international

In Viet Nam, the process of tackling the consequences of catastrophes has been incorporated into life of the whole country and reflected in government policies. Viet Nam considers disaster prevention and mitigation a vital issue when planning socioeconomic policies and has tended to mobilize various sources for preventing and mitigating catastrophic losses after disasters.

In practice, the Government of Viet Nam (2001) has issued a series of specific policies and measures as follows:

- An Ordinance specialized in flood and storm prevention regulating articles necessary for natural disaster prevention. A system of standing committees of flood and storm prevention from the higher government authority (the Center) to communities has been set up (Standing Committee of the National Assembly of Viet Nam 2000).
- Financially, besides contingency state budget lines and emergency relief resources, the government has allowed local flood and storm prevention funds to be established by raising compulsory contributions for flood and storm prevention activities.
- A national action plan and strategy on preventing and mitigating natural disasters covering 12 sorts of natural disasters comprising nature-related, water-related, and human-related natural disasters.
- The government has incorporated natural disaster management in planning development projects in natural disaster-prone areas nationwide.
- The government has applied a series of financial, credit, and insurance policies to mitigate consequences after natural perils for quickly recovered livelihoods of households suffering from natural disasters.
- The government has issued a policy of "co-staying with floods" by regulating residential areas, transport networks, and fruit gardens to mitigate losses after floods.
- Heightening the community's awareness of natural disaster prevention and mitigation.

Besides various (infrastructure and non-infrastructure) natural disaster mitigation measures, Viet Nam has carried out many policies and mobilized resources to take the initiative in mitigating consequences of natural disasters. Simultaneously, the government looks for support from international organizations,

---

specialists and funded by the UNDP Viet Nam via project VIE/01/014 "Capacity Building for Disaster Mitigation in Viet Nam."

such as UNDP and the World Bank, as well as other donors for natural disaster prevention and mitigation.

20.2 Analysis of Financial Mechanisms and Resources

Recent research and surveys (Institute of Financial Science 2005a, 2005b) have identified financial resources available for natural disaster prevention in Viet Nam. The resources include:

- state budget including the central budget and local budgets
- Fund for Flood and Storm Prevention (FFSP) of localities
- funds from relief, support of domestic individuals, organizations; charitable aid from individuals, nongovernmental organizations, foreign governments, and international organizations
- credit (formal and informal, preferential and un-preferential schemes)
- insurance
- contingency liability included in the state budget, including the government budget and the local budgets,[2] so that local authorities, ministries and sectors have contingent equipment (such as lifebuoys, jolly-boats, clothes, blankets, mosquito nets, food) for disaster recovery

Most responses (69.2%) in the survey stated that locally raised financial sources meet below 50% of needs, of which the main proportion is spent on natural disaster settlement, followed by natural disaster prevention. The ability to deal with catastrophes from local budgets is low. This indicates that other financial sources outside localities (from the Center, donors, enterprises, and households in other localities) should be mobilized for natural disaster mitigation. Locally mobilized resources are often seen in the form of human assistance, contributions in kind under community solidarity, and mutual assistance to settle short-term problems.

Regarding the stability degree, localities consider budget resources the most stable and important. The rankings are (i) the central budget and the FFSP are considered the most stable while the provincial budget contributes the most; (ii) contributions from enterprises, oversea aid, emergency relief, voluntary, informal credit are still very low and unstable; and (iii) contributions from households are considered stable and at the normal level.

Regarding the importance of each source, the rankings are: (i) the central budget being the most important, (ii) local budget lines, (iii) funds mobilized from

---

[2] In Viet Nam, the state budget includes the government budget, the provincial budget level, the district budget level and the municipal budget level.

residential communities, *(iv)* domestic loans, *(v)* support from other localities, *(vi)* international loans and relief, and *(vii)* and donations from charitable organizations. This result demonstrates that localities primarily rely on support from the center for natural disaster prevention but they are not active in looking for local funds and enterprises' contributions. This is a weakness in mobilizing financial funds for natural disaster prevention.

Regarding the credit source, most feedback (43.8%) stated that credit plays an extremely important role in mitigating ex-post consequences. Few survey respondents argued that credit could provide financial capacity for urgent on-the-spot settlement of natural disaster settlement. In other words, when a natural disaster occurs, households do not rely on credit. Regarding the degree of contribution from credit sources, most responses (84.1%) said that credit meets less than 50% of the needs of loans for natural disaster prevention; nearly half (49.2%) said below 20% of needs is met. Credit sources are mainly spent on ex-post natural disaster mitigation, followed by prevention and ex-ante mitigation.

Most localities have collected fees for the FFSP under Decree 50/CP, but some have not yet (mainly those in the east and west of southern Viet Nam, Tay Nguyen province, and the central part).

Assessment of the timely degree for natural disaster mitigation. The questionnaire-based results indicated that although the allocation of the state budget, the central budget, various local budgets, and locally raised funds are sometimes delayed, they are still efficient at an acceptable degree. The most delayed funds are from overseas aid. Delay exists in all phases of the process to finance natural disaster prevention. The appraisal procedures phase is considered the slowest due to the complexity and involvement of many same-level competent intermediary agencies. Besides, bureaucracy and irresponsibility existing at various levels of local government are also causes of delaying the financing of natural disaster prevention.

Regarding the legal framework, at present the Ordinance on Flood and Storm Prevention is the highest legal document that directly stipulates the special and financial issues referring to flood and storm prevention (the FFSP). Apart from the regulations on the FFSP, other regulations on financing schemes relating to natural disaster prevention are found scattered in various documents on financial, budget, taxation, credit, and insurance issues. In addition, policies and levels to finance natural disaster prevention are inconsistent among localities. The center only stipulates the maximum or minimum financing levels as general instructions while provinces issue specific documents. Thus, differences in natural disaster prevention expenditures exist among localities where those expenditures depend a lot on how much can be mobilized for natural disaster prevention. Clearly, this shows the complexity of the regulations on finance, credit, insurance, among others, with reference to natural disaster prevention. This complexity should be simplified and integrated.

There are policies and spending degrees for natural disaster prevention. For the state budget's contingency sources, the decentralization mechanism was implemented under the 2002 state budget law. The budget contingency degree at each government authority level is decided by the level itself (within the general regulation framework of the center about the extraction ratio, spending content, etc.). When natural disasters happen, the government authority level can decide how much is spent from its contingency liability (a budget line for annual contingency). If their contingency liabilities are inadequate they can ask for a higher budget for extra capital support.

The biggest feature of the current budget, relief, and the FFSP system is "non-refundable" when those resources are spent. That means money paid out never comes back. In contrast, credit and insurance mechanisms show "refundable" when funds are reasonably spent. Insurance mechanisms will also help ensure the sustainability of the Self Reliant Fund (SRF) in case of claims for losses after catastrophic disasters.

At present, there is no mechanism to promote self reliance and input–output balance. With the developing market economy, self reliance will be ensured by the use of insurance, loans, or financial investments schemes. If temporarily unspent, money would be invested, they will make profits and help maintain the capital for the SRF. If fund allocation, support, and relief for natural disasters are well-integrated with schemes of financial investments or credit, insurance for the purpose of natural disaster mitigation, the requirements of an SRF will be ensured.

Regarding the organization system, in Viet Nam many organizations are getting involved in activities of natural disaster prevention. They include people's committees, people's councils and government agencies such as financial and planning departments, agriculture and rural development departments, the flood and storm prevention organization, fatherland fronts, women's unions, farmers' unions, the Red Cross, the Bank of Social Policy, and the Bank of Agriculture and Rural Development. All organizations actively contribute money but at different degrees for natural disaster mitigation. Of those organizations, the PC always has a decisive role in allocating money for natural disaster prevention.

Apart from the flood and storm prevention organization, all 63 provinces have their own executive committees of flood and storm prevention that are run by vice chairpersons of equivalent-level people's committees. Permanent chiefs of executive committees of flood and storm prevention are representatives of agriculture and rural development organizations. Other coordinators are representatives from financial organizations, banks, and social insurance, among others. Overall, the system of executive committees of flood and storm prevention, well organized at all intergovernmental levels, has played a decisive role in promoting natural disaster prevention activities. However, most of the officials are part-time with full-time positions with and paid by other agencies.

The state budget is mainly spent on restoring and repairing damaged infrastructure (including traffic systems, irrigational systems, schools, and medical centers). In case of big natural disasters, besides food aid for hunger mitigation within a locality, part of the state budget is spent on supporting plant varieties, domestic animals, and households whose family members die and properties are imputed after a natural disaster. The support level is quite small in comparison with the money spent on public infrastructure.

Regarding the processes of allocation, payment, and administrative procedures, results indicated that it is random and very complex administratively. Therefore, it is difficult and sometimes slow for households to get access to existing funds for natural disaster prevention and mitigation to timely finance their emergency needs.

Regarding the involvement of households in supervising the process of allocation and usage of budgets for natural disaster prevention, most respondents said that households supervise the process above mainly through their representatives. Very few localities allow households to supervise the process directly except for contributions paid by households.

The research on the coordination in supplying funds for natural disaster prevention and mitigation concluded that the coordination between competent state agencies themselves and with unions is generally considered proper and close. However, around 25% replied that the current combination was not effective. Loose coordination is a cause of the inefficiency in allocating, managing, and spending funds for natural disaster prevention and gaps for some mercenary actions.

In Viet Nam, many financial mechanisms and resources are used for natural disaster prevention. However, their subjects are insufficient, likely to overlap, their resources are scattered and their implementation is inconsistent. The main reason for those weaknesses is the existence of many intermediaries in managing and distributing resources for natural disaster prevention. The highly self-reliant unification and long-term nature of natural disaster financing mechanisms are still low and there is a lack of an integrated, capable, and financially powerful mechanism for timely natural disaster mitigation for households. Also, due to the existing scattering problem, the current society and government funds raised for natural disaster prevention and mitigation meet only a small portion of the actual needs and losses. Thus, a study is needed to unify mechanisms and organizations and instantly remove the scatter, overlap, and multi-partners in financing affected households after natural disasters. Practical, efficient mechanisms and policies are needed to unify various financial sources (with proper schemes) for natural disaster prevention and mitigation.

Local budgets, FFSP, household contributions, and the central budget's support are the most crucial financial resources presently playing a key role in financing natural disaster prevention and mitigation. Besides the key resources described above, other resources are being mobilized from individuals and organizations inside and outside localities for support and relief of inhabitants. Nevertheless, those (irregular and infrequent) resources only arise in case of natural disasters and focus on emergency cases but not the general economic effectiveness of the coordination for sustained natural disaster prevention and mitigation. The practically of relief, especially relief in kind, is obvious but in many cases problems arise as relief does not meet households' exact needs.

Regarding efficiency, existing financial resources are untimely. The fund's activities tend to be of subsidization, lack self-reliance, rely on mutual help and community solidarity, and settle the short term rather than the long term. They mainly depend on the state budget, relief efforts, urgent support, and credit sources. Though an insurance scheme has been carried out, its efficiency is very low. The presence of various financial mechanisms and sources reflect the active participation of the state; residential communities; nongovernmental organizations; enterprises; financial, credit, and insurance institutions; individuals; public union; and religious bodies inside and outside natural disaster-affected areas. Nevertheless, the degrees and efficiency of financial contributions, management, and usage are different.

The data indicated that the community level lacks on-the-spot financial capacity to settle natural disaster-related matters for households. Although community-level contingent liability of the annual budget is regulated by state budget law, it has not been implemented in practice mainly because of a small community budget scale. Meanwhile, households' financial contingency against natural disaster risks has not been concerned due to their low incomes and poor economic conditions. When natural disasters occur, most households are affected and it is difficult to mobilize emergency funds from communities. Hence, a mechanism is needed to overcome that problem. That mechanism, on the one hand, should create conditions for community-level government authorities in raising financial resources locally available, decentralized and spent to take the initiative in settling matters following natural disasters. On the other hand, that mechanism shall be a shield that households can rely on in case of natural catastrophe. At present, as current FFSPs are not run by communities and contributions are low, those funds are unable to provide an on-the-spot financial capacity to mitigate natural disasters for households.

Schemes of credit, microcredit, and savings credit in rural areas have had a lot of specific, practical contributions to households' needs but they have not been well incorporated into natural disaster prevention. Overall, ex-ante prevention activities based on credit are still immature and lack the combination of current financial and credit tools. Meanwhile, all financial and microcredit tools serve the same purpose of strengthening or increasing incomes for better natural disaster settlement and mitigation for households.

Insurance is also considered one of the efficient financial mechanisms for natural disaster prevention but it is not common in Viet Nam yet. In fact, a lot of issues pertaining to natural disaster and agriculture insurance schemes need to be studied and settled. Though insurance institutions have played an increasing role in sharing risks among communities and directly or indirectly mitigating consequences of natural disasters, a challenge here is how to solve the requirement of "a profitable business" to an insurance company. Therefore, an emerging issue is how to study a mechanism that simultaneously satisfies all stakeholders: natural disaster victims, insurance companies, government, and residential communities when the insurance mechanism is applied for natural disaster prevention.

International loans and aid also contribute to natural disaster prevention but they are random. Some provinces (such as Tay Nguyen) do not receive international loans and aid.

Although there are many financial mechanisms for natural disaster prevention, money directly coming to the hands of households, in terms of both scale and value, is still very low. At present, mechanisms making cash available to natural disaster sufferers are mainly credit, relief, and emergency aid. Yet, that money accounts for a small proportion of the total value spent on natural disaster prevention as well as financial emergency needs of households. The biggest expenditures on natural disaster prevention are for public infrastructure, equipment purchase, and current expenditures on natural disaster prevention such as rehearsing and patrolling. Those expenditures are important and necessary, but do not give cash directly to households.

Apart from relief, emergency aid sporadically arising to "immediately save"[3] households following natural disasters, households do need support in the form of capital to consolidate their financial capacity for ex-ante natural disaster prevention such as for sustainable livelihoods, increased incomes, and strengthened homes. This can substantially increase ex-ante and ex-post ability to mitigate consequences following a natural disaster. The concepts of maintenance and development of funds are unavailable. Also, there are no mechanisms available to use unspent resources when natural disasters do not occur. Natural disaster prevention expenditures are mainly thought to be in the form of subsidies or allocations for free. Concepts related to self reliance and new approaches in modeling and organizing advanced financial measures and mechanisms to actively maintain and develop resources for natural disaster prevention are generally unavailable.

Some key findings of the research conducted by the Institute of Financial Science (2005b) are:

- At present, there is no financial scheme or fund worldwide detailing government support for low income households for ex-ante and ex-post risk management activities against natural disasters.

---

[3] When disasters occur, households need emergency aid such as rice and noodles. This aid is very useful and helps them to survive. But, after a disaster, households need capital support to deal with ex-post natural disaster. In the long term, households also need financial measures for ex-ante natural disaster prevention.

- No financial scheme or fund specifically describes a scheme, albeit some small features like the self-sufficient fund for natural disaster prevention and mitigation for households. Furthermore, no schemes have proposed, approached, and incorporated activities of preventing and mitigating losses after natural disasters for poor households.
- The current insurance schemes in other countries, only focus on agriculture insurance but not a scheme purely serving the prevention and mitigation of impacts of natural disasters on households.
- Ongoing sponsorship for post-disaster losses could be more efficient and practical if focusing on ex-ante schemes (aimed at financing the prevention stage through insurance, reinsurance, and small-value loans to enhance the financial ability for natural disaster prevention and mitigation). Ex-post assistance (donor relief and aid after natural disasters), despite being traditionally common, is not as effective and practical as ex-ante assistance.
- Ex-post (both domestic and international) assistance, though valuable, is often late and insecure, thus this help cannot take the financial initiative of localities forward. If a local financial mechanism or fund exists in the hands of localities, they can take the much better initiative in preventing and settling natural disasters as well minimizing consequences following natural disasters.
- Goes and Skees (2003) found out the relations between the natural disaster frequency, their damage degrees, and suitable solutions in terms of insurance techniques. The details are:

  o risk of high-frequency natural disasters (regular events) with insignificant losses needs financing from the government, communities, and households
  o risk of medium-frequency natural disasters with medium losses should be insured
  o risk of low-frequency natural disasters but with huge losses should be reinsured
  o catastrophe CAT bonds should be applied at a very low-frequency when rare and extremely heavy losses natural disasters happen

20.3 Design Solutions to Establish the Self Reliant Fund

20.3.1 Defining the Concept of the Self Reliant Fund

The Self Reliant Fund (SRF) for natural disaster mitigation for households is a financial fund, which is highly communal, not-for-profit, but a self-reliant, sustainable, maintained, and developed financial fund. The SRF is formed by the government and is entitled to raise compulsory and voluntary funds from households, enterprises, political, and religious organizations internationally and domestically to mitigate consequences before, during, and after natural disasters for households, and to ensure its self reliance and sustainability.

The concept of the SRF for natural disaster mitigation for households needs to be understood correctly and comprehensively, as this is a new type of fund in Viet Nam and the world. Therefore, the true comprehension about the concept of the SRF will be very crucial in deciding its modality, purposes, contents, operational mechanisms, management, and operation. Particularly, the agreement on the understanding of "financial self reliance" and "natural disaster mitigation" for households should be integrated.

20.3.2 Key Characteristics of the Self Reliant Fund

During operation, the SRF applies techniques of market economy rules to maintain and develop resources raised. This is presented by measures the fund will employ (insurance, credit, finance, allocation, and investments in financial markets). These are technical factors that can ensure the fund's self reliance as well as maintain and develop its capital and assets.

The SRF is self reliant in terms of the right to make decisions on mobilizing and spending the funds. The center (or the higher government authority) only issues the legal framework, regulation, and guideline documents but the documents shall be issued in advance and be publicly available. Specific management is decentralized to each government authority to promote responsibility, resilience, and creativeness of the authority as well as various levels of fund management.

The feature of "financial self reliance" makes the SRF quite different from existing financial funds (contingency, support, relief, or quick response funds) for natural disaster mitigation in terms of all processes from mobilization, spending, and operation as well as depletion and/or non-depletion features of the SRF's capital.

The SRF feature of natural disaster mitigation for household bears resemblance to a communal fund mobilizing contributions from many organizations and individuals for communal purposes (Figure 20.1).

20.3.3 Self Reliant Fund Instruments

Key financial and credit instruments for actively mobilizing, preserving, and developing the SRF's capital include *(i)* financial subsidies (to support losses after natural disaster), *(ii)* microcredit (at the community level), *(iii)* insurance, reinsurance, mutual insurance to share risk and to compensate natural disaster losses, and *(iv)* financial investments (from temporarily unspent capital).

Figure
20.1

Schematic View of the Self Reliant Fund

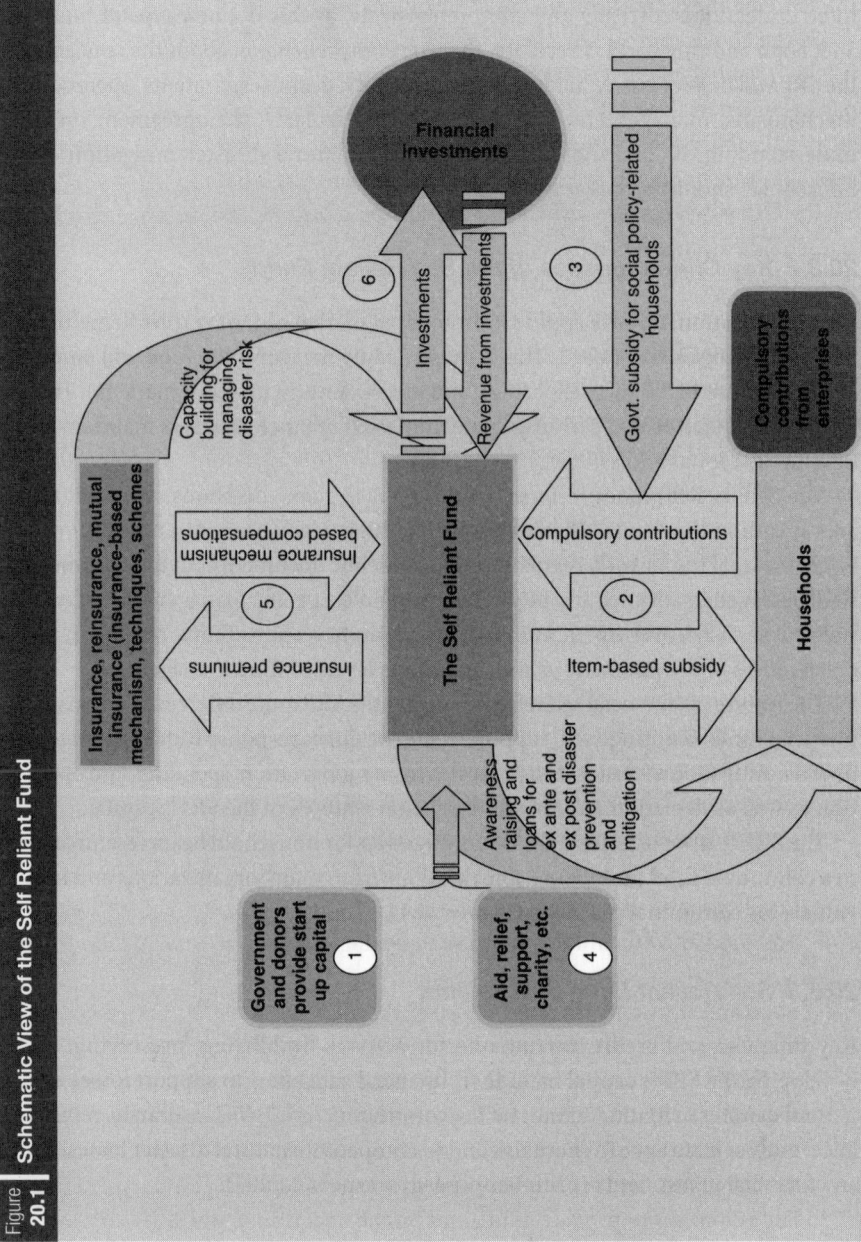

Source: Prepared by author.

20.4 Conclusion and Recommendations

There are three approaches to financing catastrophic damages: *(i)* ex-post (such as government assistance and emergency donor relief; *(ii)* ex-ante (in the form of insurance, microfinance and risk sharing in the stock market); and *(iii)* financing the settlement expenditures that arise during natural disasters.

Of the above approaches, the ex-post assistance is socially inefficient, because it is invariably generated under conditions of haste, disorder, and disarray after natural disasters. It is, too little, too late. In addition, ex-post social relief for losses after natural disasters is much more costly than ex-ante prevention. Therefore, an active ex-ante financing scheme should be considered.

The SRF is a necessary instrument to transfer, diversify, and minimize risks for households' strengthened livelihoods, sustainable poverty reduction, and sustainable development to mitigate catastrophic disasters for households within localities. Bearing the nature of a mutual help fund, the SRF can share risk within localities by insuring natural disaster risk through a supportive system. Households in areas not prone to natural disasters can support those in highly disaster-prone areas by contributing money to the SRF. As being highly communal and supported by the majority of households and various government authorities, the SRF will be feasible and sustainable.

The proposed SRF in Viet Nam is organized on a four-level model parallel with the four-level government authority system, the four-level state budget system, and the current four-level executive board system of Flood and Storm Prevention, with techniques, skills, and instruments subject to mechanisms and schemes presented above. The SRF's tools allow it to comprehensively and flexibly use financial, insurance, reinsurance, mutual insurance, financial subsidy, and microcredit tools to make its capital stable. The SRF's tools are also useful for households' improved risk mitigation, decreased vulnerability, increased ex ante and ex post natural disaster risk management, better recovered livelihoods and sustained development.

The key proposals to take the SRF forward include:

- Legal bases shall be improved for the SRF's introduction: The Ordinance on Flood and Storm Prevention shall be upgraded to the Law on Flood and Storm Prevention. In the Law on Flood and Storm Prevention, the scope of flood and storm prevention shall be upgraded to natural disaster prevention for households. Regulations on the FFSP shall be replaced by regulations on the SRF. The government shall also issue a decree providing guidelines on the implementation of the law on disaster prevention and decisions to set up and regulate the SRF's organization and operation. The Ministry of Finance, the State Bank, and the Ministry of Agricultural and Rural Development shall issue circulars to provide guidelines on the operation of the SRF.

- Start up capital provided for the SRF: To make the goals of sustainable development, hunger eradication, and poverty reduction viable in Viet Nam, the government and donors shall have integrated agreements and conclusions in providing the SRF's start up capital (based on a proposed 50-50 share). The government shall play a crucial role in deciding the establishment and development of the SRF; providing legal foundations, mechanisms, policies, start up capital, even part of operational expenditures, and insuring the SRF in case of financial risk due to catastrophic events.
- Setting up, training, transferring, and operating the SRF within localities: As a new financial scheme (not only in Viet Nam but also the world), the SRF targets natural disaster prevention (ex ante) as well as natural disaster mitigation (ex post). The SRF simultaneously uses subsidy-based schemes (to subsidize listed items) and market-based schemes (financial investments, microcredit, insurance, and reinsurance). At the same time, the SRF performs two missions: "humane relief" for losses of affected households after natural disasters and "capital maintenance and development". Good preparation, from various government authorities from the center to communities, in terms of thoughts, human resources, officials, and training, to put the SRF into effect.
- Start up period of the SRF: To make the SRF's introduction and operation "smooth" in the beginning period of 1–3 years, the center shall provide the following regular and systematic technical assistance to localities:

  o Guiding the establishment of the SRF.
  o Compiling training materials of operational skills of the SRF.
  o Training the SRF's management skills.
  o Setting up a website for information flow about the SRF.

All the above issues, both in the short and long term, shall be licensed by competent organizations and supported by various government authorities, managers, and experts from the center down to communities. Especially, there shall be support of international donors led by the UNDP, the World Bank, and Asian Development Bank. Finally, the SRF shall be supported by households, enterprises, and multilateral donors as most crucial factors to make its development sustainable.

References

Goes, A. and Skees, J. R. 2003. *Financing Natural Disaster Risk Using Charity Contributions and Ex Ante Index Insurance*. Paper prepared for the American Agricultural Economics Association Annual Meeting, Montreal, Canada, July, 27–30.

Government of Viet Nam. 2001. *Second National Strategy and Action Plan for Disaster Mitigation and Management in Viet Nam–2001 to 2020*, Government of Viet Nam, Ministry of Agriculture and Rural Development, Central Committee for Flood and Storm Control, Hanoi.

Institute of Financial Science. 2005a. *Questionnaire-based Survey Report*. Hanoi.

———. 2005b. *Report of International Experience on Financial Measures for Disaster Mitigation*. Hanoi: Institute of Financial Science.

Chapter 21

Economic Analysis of Climate Change Impacts on Agriculture at Farm Level

Kuppannan Palanisami, Coimbatore Ramarao Ranganathan,
Samiappan Senthilnathan, and Sevi Govindaraj

21.1 Introduction

Climate change has a multidimensional effect on human activities. Of these, agriculture is prone to be the most affected because the very important input variables, that is, precipitation and temperature, are mainly climate related. It puts agriculture at great risk. If population growth remains high and economic growth in the developing world is low, the impacts of climate change will add to the number of undernourished people in the world (IIASA 2002). Since agriculture continues to be the main sector of economy in many countries, it is important to study the economic impact of climate change on agriculture. Globally several studies focused on the effect of climate change on agricultural production (Carraro and Sgobbi 2008; Kameyama et al. 2008; Kurukulasuriya and Rosenthal 2003). There are some studies on the economic impact of climate change on Indian agriculture. If temperatures rose by 4 degrees centigrade (°C), grain yield would fall 25%–40%, rice yield would fall by 15%–25%, and wheat yield by 30%–35% (Kumar and Parikh 1998b). Mendelsohn et al. (1994) indicated that the global warming would decrease net income by 8%.

Regional-level economic impacts have been examined in a few studies (for example, Palanisami et al. 2009; Ranganathan 2009). These studies will be helpful in formulating adaptation and mitigation strategies applicable to the concerned regions. This will help greater reduction in the loss due to climate change. The objective of this chapter is to use the Ricardian model to study the economic impact of climate change on agricultural production in Tamil Nadu State, India. This model is preferred to the traditional estimation methods, given that instead of ad hoc adjustments of parameters that are characteristic of the traditional approach, the Ricardian technique automatically incorporates efficient adaptations by farmers to climate change (Palanisami et al. 2009).

21.2 Overview of Agriculture in Tamil Nadu

Tamil Nadu state has historically been an agricultural state and it is India's second biggest producer of rice (Government of Tamil Nadu 2008). In all, 90% of farmers belong to the small and marginal category and their operational holdings account for 56% of the total area. The gross cropped area is around 5,843 thousand hectares of which the gross irrigated area is 3,309 thousand hectares (57% of the total cropped area). The balance of the area (43%) is under rain-fed cultivation. Paddy is the major crop grown in all districts with a share of 44% of the total cropped area followed by groundnut and pulses each at 15% of the cropped area. The total cropped area has declined gradually over years. Except for maize, sugarcane, and bananas, the area under all other crops has decreased and the total cropped area has also decreased from 4,808 thousand hectares to 3,922 thousand hectares, a decrease of 18.4% from 2001–2002 to 2005–2006. The main causes of this decline are the failure of the monsoon, decrease in labor availability, and urbanization. In spite of this, there is an increase in productivity of major crops due to the usage of modern technologies and varieties. The decrease in cropped area coupled with demand for higher productivity and production due to population explosion, imply that greater depth of studies are required to overcome the negative impacts of climate change on agriculture.

21.3 Brief Review of the Literature

There are many studies to estimate the effects of environmental changes on crop productivity levels using agro-economic models or regression analysis. Assuming that climate change affects the area and productivity of crops, three approaches have been widely used in the literature to measure the sensitivity of agricultural production to climate change: agronomic–economic models, cross-sectional models, and AEZ models.

The agronomic–economic method begins with a crop model that has been calibrated from carefully controlled agronomic experiments (FAO 2000; Kumar and Parikh 1998a). Crops are grown in field or laboratory settings under different possible future climates and carbon dioxide levels keeping all farming methods across experimental conditions fixed so that all differences in outcomes can be attributed to the climate variables, that is, temperature, precipitation, or carbon dioxide.

In a cross-sectional approach, also known as the Ricardian method, farm performances are examined across climate zones (Kumar and Parikh 1998b; Mendelsohn et al. 1994, 1996). Ricardo observed that land values would reflect land productivity at a site (under competition). In this approach land value is regressed on a set of environmental inputs to measure the marginal contribution of each input to farm income. The approach has been applied to the United States (Mendelsohn et al. 1994, 1996). In the Ricardian analysis, prices of both inputs and outputs are assumed to remain proportionately constant. Climate parameters are precipitation and minimum, maximum, and diurnal temperatures. Usually climate normals, based on time-series averages over a fairly long period of time, are considered.

The third approach to measure the impacts of climate change utilizes AEZ (FAO 1996). The main advantage associated with the AEZ is that they have been measured and published for all developing countries (FAO 1992). Detailed information is available about the climate and soil conditions, crops, and technologies being used throughout the tropical zone. The AEZ model develops a detailed eco-physiological process model. Factors such as length of growing cycle, yield formation period, leaf area index, and harvest index that explain plant growth are inputs to the model. Existing technology, soil, and climate are combined to predict land utilization types (LUT). Combining these variables, the model determines which crops are suitable for each cell. The impact of changes in climate variables on potential agricultural output and cropping patterns are thus simulated.

Palanisami et al. (2009) employed Ricardian type regression equations to study the effect of climate change on area and productivity of three crops in Tamil Nadu. The time series data for their study included three major crops, that is, paddy, groundnut, and sugarcane grown in 11 districts. These crops account for the major cultivated area of the state besides being grown in almost all districts. Based on data availability for various variables considered for the analysis, the panel data set was constructed for the period from 1990–1991 to 2000–2001. Dependent variables considered for analysis were area and yield of the three crops. Thirty-year averages of rainfall, minimum and maximum temperatures, and diurnal variations obtained from the Indian Meteorological Department, Pune, India were included as the independent variables. Finally, data from HADCM3 climate change projections for Tamil Nadu region downloaded and extracted from the global circulating model (GCM) outputs of IPCC scenarios (http://www.ipcc-data.org/sres/hadcm3_download.Html) was used in the Ricardian model regressions to estimate the impacts of climate change on the crop area, yield, and production levels. Using the HADCM3 scenario predictions, that is, an average temperature increase of 0.8°C in 2020 and an average increase of 38 mm of rainfall in Tamil Nadu (Ranganathan 2009), the Ricardian model predicted reduction in crop area and yields by about 3.5% to 12.5% due to impact of climate change. Consequently overall crop production is expected to decrease from 9 to 22% in the state. The Ricardian approach is a cross-sectional model applied to agricultural production. It takes into account how variations in climate change affect net revenue or land value. Following Mendelsohn, Nordhaus, and Shaw (1994), the approach involves specifying a net productivity function of the form

$$R = \sum p_i q_i (x, f, z, g) - \sum p_x x \qquad (1)$$

where R is net revenue per hectare, p_i is the market price of crop i, q_i is output of crop i, x is a vector of purchased inputs (other than land), f is a vector of climate variables, z is a set of soil variables, g is a set of economic variables such as market access, literacy, and population density, and p_x is a vector of input prices. The farmer is assumed to choose x to maximize net revenues given the characteristics of the farm and market prices. Assuming a quadratic function for crop output, the standard Ricardian model is specified by the quadratic function

$$R = \beta_0 + \beta_1 f + \beta_2 f^2 + \beta_5 z + \beta_6 g + u \qquad (2)$$

where u is an error term and f and f^2 are levels and quadratic terms for temperature and precipitation. The inclusion of quadratic terms for temperature and precipitation ensures non-linear shape of the response function between net revenues and climate. Normally we expect that farm revenues will have a concave relationship with temperature. When the quadratic term has positive sign, the net revenue function is U-shaped, but when the quadratic term is negative, the function is hill-shaped. Since for each crop there is an optimal temperature at which it has maximum growth, the function is expected to have a hill shape. From the fitted equation, we can find the marginal impact of a climate variable on farm revenue. The marginal impacts are usually found at the mean level of the climate variable. Thus from equation (2) we have

$$\frac{\partial R}{\partial f} = \beta_1 + 2\beta_2 \bar{f} \qquad (3)$$

where $\bar{f}$ is the mean of the climate variable. This shows that the marginal effect of a particular climate variable is equal to the sum of (i) coefficient of the linear term and (ii) twice the product of the coefficient of the quadratic term multiplied by the mean level of the climate variable. The climate variables included in the model are season temperatures (winter, spring, summer, and fall) and their squares and season precipitations and their squares. The model described above was fitted to cross-sectional data using MATLAB software package.

21.3.1 Data and Variables

The cross-sectional data for the present study was collected from a farm-level survey of 450 farmers spread over 10 districts of Tamil Nadu. These 10 districts were chosen to represent various agro-climatic zones of Tamil Nadu. In all the 10 districts, the majority of the farmers grow rice. Other crops grown are cotton, sugarcane, and turmeric. In the dry districts, that is, Perambalur, Dharmapuri, and Ramnad, cotton is the major crop followed by groundnut although rice is also grown as rain-fed crop. In Ramnad district rice is grown both as rain-fed and irrigated crops.

The net-revenue per acre for each farmer was approximated using the equation

$$NR_F = \frac{\left(\sum\limits_{i=1}^{i=n} P_i Q_i - \left(\sum\limits_{j=1}^{m} p_j X_{ij} \right) \right)}{A_F} \qquad (4)$$

where

NR_F = Net farm revenue per acre for farmer F
P_i = Unit price of crop i
Q_i = Quantity produced of crop i
P_j = Unit price of input j
X_{ij} = Quantity of input j used in crop i and
A_f = Total area planted by farmer F in a year

As suggested by equation (2), the net revenue per acre was regressed on climate, soil, and socioeconomic variables. The climate variables included were *(i)* maximum temperature *(ii)* minimum temperature and *(iii)* rainfall. The squares of these variables were also included. Percentage of wetlands was used as a proxy to represent soil wetness. The socioeconomic variables used in the study are *(i)* age, *(ii)* education in years, *(iii)* family size, *(iv)* livestock income (in '000 rupees), and a dummy variable for adopting crop insurance by the farmer.

21.3.2 Results and Discussions

21.3.2.1 Mean Values

The mean values of the variables for each district are provided in Table 21.1. It shows that the average net revenue per acre is the greatest in Coimbatore district followed by Kanyakumari district. Maximum temperature varies between 31.17°C and 34.64°C, excluding Nilgris. Rainfall has a high fluctuation between 687 millimeters (mm) to 1,433 mm while minimum temperatures range between 21.79°C to 29.34°C. The soil wetness index also has very high fluctuation. The average age of the farmer is about 44 years with an education of about 6 years. Family size is almost a constant with four members. Farmers have good livestock income in all the districts. The percentage of farmers who had opted for crop insurance ranges between 36% and 71%. The crop insurance dummy was included in the model as it represents an adaptation strategy by the farmers against climate change.

21.3.2.2 Fitted Model Parameters

Table 21.2 gives the statistics on fitted model parameters. It shows that maximum temperature, its square, and rainfall are important climate variables affecting net

Table 21.1 | **Mean Values of the Variables Used in the Ricardian Model**

District	Net revenue/ unit area (in '000 Rs)	Maximum temperature (°C)	Rainfall (mm)	Minimum temperature (°C)	Wetlands (%)	Age (in years)	Education (in years)	Family size (number)	Livestock income ('000s)	Crop insurance (dummy)
Vellore	6.45	32.84	913.92	22.96	69.58	45.53	6.16	4.44	10.24	0.53
Dharmapuri	12.60	32.93	891.12	22.13	0.00	39.73	5.93	4.24	38.84	0.60
Perambalur	11.29	34.64	952.80	29.34	46.64	39.71	5.47	3.82	6.46	0.64
Ramnad	3.14	31.84	1075.00	26.19	99.68	52.44	5.33	4.16	15.36	0.62
Nilgris	5.35	19.54	1433.40	10.09	58.42	40.16	5.40	4.16	16.35	0.58
Cuddalore	6.66	33.36	1306.20	24.27	20.10	50.44	5.69	4.11	12.74	0.49
Kanyakumari	16.05	31.17	867.92	24.63	50.55	41.58	5.38	4.00	5.85	0.36
Tanjavur	11.92	32.98	1206.08	23.90	95.83	39.31	7.44	4.49	26.47	0.71
Coimbatore	23.13	32.92	687.16	21.79	19.93	43.89	6.49	4.42	29.50	0.42
Trichy	4.34	34.35	941.86	24.31	81.88	48.04	5.18	4.38	6.45	0.62
Total	10.09	31.66	1027.55	22.96	54.26	44.08	5.85	4.22	16.83	0.56

Source: Field survey.

Table 21.2	Estimated Regression Coefficients of the Ricardian Model			
Variable	Coefficient	Standard error	t-stat	P-value
Intercept	−256.87	223.79	−1.15	0.25
Maximum temperature	38.27	22.36	1.71*	0.09
Rainfall	−0.24	0.07	−3.57***	0.00
Minimum temperature	−16.53	11.80	−1.40	0.16
Maximum temperature-squared	−0.61	0.35	−1.74*	0.08
Rainfall-squared	0.00	0.00	3.45***	0.00
Minimum temperature-squared	0.33	0.23	1.45	0.15
Percentage of wetlands	0.04	0.02	1.57	0.12
Age	−0.26	0.10	−2.56***	0.01
Education in years	−0.11	0.31	−0.34	0.74
Family size	1.12	0.91	1.23	0.22
Livestock income ('000s)	−0.04	0.05	−0.95	0.34
Crop insurance	−1.61	1.91	−0.84	0.40

Source: Estimated model using survey data.
Notes: ***Significant at 1% level; *Significant at 10% level.

revenue. Age of the farmer is the significant socioeconomic variable having a negative influence on net revenue, that is, the higher the age, the lower the income of the farmer. This conclusion is quite reasonable as generally productivity of humans decrease with age. Other socioeconomic variables are found to be non-significant.

21.3.2.3 Overall Effect of Climate Variables

A statistical analysis was performed to test the combined effect of climate variables on the net revenue per acre. For this a separate regression with net revenue as dependent variable and socioeconomic variables as explanatory variables was performed and the resulting mean sum of squares from the two regression equations were tested using F-test. The results of the analysis are provided in Table 21.3.

The analysis clearly shows that the climate variables do have significant effect on net revenue as the F-ratio is significant at 1% level.

21.3.2.4 Marginal Analysis

Next, marginal impact analysis was undertaken to observe the effect of an infinitesimal change in temperature and rainfall on Tamil Nadu farming. This was done by making use of equation (3).The computed marginal impacts of climate variables are given in Table 21.4.

Table 21.4 shows that except in Nilgris district (which is a hilly district), in all the other districts, climate variables generally have negative effect on net revenue. This means that the higher the climate variable the lower will be the net revenue.

Table 21.3	Statistical Test on Effect of Climate Variables on Net Revenue
Sum of squares with climate variables	15,285.52
Sum of squares without climate variables	3,942.85
Increase in sum of squares	11,342.67
Number of climate variables	6
Increase in mean sum of square	1,890.45
Residual sum of squares with climate variables	138,667.78
Error degrees of freedom with climate variables	437.00
Residual mean sum of squares with climate variables	317.32
F-Ratio: Numerator	1,890.45
F-Ratio: Denominator	317.32
Calculated-F-Ratio	5.96
F-Ratio-table value at 1%	2.84

Source: Estimated from the field survey data.

Table 21.4	Marginal Effects of Climate Variables on Net Revenue		
Region	Maximum temperature	Rainfall	Minimum temperature
Vellore	−1.60	−0.04	−1.18
Dharmapuri	−1.70	−0.04	−1.74
Perambalur	−3.78	−0.03	3.08
Ramnad	−0.38	0.00	0.98
Nilgris	14.54	0.08	−9.78
Cuddalore	−2.24	0.05	−0.30
Kanyakumari	0.43	−0.05	−0.07
Tanjavur	−1.77	0.03	−0.55
Coimbatore	−1.69	−0.09	−1.96
Trichy	−3.43	−0.03	−0.28
Total	−0.16	−0.01	−1.18

Source: Derived from the estimated model.
Note: R = rupee

For example, in Cuddalore district, one degree increase in maximum temperature will decrease the net revenue of the farmer by 2,240 rupees. Similarly in Dharmapuri district farmers will experience a loss of 1,740 rupees for one degree increase in minimum temperature. The percentages of losses were also worked out. The analysis showed that Trichy, Cuddalore, and Perambalur districts will suffer a loss of 79%, 34%, and 33% in net revenue per one degree increase in maximum temperature. Similarly Vellore, Dharmapuri, and Coimbatore districts may experience respectively 18%, 14%, and 9% decrease in net revenue per acre for one degree rise in minimum temperature.

21.3.2.5 Effect of HADCM3 Scenario

Using HADCM3 scenario predictions for Tamil Nadu and the marginal effects presented in Table 21.4, it is now possible to project the HADCM3 projected net losses in revenue per acre for different districts, as presented in Table 21.5.

Table 21.5 shows that HADCM3 scenario will have maximum effect on Perambalur farmers with a loss of about 3,000 rupees per acre followed by Trichy farmers whose losses will be around 2,740 rupees per acre.

21.4 Conclusion

The impact of climate change on crop productivity and income is clearly indicated by the results of the analysis. Even though several models are used to capture the impact at macro level, the Ricardian type of the model applied in this study with cross section data captures the impact of climate change at farm level. The marginal impacts can be used in developing adaptation strategies to address the most significant impacts both in the short and long term perspectives. Hence it is important to develop and field test the adaptation strategies that will help minimize the impacts of the climate change. Also it is recommended to study the transaction cost of implementing and adopting these strategies as adaptation measures by the farmers. Hence the future line of research in this area will be studying the economics of different adaptation strategies being followed by farmers as well as new strategies that are being field tested by the research. A study of farmers' responses to these strategies is also warranted as it is not uncommon to see that most efficient strategies

Table 21.5	HADCM3 Projections for Losses in Net Revenue Per Acre in Different Districts		
Region	Maximum temperature	Rainfall	Minimum temperature
Vellore	−1.28	−0.02	−0.94
Dharmapuri	−1.36	−0.02	−1.39
Perambalur	−3.02	−0.01	2.46
Ramnad	−0.30	0.00	0.78
Nilgris	11.63	0.03	−7.82
Cuddalore	−1.79	0.02	−0.24
Kanyakumari	0.34	−0.02	−0.06
Tanjavur	−1.42	0.01	−0.44
Coimbatore	−1.35	−0.03	−1.57
Trichy	−2.74	−0.01	−0.22
Total (Except Nilgris)	−1.44	−0.01	−0.18

Source: Prepared by the authors.
Note: R = rupee

in terms of profit are not followed by farmers as there are operating constraints in implementing these profit maximization strategies.

At government level, several interventions have been made to address the possible impacts of climate change in agriculture. The major interventions include: subsidies for electricity for irrigation water pumping, subsidies for drip and sprinkler irrigation systems, advanced crop production practices like system of rice intensification, precision farming practices, waiver of agricultural loans particularly during droughts, and weather based crop insurance programs mainly to benefit the rain-fed farmers.

References

Carraro, C. and A. Sgobbi. 2008. *Climate Change Impacts and Adaptation Strategies in Italy.* An Economic Assessment. http://ssrn.com/ abstract= 1086627

Food and Agriculture Organisation (FAO) 1992. *Agrostat.* Rome: FAO.

———. 1996. *Agro-ecological Zoning: Guidelines.* FAO Soils Bulletin No. 73. Rome.

———. 2000. *Two Essays on Climate Change and Agriculture.* FAO Economic and Social Development Paper No. 145. Rome: FAO.

Government of Tamil Nadu. 2008. *Policy Notes for Discussion.* Chennai: Agriculture Department.

International Institute for Applied Systems Analysis (IIASA). 2002. *Press Release on Climate Change and Agricultural Vulnerability.*

Kameyama, K., A. Sari, M. Soejachmoen, and K. Norichika, eds. 2008. *Climate Change in Asia: Perspectives on the Future Climate Regime.* Tokyo: United Nations University Press.

Kumar, K. and J. Parikh. 1998a. Climate Change Impacts on Indian Agriculture: Results From a Crop Modelling Approach. In A. Dinar, R. Mendelsohn, R. Evenson, J. Parikh, A. Sanghi, K. Kumar, J. McKinsey, and S. Lonergan, eds. *Measuring the Impact of Climate Change on Indian Agriculture: World Bank Technical Paper No. 402.* Washington, DC: World Bank.

———. 1998b. Climate Change Impacts on Indian Agriculture: The Ricardian Approach. In A. Dinar, R. Mendelsohn, R. Evenson, J. Parikh, A. Sanghi, K. Kumar, J. McKinsey, S. Lonergan, eds. *Measuring the Impact of Climate Change on Indian Agriculture: World Bank Technical Paper No. 402.* Washington, DC: World Bank.

Kurukulasuriya, P. and S. Rosenthal. 2003. Climate Change and Agriculture: A Review of Impacts and Adaptations. *The World Bank Environment Department Climate Change Series Paper 91.* Washington, DC: World Bank.

Mendelsohn, R., W. D. Nordhaus, and D. Shaw. 1994. The Impact of Global Warming on Agriculture: A Ricardian Analysis. *American Economic Review.* 84 (4). pp. 753–771.

———. 1996. Climate Impacts on Aggregate Farm Values: Accounting for Adaptation. *Agriculture and Forest Meteorology.* 80 (1). pp. 55–67.

Palanisami, K., P. Paramasivam, C. R. Ranganathan, P. K. Aggarwal, and S. Senthilnathan. 2009. Quantifying Vulnerability and Impact of Climate Change on Production of Major Crops in Tamil Nadu, India. In M. Taniguchi, W.C. Burnett, Y. Fukushima, M. Haigh, and Y. Umezawa, eds. *Headwaters to the Ocean: Hydrological Changes and Watershed Management*, pp. 509–551. London: Taylor and Francis.

Ranganathan, C. R. 2009. *Quantifying the Impact of Climatic Change on Yields and Yield Variability of Major Crops and Optimal Land Allocation for Maximizing Food Production in Different Agro-climatic Zones of Tamil Nadu, India*. An Econometric Approach Working Paper No. 2009–008. Kyoto, Japan: Research Institute for Humanity and Nature.

Chapter 22

Supporting Climate Action Plans: The Role of the Adaptation Knowledge Platform

Serena Fortuna

The Regional Climate Change Adaptation Knowledge Platform for Asia (hereafter Adaptation Knowledge Platform) was developed to respond to demand for effective mechanisms for sharing information on climate change adaptation and developing adaptive capacities in Asian countries, many of which are the most vulnerable to the effects of climate change. Discussions on this initiative started in 2008 and the first phase was officially launched in 2009 by the Swedish Environmental Secretariat for Asia (SENSA), the Stockholm Environment Institute (SEI), the United Nations Environment Programme (UNEP) and the Asian Institute of Technology (AIT)/UNEP Regional Resource Centre for Asia and the Pacific (AIT/UNEP RRC.AP)—which also hosts its secretariat. The Adaptation Knowledge Platform supports research and capacity building, policy making, and information sharing to help countries adapt to the challenges of climate change. It also facilitates climate change adaptation at local, national, and regional levels and strengthens adaptive capacity of countries in the region, while working with existing and emerging networks and initiatives.

The Adaptation Knowledge Platform is working toward building bridges between current knowledge on adaptation to climate change and governments, agencies, and communities that need this knowledge to inform their responses to the challenges that climate change presents them. The goal is to facilitate climate change adaptation in Asia at local, national, and regional levels and to strengthen adaptive capacity.

The specific purpose of the initiative was to establish a mechanism that facilitates the integration of climate change adaptation into national and regional economic and development policies, processes and plans; strengthens linkages between adaptation and the sustainable development agenda in the region; and enhances institutional and research capacity, in collaboration with national and regional partners.

The Adaptation Knowledge Platform brings together policymakers, adaptation researchers, practitioners, and business leaders who will work through activities to achieve the following three components:

1. Regional knowledge-sharing system: A regionally and nationally owned mechanism to promote dialogue and improve the exchange of knowledge, information, and methods within and between countries on climate change adaptation and to link existing and emerging networks and initiatives.
2. Generation of new knowledge: To facilitate the generation of new climate change adaptation knowledge, promoting understanding and providing guidance relevant to the development and implementation of national and regional climate change adaptation policies, plans, and processes focused on climate change adaptation.
3. Application of existing and new knowledge: Synthesis of existing and new climate change adaptation knowledge to facilitate its application in sustainable development practices at the local, national, and regional levels.

The need for such an initiative is clear. The inception year, 2009, was used to define the form the initiative should take, to establish the management and implementation modalities, to develop contacts with and the ownership of stakeholders at both national and regional levels, to assess needs for knowledge generation and sharing and capacity building, and to prepare plans for implementing the Adaptation Knowledge Platform in 2010–2011.

Overall, the activities implemented in 2009 achieved these aims. Activities were initiated in the five pilot countries, Bangladesh, Cambodia, Nepal, Thailand, and Viet Nam, with local partners mobilized and key knowledge and capacity gaps identified. The management arrangements for the long-term development of the Adaptation Knowledge Platform are in place and the structure of the regional knowledge sharing mechanism has been defined. Effective communications have been initiated, leading to awareness of the Adaptation Knowledge Platform's development culminating in its high profile launch on October 3, 2009, together with the Asia Pacific Climate Change Adaptation Network, in Bangkok, Thailand. Capacity-development activities include training for officials and researchers from across the region, with substantial progress in the inventorying of existing and generation of new knowledge products. Sharing of knowledge on climate change adaptation has been initiated, focusing on the impacts of climate change on high-altitude ecosystems. Linkages and collaboration with other relevant initiatives have been initiated, with the agreement reached with the Adaptation Network and the Southeast Asia Network of Climate Change focal points for delivery of country needs on climate change adaptation in South and Southeast Asia.

A significant outcome of the inception phase in 2009 was the strategy for the future development of the Adaptation Knowledge Platform. This strategy details the activities to be undertaken for the three components identified in the program framework, along with a number of specific communications activities. These three components and identified activities are:

1. Regional knowledge sharing system: This includes the continued development of the Adaptation Knowledge Platform website and communications products to reach stakeholders across the region, the conduct of an Asian Climate Change Adaptation Forum (both to be implemented in collaboration with the Adaptation Network), training and capacity development activities, the synthesis and dissemination of information and global experiences on adaptation actions, and the development of national-level knowledge sharing and capacity development activities. The first Adaptation Forum was on October 21–22, 2010 in Bangkok, with very successful results and a large participation of approximately 500 participants actively debated on issues related to mainstreaming climate change adaptation into development planning in Asia and the Pacific. Building from the Adaptation Forum 2010, the 2011 edition (October, 27–28 2011) will aim to focus on "Adaptation in Action" signifying a shift from deliberations to decisions, plans to policies and policies to practices. More information is available on the Adaptation Forum website at http://www.asiapacificadapt.net/adaptationforum2011/

2. Generation of new knowledge. This includes the development of generic knowledge products focused on the analysis of resilience and vulnerability, understanding the links between disaster risk reduction and climate change adaptation and downscaling of climate change and impact assessment data. It will also include four new studies that address key gaps in knowledge and understanding for the mainstreaming of adaptation into development planning. These four studies are: *(i)* Understanding planning; *(ii)* Perceived and actual knowledge gaps; *(iii)* Comparing adaptation and development; and *(iv)* How "autonomous" are autonomous responses?

3. Application of existing and new knowledge: The focus here is where knowledge is applied within the countries of the region through mainstreaming adaptation into development planning. Follow up activities were planned for the five pilot countries listed above and, in addition, in 2010 and 2011 the Adaptation Knowledge Platform activities initiated in other focal countries, such as: Bhutan, Sri Lanka, the Philippines, Myanmar, Lao People's Democratic Republic, and Malaysia. Following its approach, also in these countries, the Adaptation Knowledge Platform worked through partnerships with local institutions and 30% of the budget for 2010–2011 was planned to be dedicated to these partners. There will also be activities to develop generic knowledge-to-practice products at the regional level.

Together with the dedicated communications activities, these components will achieve the objectives of this phase of the Adaptation Knowledge Platform. They will also build a base for the long-term development of the initiative as a knowledge-based, demand-driven structure through which planning for and capacities to address climate change adaptation as a core challenge for the future development of Asia. It is anticipated that this legacy will be carried forward through new phases of the Adaptation Knowledge Platform if and when there is demand from the countries of Asia for the services the Adaptation Knowledge Platform provides.

PART V

Capacity Building Strategies
for Mainstreaming
Climate Change Adaptation

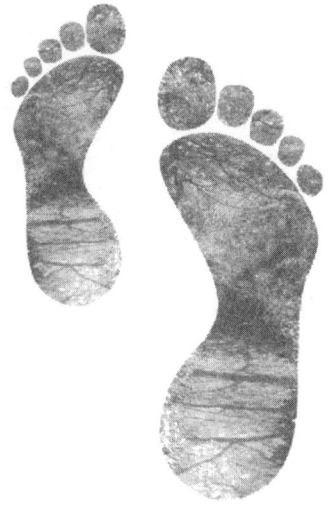

PART V

Capacity Building Strategies for Mainstreaming Climate Change Adaptation

Key Messages

Incorporating or integrating adaption to climate change into the planning process is a necessary strategy for sustainable development of the region. Climate change impact does not happen in isolation; the impact on one sector can adversely or positively affect another. There are difficulties in mainstreaming adaptation concerns into development planning due to low staff capacity for monitoring, poor data on adaptation options, lack of mechanisms for information sharing and management, and limited funding. Countries in the region need new funding and improved access to funding to provide effective technical and financial support and capacity building opportunities.

Given that many countries experience similar effects from climate change, sharing experiences can broaden the knowledge on how to address adaptation challenges. International collaboration mechanisms fostered by the UNFCCC process include the assessment of vulnerability and risk associated with climate change such as funding for national communications and national adaptation programs of action (NAPAs), public education and outreach, data and observation, decision support, adaption planning, and implementation. Operational guidelines could be prepared to help integrate adaptation into various sectors from national to local level and from local to national level, and to encourage countries in the region to implement more pilot projects and provide funding for such projects.

Effective implementation of climate change adaptation is complicated because of different scales involved; the level at which action leading to change occurs is often different from the level at which decisions regulating such actions are taken. Lack of cooperation among ministries is highlighted as a major barrier to progress on adaptation. In order that real progress can be made, key government agencies such as ministries of finance need to be informed of the relevant outputs of impact and vulnerability assessments. Sectoral institutions need to be strengthened in order to address the complexities of coordinating the implementation of adaptation action.

Capacity building at local level (for example, strengthening coping strategies and feedback to national policies), national level (for example, inter-agency policy coordination in the water sector and legal provisions for mainstreaming), and regional level (for example, incorporating climate risks in projects of development agencies) is vital to enable developing countries to adapt to climate change. Stakeholders and development partners must recognize the role of university and knowledge institutes. Enhanced support is needed for institutional capacity building, including establishing and/or strengthening centers of excellence, so that they can resolve the complexity of addressing and coordinating the planning and implementation of adaptation actions. Effective regional cooperation among countries will also help disseminate internationally and nationally the best practices, share climate information, support institutional coordination, and generate additional resources for enhancing the adaptive capacity at local level.

Chapter 23

Enhancing the Adaptive Capacity in the Asia and Pacific Region: Opportunities for Innovation and Experimentation

Venkatachalam Anbumozhi

23.1 Introduction

Developing countries of the Asia and Pacific region, as a group, are the ones most threatened by climate change. Although climate change will have impacts that can be positive for other regions, the most significant impacts are expected to be negative for the region (ADB 2009, 2010). This is because many of these countries lie in areas where climate change related effects including flooding, drought, tropical cyclones, and ecosystem destruction will be more damaging affecting the production activities within natural resources management sector. Furthermore, the region is more vulnerable to climate change due to weak institutions and their ability to respond to adverse impacts is limited. If not addressed immediately, this has the potential to undermine economic growth.

Enhancing the adaptive capacity is therefore essential to reduce vulnerability to adverse effects. An adaptive capacity building framework was scoped at the conference of parties (COP 7) (Marrakesh 2001) where developed and developing nations designed a roadmap for adaptive capacity building actions. The framework stressed that capacity building is a country-driven and results-oriented process that specifically addresses countries' needs and reflects their national strategies for sustainable development (UNFCCC 2001). It also underlined the point that capacity building should be implemented in a flexible manner to encourage its cost-effectiveness evaluation.

However, adaptation to climate change has not yet become a high-priority policy issue in most parts of Asia and the Pacific, as policymakers are preoccupied with other developmental priorities. Improbability is often cited as a reason for inaction and could be interpreted as the case of limited knowledge on cost effective best practices. Adaptation strategies are also largely being dealt in isolation from other development issues. In order to mainstream adaptation into development planning, policies should be integrated at sectoral and local levels, rather than designing separate ones. Efforts to mainstream adaptation may find resistance, particularly in the sectors directly related to economic and social development, as commonly there are tradeoffs between climate change adaptation and economic development. Resilience of the region to climate change needs to be enhanced through building regional, national, and institutional commitment, as well as technical and scientific capacity.

This chapter provides a framework for adaptive capacity building from the perspective of the natural resources management sector. Based on country experiences, it provides a broad overview of issues in strengthening adaptive capacity, describing its key dimensions and suggesting promising interventions for further exploration. It also serves as a basis for planning for mainstreaming climate change adaptation into sectoral planning.

23.2 Vulnerability and Adaptive Capacity

In order to design adaptive strategies, it is necessary to assess the vulnerability of the natural resources sector to climate change. Unfortunately, there is not a single universally accepted definition of vulnerability. Chapters 4, 8, 12, and 20 in this book describe various approaches to vulnerability assessment and adaptive capacity. A general framework that has emerged from the regional capacity building workshops conducted by ADBI on climate change adaptation is shown in Figure 23.1. This framework links human resources development to climate change through the key concept of adaptive capacity.

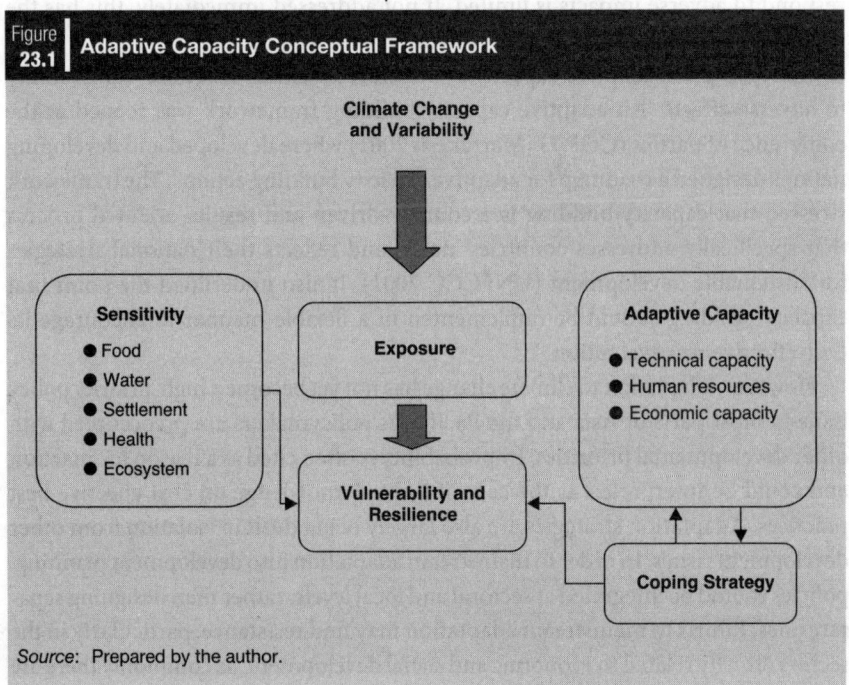

Figure 23.1 | Adaptive Capacity Conceptual Framework

Climate Change and Variability

Sensitivity
● Food
● Water
● Settlement
● Health
● Ecosystem

Exposure

Adaptive Capacity
● Technical capacity
● Human resources
● Economic capacity

Vulnerability and Resilience

Coping Strategy

Source: Prepared by the author.

According to the UN Conference on Environment and Development (UNCED), capacity building refers to the ability to plan, evaluate, and implement strategies and measures towards sustainable development, based on an understanding of environmental potentials and limits and needs of local communities (UNCED 1992). Thus, capacity building supports country-led initiatives, such as national communications, national capacity self-assessments, national adaptation programs of action, and policy-gap assessments.

Adaptive capacity is the ability to adapt. It is a function of a country's stock of infrastructure, human resources, technology base, educational system, research capacity, natural resources base, the structure of its economy, and many other determinants. But this is the key intervention point for development planning. Strengthening adaptive capacity to deal with the variability of climate change induced negative impacts comprises the sum of actions taken to change behaviors, shift priorities, produces necessary goods and services, and to plan and respond in ways that reduce harmful climate change impacts or transform them into "no regret" economic opportunities. No regret measures include strategies to reduce GHG emissions that would have a positive impact on the economic development, even in the absence of climate change.

23.3 Adaptive Capacity in the Region: Past, Present, and Future

Given the risks and the scale of potential impacts of climate change, serious efforts are underway to address impacts of climate change in most parts of the Asia and Pacific region. Actions are classified as structural and nonstructural measures. Structural measures (often referred as hard measures) include physical construction of infrastructure or application of engineering techniques to reduce the intensity of possible climate change derived hazards. Nonstructural measures (or soft measures), on the other hand, aim at changing behaviors, methods, and practices to cope with risks and impacts of climate change. Table 23.1 summarizes various structural and nonstructural measures of adaptation being implemented in different countries in the Asia and Pacific region. Among these approaches, a key issue is the identification of successful cost effective adaptation practices. It is necessary to distinguish adaptation by who is undertaking it and the interests of the diverse stakeholders involved. It is clear that countries and communities will adapt and have been adapting to climate change over the course of human history. Vulnerability to climate change can act as a driver for adaptive resource management. There are various scales and actors involved in adaptation. Some adaptation by individual communities is undertaken in response to climate threats, often triggered by individual extreme events (Ribot, Najam, and Watson 1996). Other adaptation measures are undertaken by governments sometimes in anticipation of change but again, often in response to individual events.

Table 23.1	Overview of Ongoing Adaptation Related Actions and Policies in Country Plans
Country	Adaptation actions and policies
Bangladesh	**Structural:** Flood management schemes, irrigation schemes.
	Nonstructural: Climate Change Action Plan (2009–2018); established a climate change cell; developed a network of 34 focal points in various government agencies, research and other organizations; created flood-warning systems; expanding community-based disaster preparedness; established agricultural research programs to develop saline, drought, and flood-adapted high-yielding crops.
Cambodia	**Structural:** River bank modification, crop diversification.
	Nonstructural: Released the National Adaptation Programmes of Action (NAPA) in 2006 that identifies water management, and vulnerability assessments and adaptation measures in coastal areas as priorities. Submitted its Initial National Communication (INC) to United Nations Framework Convention on Climate Change (UNFCCC) in 2002. Its 2nd Socio-Economic Development Plans (SEDP II) briefly acknowledge the negative impacts of climate change. The Royal Decree on the Creation and Designation of Protected Areas (1993), the Law on Environmental Protection and Natural Resource Management (1996), and the Forestry Law (2002) are relevant to environment and sustainable development but do not explicitly mention climate change.
Fiji Islands	**Structural:** Watershed management project for drought-prone sugarcane growth regions, integrated coastal zone management program.
	Nonstructural: Fiji Islands Climate Change Response report (2005), national vulnerability assessment study, community based adaptation strategy actions (LMMA) to support the survival of local communities and protect marine resources, adaptation project (ongoing) to reduce the vulnerability of the tourism industry.
Indonesia	**Nonstructural:** 2007 National Development Planning Response to Climate Change, 2007 National Action Plan on Climate Change (NAPCC) and the Climate Change Roadmap (March 2010)
India	**Structural:** Crop improvement, drought proofing, livelihood preservation, disaster management programs.
	Nonstructural: First National Action Plan on Climate Change (NAPCC) report (2008), increasing awareness and education, risk financing, health, National Environment Policy 2006.
Kazakhstan	**Structural:** Integrated water management; community-based adaptation measures regarding rehabilitation of natural rangeland ecosystems, stabilization of slope-wash, introduction of pasture-rotation methods and changing crop patterns, reduction of land-degradation pressures.
	Nonstructural: Awareness-raising campaigns; fostering local institutions for cooperative community management; designing multidisciplinary pro-sustainable development policies, mainly in agriculture, forestry, fishery, and water sectors.

Lao People's Democratic Republic	**Structural:** Flood and drought mitigation programs, construction of irrigation systems. **Nonstructural:** NAPCC (2009) Report to address immediate and urgent needs related to current and projected adverse effects of climate change in key sectors (agriculture, forestry, water and water resources, and human health). Disaster Management Strategic Plan (2003).
Malaysia	**Structural:** Enlarging reservoir capacities, improving hydrological forecasting, promoting widespread use of groundwater; changing land-use practices, developing demand-side management for water resources, creating buffer zones in agriculture and forestry industries to minimize erosion and sedimentation, constructed the multi-purpose smart tunnel that is used as both a motorway and flood-diversion channel. **Nonstructural:** INC (2000) and Second national communication (SNC) to the UNFCCC (2011), reporting the national GHG emissions and measures taken to address climate change. The 2009 national climate change policy, formulating Clean Air Action Plan, establishing technical secretariat for Clean Development Mechanisms (CDM), incorporated climate change projects into the 9th Malaysia Plan, establishing an inventory of agricultural Greenhouse Gas (GHG) emissions, conducting lifecycle assessments and renewable energy research.
Maldives	**Structural:** Developing coastal protection of designated safer islands and the Malé International Airport, flood control measures. **Nonstructural:** Maldives Climate Change Strategy, NAPA, Population and Development Consolidation program, carbon neutral policy, established a multidisciplinary National Climate Change Technical Team, strengthening health care capacities, improving education and awareness.
Mongolia	**Nonstructural:** National Action Programme on Climate Change, focusing on pasture land, animal husbandry, arable farming, water resources, forests, soil degradation, and desertification.
Myanmar	**Nonstructural:** Hydrological research study and field survey of the 2008 Nargis Cyclone, including historical analysis of the magnitude and frequency of cyclones and cyclone tracks over time.
Nepal	**Structural:** Water saving irrigation methods, upland land use changes. **Nonstructural:** Preparing a NAPA, established a Climate Change Network and establishing a Himalayan Research Center, the Ministry of Home Affairs has drafted National Strategies for Disaster Risk Management, introduced mandatory Environmental Impact Assessment (EIA), developed a three-year interim plan (2008 to 2010) to prioritize policies and strategies related to climate change in the development agenda, monitoring glaciers.
Pakistan	**Structural:** Water Resources Development Plan to improve flood control and protection, resource conservation cultivation, and high efficiency irrigation systems. **Nonstructural:** INC (2006) that presents the national greenhouse gas (GHG) inventory and identifies key sources and sinks of direct and indirect GHGs.

(Continued)

(Continued)

Country	Adaptation actions and policies
Philippines	**Nonstructural:** 2000 Initial National Communication on climate change to UNFCCC, established the Presidential Task Force on Climate Change and other task groups (fisheries, watershed protection, water recycling, rainwater conservation, atmospheric activities, Conservation, protection and restoration (CPR) economics, fossil fuels, information), progressing a climate change bill for mainstreaming climate change adaptation into all government policies and programs, established a Philippine government-UN joint program. Strengthening the Philippines' institutional capacity to adapt to climate change, that has already implemented five adaptation demonstration projects across the country.
People's Republic of China	**Structural:** Enhancing technology development and transfer in agriculture sector, improving livestock management, intensifying ecological agriculture in high-intensive production areas, enhancing water resource management. **Nonstructural:** Implementing regulations for improved agricultural production and increased agricultural ecosystem carbon storage, developing farmland and pasture protection construction plans.
Sri Lanka	**Structural:** Programs to improve crop and water management, distribution of flood-resistant crop varieties (2005–2008), post-tsunami coastal rehabilitation and resource management program, improving fisheries. **Nonstructural:** Enhancing training capacities, implementing the Soil Conservation Act, National Rain Water Policy.
Thailand	**Structural:** Improving crop resilience, local-community water resource management and farming practices, and alternative livelihood and tourism activities. **Nonstructural:** 2000 Initial National Communication (INC) on climate change to UNFCCC and is currently in the middle stages of completing its Second National Communication, launched Thailand's Strategic Plan on Climate Change 2008–2012, established the Project Steering Committee, undertaking climate scenarios modeling, strengthening human resources and learning processes, social protection systems and empowering local communities, integrating adaptation measures with natural hazard reduction and disaster prevention programs, established early warning and preparedness systems.

Source: ADBI (2009); PICCAP (2005).

But these levels of decision making are not independent—they are embedded in the planning or social development processes that reflect the relationship between institutes, individuals, their networks, capabilities, and social capital (Adger 2001). Sometimes a distinction is drawn between planned adaptation assumed to be undertaken by governments and autonomous adaptation by communities.

Realising that action is required to enhance the adaptive capacity of the most vulnerable sectors, efforts should focus on identifying generic determinants of adaptive capacity that may vary from country to country. These determinants include the social capital of societies, the flexibility and innovation in government institutions and the private sector to grasp opportunities associated with climate change.

23.4 Progress in Strengthening the Adaptive Capacity at Sectoral Level

As a part of the capacity building workshops, ADBI conducted a survey to understand the determinants of adaptive capacity of selected countries in the Asia and Pacific region. Participants at the ADBI workshop on Mainstreaming Climate Change Adaptation into Development Planning held from April 14 to 17, 2009 completed the survey. The questionnaire focused on raising awareness of climate change vulnerabilities, policy endorsement, climate change impacts, operational measures on mainstreaming adaptation, and regional cooperation. The responses appearing in this report are not official responses, but rather the professional judgment of officials who are directly in charge of climate change adaptation related policies and projects.

23.4.1 Institutional Awareness on the Risks Posed by Climate Change

Any operational measures to integrate adaptation that can be developed and put in place by sectoral agencies require a certain degree of awareness of climate change and the risks it poses to development. It is therefore important that sectoral agencies conduct awareness-raising activities on the risks posed by climate change, both internally and as part of their interaction with their stakeholders.

The activities however, vary considerably across countries in terms of emphasis, specificity, scope, and whether they are once or recurrent. Almost all the fifteen survey respondents, who are senior level policymakers, indicated that their agency had undertaken internal awareness-raising activities on climate change. Internal awareness-raising initiatives generally rely on a combination of written material and training seminars, while policy dialogues are used in conjunction with written material and training courses to raise awareness on climate risks (Figure 23.2).

All respondents reported that similar initiatives were being undertaken in partnership with other agencies.

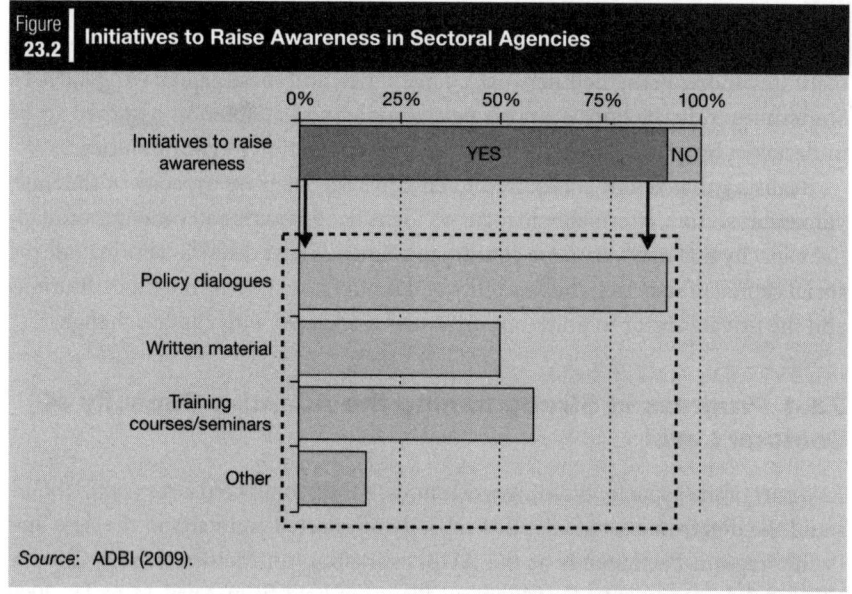

Figure 23.2   Initiatives to Raise Awareness in Sectoral Agencies

Source: ADBI (2009).

The written materials used for raising awareness are produced by the research community, international organizations, or the agency itself. They include brochures, flyers, posters, website content, communication briefs, and e-mail distribution of news items on climate change. Some agencies disseminate information on broad themes such as sustainable development.

Besides written materials, some agencies conduct training courses and seminars to raise awareness. International institutions like the Japan International Cooperation Agency (JICA), and GIZ, regional development banks like ADB and its institute (ADBI), and regional networks such as the Network of Asian River Basin Organizations (NARBO) are helping them to advance in this area. They conduct training courses for planning professionals on disaster risk reduction and climate change adaptation issues, as well as capacity building seminars for relevant developing country government agencies and other related stakeholders (Box 23.1).

23.4.2 Multi-sectoral Initiatives

Almost all the survey respondents reported that their agencies hold discussions on climate change in regular policy dialogues with other sectoral authorities. The relative emphasis on joint actions depends on several factors, including the level of sectoral interests of policymakers, as well as contextual issues. In general, nonstructural measures of adaptation tend to dominate in dialogues with middle income countries, while structural measures are a priority focus in the most vulnerable

Box 23.1	NARBO Capacity Building Activities

The Network of Asian River Basin Organizations (NARBOs) was established in 2004 by ADB, ADBI, and the Japan Water Agency. NARBO aims at strengthening the capacity and effectiveness of River Basin Organizations (RBOs) in promoting integrated water resources management (IWRM) and improving water governance, through training and exchange of information and experiences.

Addressing the needs of member organizations, NARBO has organized several regional capacity building events to raise awareness for IWRM, and share good practices and lessons learnt for IWRM among RBOs. The Regional Workshop on Developing Partnerships for Water and Climate Change Adaptation was organized in Malaysia in 2008. The objective of the workshop was to increase the understanding of the impacts of climate change on water management and develop partnerships for better results in climate change projections, impact assessments, and adaptation strategies. River improvement activities to adapt to the impact of climate change and state-of-the-art forecasting rainfall systems such as down scaling model by using Geographic Information System (GIS) and satellite information systems were also discussed.

ADB and ADBI also introduced GIS and satellite information systems to predict rainfall patterns in related workshops.

Source: NARBO (2010).

countries such as small island states. In addition, several respondents indicated that awareness in other sectors when adaptation issues are included in country programming funded by external agencies. UN agencies, for example, support partner countries in preparing national adaptation programs of action (NAPAs) that identify priority responses to most urgent immediate adaptation needs.

However, in terms of using instruments such as risk-screening tools and guidelines for mainstreaming adaptation, only 8% reported using the instruments developed by other sectoral agencies and international organizations (Figure 23.3).

The result implies that the initiatives to achieve sector-wide commitments to work together have been taken in the absence of shared instruments.

23.4.3 High-level Policy Endorsement

There is an interdependent relationship between the degree of internal awareness amongst agency staff of the challenges posed by climate change and the level of policy endorsement at senior levels within the agency for the need to integrate adaptation into development activities. On the one hand, a certain level of climate change is often a prerequisite before the issue reaches the high-level policy agenda. On the other hand, high-level policy endorsement of the need to take climate risks into account can, in turn, further enhance the level of awareness across the agency

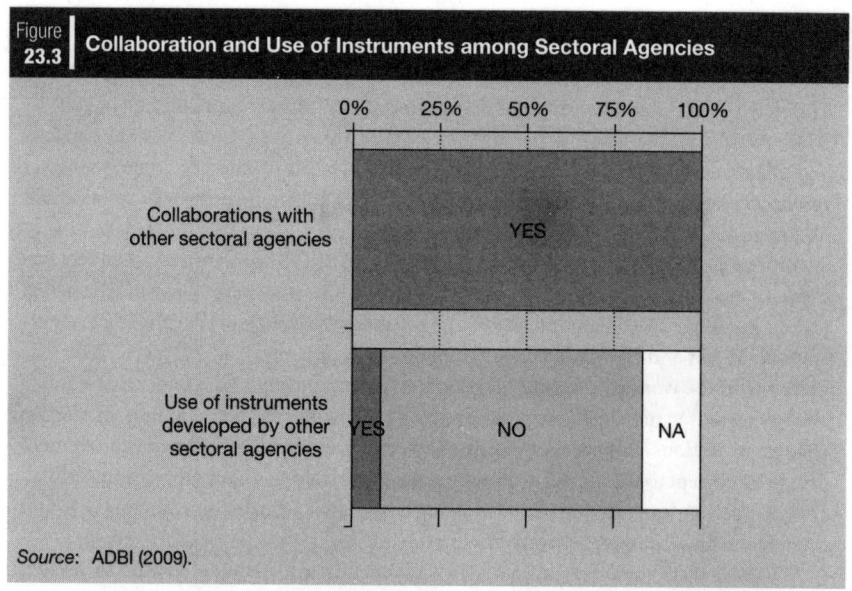

Figure 23.3 | Collaboration and Use of Instruments among Sectoral Agencies

Source: ADBI (2009).

and provide both the impetus and the enabling environment for operationalizing adaptation activities. Therefore, high-level policy endorsement is an important criterion for monitoring progress in this area.

A majority of the surveyed officials reported to having such high-level policy endorsement at the agency and/or national level. There are a number of high-level national policy initiatives with broad environmental objectives and developmental priorities. Climate change adaptation is explicitly or implicitly contained within these broader mandates. Some countries like the People's Republic of China (PRC) and India have medium-term policies to address urgent adaption issues. Those documents outline, among other issues, cross sectoral efforts in dealing with adaptation challenges. With the goal to advancing the adaptation agenda, environment ministries have established expert committees for building and enhancing adaptive capacity, and enhancing collaboration among the sectoral agencies. These initiatives range from agreements with broader environmental and development objectives, to climate change initiatives comprising both mitigation and adaptation issues, to specific agreements aimed at integrating climate change adaptation into developmental planning. Major areas of action were referred to be agriculture (34%), disaster prevention (17%), and overall development goals (17%), as illustrated in Figure 23.4.

23.4.4 Mainstreaming Climate Change Adaptation

Even though all the countries are aware of the consequences of climate change, they have limited capacity to design and implement adaptation programs to avoid the

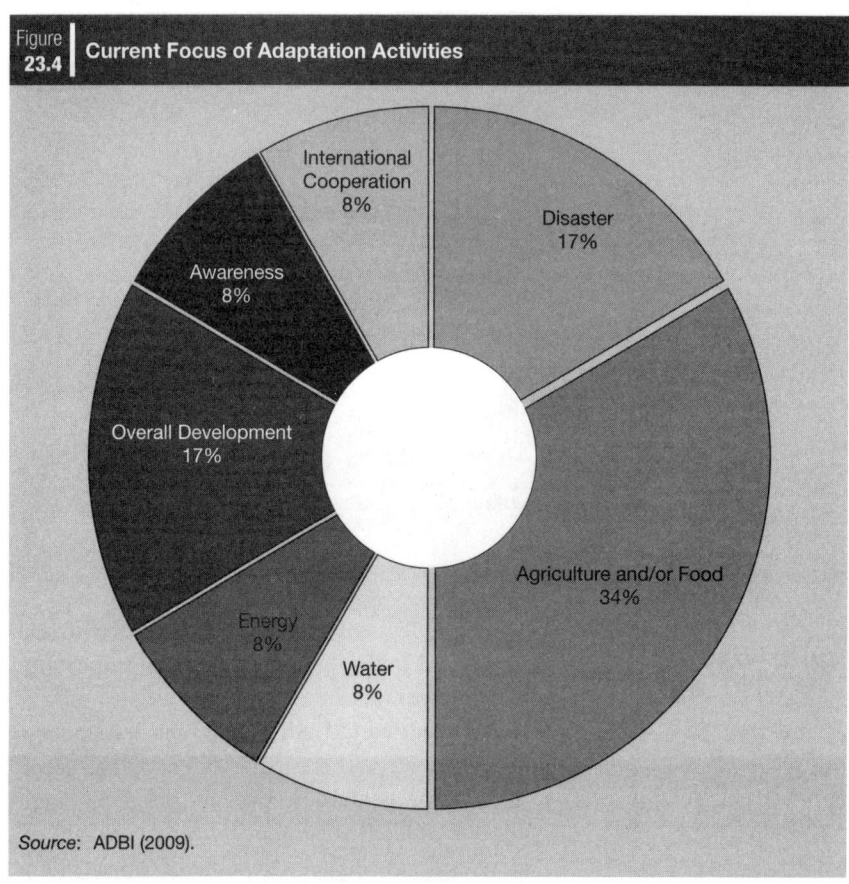

Figure 23.4 | **Current Focus of Adaptation Activities**

Source: ADBI (2009).

risk of climate change at a scale required. When inquired about priorities in main-streaming adaptation, respondents stressed the urgency of taking actions related to scientific and technical skills enhancement (for example, analytical studies and new technologies) (24%), inventory of baseline data (17%), increasing awareness (17%), and promoting cross-border cooperation (17%) (Figure 23.5).

In all, 83% of the respondents had conducted assessments on the implications of climate change on their activities (for example, country, regional, and sectoral strategies, technical cooperation, and projects), 92% had conducted assessments on documents (for example, country strategies, policy, and project descriptions whether they make reference to climate change impacts and vulnerabilities), and 34% had conducted assessments on exposure of investments (for example, the proportion of its activities in sectors that are potentially affected by climate change). The results (Figure 23.6) highlight that the current attention of assessments are not climate change risk on financing but risk on activities and policies. However the

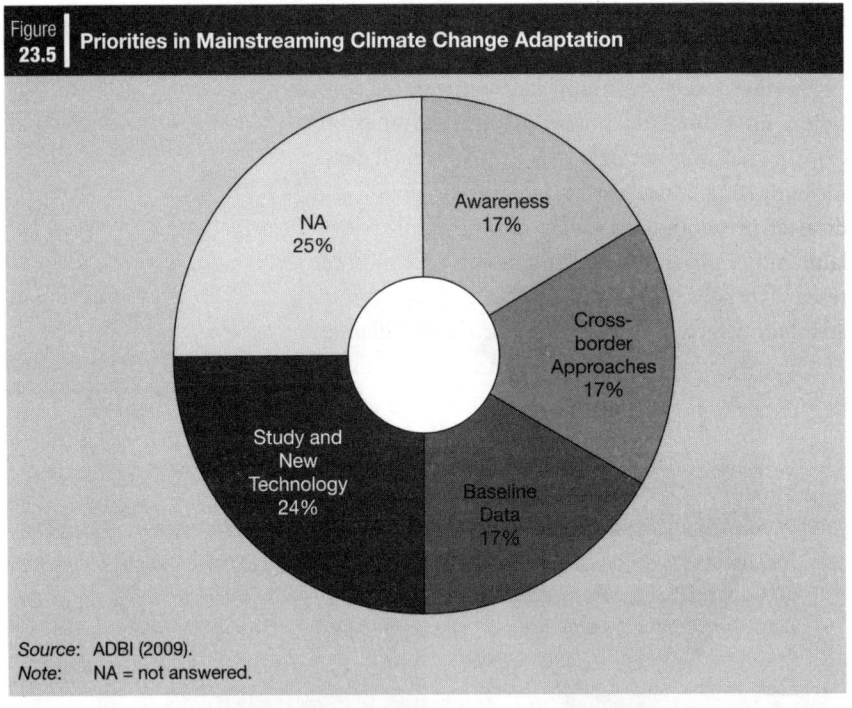

Figure 23.5 | Priorities in Mainstreaming Climate Change Adaptation

Source: ADBI (2009).
Note: NA = not answered.

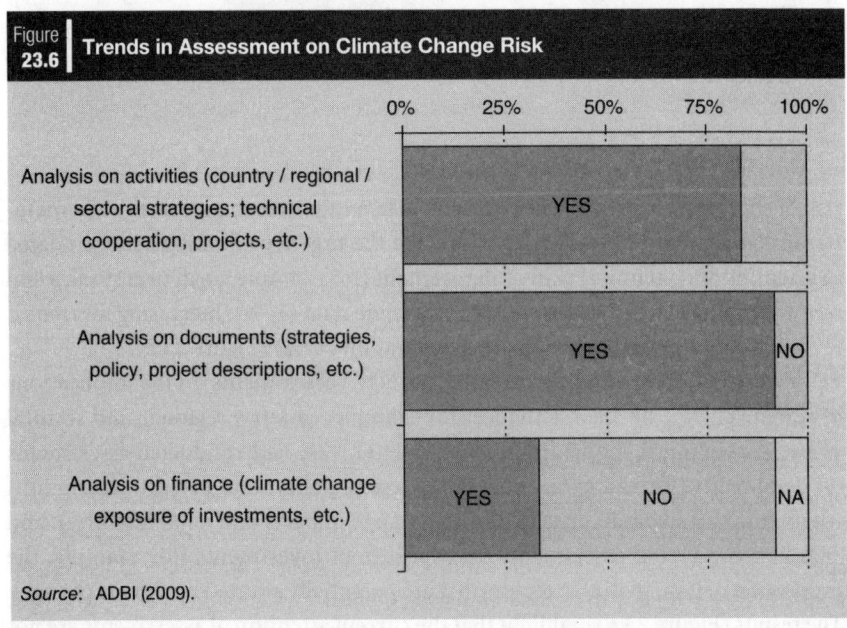

Figure 23.6 | Trends in Assessment on Climate Change Risk

Source: ADBI (2009).

impact on distribution of finance in the sectors by exposure of investment should not be underestimated.

But there are few examples of an integrated approach for climate change adaptation. Environmental policy instruments for example, do not specifically identify the integration of sectoral issues into national development planning as a priority for budgeting. More specific avenues for policy integration include *(i)* integrating disaster prevention and adaptation into all relevant development activities, *(ii)* launching a program on adaptation in agriculture including financial support to a research program, *(iii)* providing financial support to the least developed areas under adaptation agenda, and *(iv)* supporting the adaptation pillar of co-benefits.

Other barriers to mainstreaming climate change adaptation are summarized in Box 23.2.

Box 23.2	Barriers to Mainstreaming Climate Change Adaptation

- Limited understanding of the nature and extent of risks and vulnerabilities, or lack of credible climate information.
- Available climate information is often not directly relevant for development related decisions.
- Lack of information on the economics of good adaptation measures, or simply an absence of knowledge on available "no regret" strategies.
- Tradeoff between climate and development objectives exists.
- Lack of available funds or restricted access to finance.
- Segmentation within governments, no strong supportive policies, standards, and regulations.
- Differences in willingness to accept uncertainties.
- Funding modalities are not well established. Difficult for adaptation efforts to attract resources compared to more visible activities such as emergency response, disaster recovery, and reconstruction.

Source: ADBI (2009).

23.5 Strategic Approaches in Strengthening Adaptive Capacity

A country's ability to undertake actions to tackle climate risk is largely a function of its adaptive capacity. Figure 23.7 shows the grouping of countries in three tiers based on their adaptive capacity. To be most effective, adaptation must proceed with specific strategies at several levels simultaneously. Adaptation is in a fundamental way inherently local—the direct impacts of climate change are felt locally, and adaptation measures must be tailored to local circumstances. However, for these efforts to be robust they must be guided and supported by national policies and

Figure
23.7

Classification of Adaptive Capacity of Countries

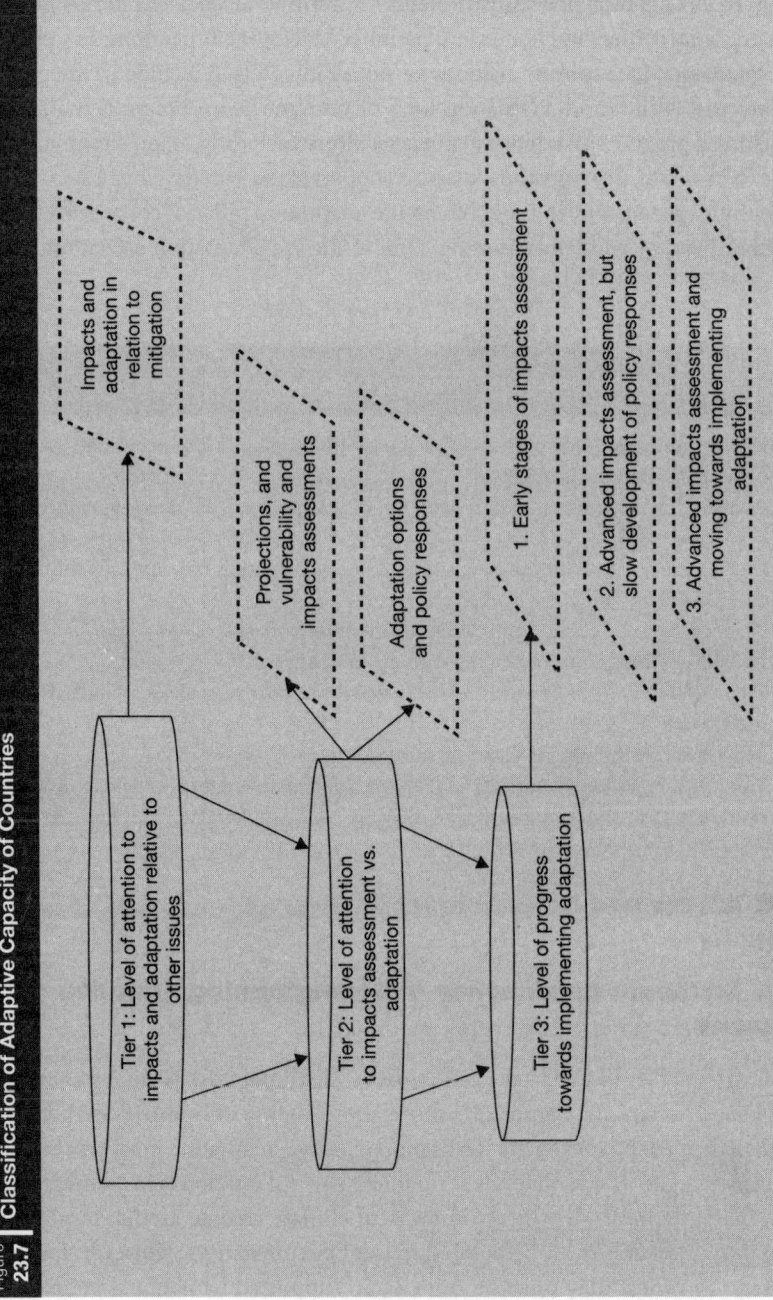

Source: Gagnon-Lebrun and Agrawala (2006).

strategies. For some countries, these need to be facilitated through international measures.

Enhancing adaptive capacity at the sectoral level could involve several stages as discussed in Chapter 1. The distinction between specific adaptations and enhanced adaptive capacity may not be clear in the initial stages. However, effective capacity building strategies must rely on the best available climate information on the nature and severity of likely impacts over different timeframes in given locales, and on the cost and efficacy of possible adaptation measure. Hence, an overriding priority is strengthening capacities in building awareness to understand potential climate impacts and devising response strategies.

Some capacity building activities are no regret, for example, resources and training to integrate adaptation considerations into development planning, expanded research into alternative crops or cropping patterns, or strengthening public health systems. Here, again, these are steps with multiple benefits beyond climate adaptation. Many specific adaptations can be effective in reducing certain risks. For example, cyclone shelters in Bangladesh have proven very effective in reducing loss of lives during climate induced disasters. However, specific adaptations deliver fewer ancillary benefits. In addition, where adaptive capacity is limited, the potential benefits of specific adaptations may be quite limited. For example, an early warning system is of limited value if the users at risk have no economic capacity to respond. Testing such measures through pilot programs is critical.

One indicator of successful adaptive capacity is to ensure that strategic adaptation actions are mainstreamed in development planning which correspondingly advances adaptive capacity. Collectively, these efforts must meet a wide range of interrelated needs. In considering how best to address these needs, sectoral planners face difficult issues stemming from the underlying institutional contexts for adaptation decision making and action, and inherent limits on available resources—all compounded by politically sensitive questions of responsibility and equity.

23.6 Actors and Stakeholders in Strengthening Adaptive Capacity

The range of actors and their specific role in enhancing adaptive capacity depends on the specific context or issue being addressed. In a generic sense, however, there are many commonalities across them. The broad groups that will be important to work with are:

1. Policymaking agencies operating across sectors at different levels
2. Scientific and educational research institutions particularly those with interdisciplinary programs
3. Private sector organizations, particularly those involved in the climate proofing infrastructure development and services

4. Civil society and community-based organizations
5. International organizations

The above mix of actors reflects the capacities required to catalyze adaptation at a scale required to avoid climate risks in a cost effective way. Local and regional governments are, however, likely to be most directly familiar with and involved in adaptation activities. Research and educational networks operating across and within sectors and interdisciplinary scientific and educational organizations are essential for cross sectoral and international learning and to link knowledge generation with major decision-making processes. Since much adaptation will occur in the future and also in a proactive way, involvement of the private sector is central to achieving impacts at scale. Finally, international organizations that combine the flexibility required for testing of innovative approaches with the explicit focus on vulnerable populations is critical. The specific roles these groups could bring to a program of adaptive capacity are discussed below.

23.6.1 Policymaking Agencies Operating at National, Subnational, and Local Levels

National governments in every country in the region are the key players in climate negotiations. However the capacity of national governments to play a significant role in climate change adaptation differs across the region. In addition to national government entities that are designated as nodal points for activities related to climate change, three types of agencies are of particular importance. First, economic planning agencies could play an important role in climate change adaptation. In India, for example, the Planning Commission has become a key player in adaptation responses and in coordinating across different sectoral agencies. Stronger involvement by economic or planning ministries to lead coordinated planning has been raised as a possible strategy in some Southeast Asian countries. In the PRC, the National Development Reform Commission, which is already the coordinating authority in climate change affairs, plays a significant role. Second, sector specific agencies, because much adaptation is likely to occur through sector specific development processes, will be of particular importance. Third, local governments play a significant role in climate change adaptation across the region. In all areas, the factors that constrain and enable both autonomous and planned adaptation will be heavily influenced by location-specific conditions. Therefore, it is particularly important to improve the capacity of local governments.

23.6.2 Scientific and Educational Research Institutions

The scientific research community in most countries across the region has primarily focused its efforts on climate change forecasting at national level. Quantitative modeling is an essential tool to assess climate change impacts, estimate systems sensitivity in response to climate change extremes, and reduce uncertainties concerning forecasts and cost-benefits of adaptation measures (OECD 2006). Developing countries with limited scientific capacity are often compelled to apply generic and global methods that do not necessarily fulfill their needs. Therefore global climate models need to be downscaled, taking into consideration regional data, increasing participation of local research centers and scientists, and including local communities' knowledge on climate change events (present and historical perspective). Inventory activities and field campaigns must be sponsored to fulfill knowledge gaps in observing networks and data collection methodologies. Although significant attempts have been carried out to develop methodologies and models to focus on Asian ecosystems, this field is still being investigated and mainly focuses on crops productivity simulation as described in Anbumozhi et al. (2003) and Reddy, Anbumozhi, and Reddy (2005). If these models expand their evaluation spectrum to include other biophysical and nonphysical variables (social and economic impacts), they will become reliable and accurate tools to support decision-making processes. Multidisciplinary teams must be formed to support these processes (Anbumozhi et al. 2001).

Scientists, who measure physical impacts of climate change and adaptation strategies, should work with economists and social scientists to include economic impacts and the perspectives of local communities. However the lack of programming skills and insufficient expertise about predicting climate events, due to short-term based activities, result in untrustworthy surveys that do not support policymaking decisions.

Regional level research networks such as System for Analysis, Research, and Training in global change science (START), Asia-Pacific Network for Global Change Research (APN), and Economy and Environment Program for Southeast Asia (EEPSEA) support regional level research. Such organizations will be among the most important for implementing research and educational programs. Building the capacity of educational and scientific institutions will be particularly important for those concerned with current policy environment; and those that produce the graduates who ultimately populate government, national sector, or policy organizations, and private business entities. Such organizations train future generations of sector specific and integrating planning and economic development experts—that is, the individuals who will ultimately "actualize" systems. Indicative systems of such capacity building programs are listed in Table 23.2.

Due to the wide array of issues involving key actors that need to be influenced, partnerships between academic, government actors, and the private sector, are likely to be particularly important in supporting adaptation at national level. International organizations and knowledge institutes are critical in strengthening the scientific capacity of local universities and policy research institutes. A closely connected network with specific targets and pragmatic time framework will greatly enhance the adaptive capacity at sectoral level. Where such networks as illustrated in Figure 23.8 do not currently exist, encouraging their formation will have greater impact than attempting to work on a one-to-one basis with individual organizations however strong they may be.

Table 23.2	Capacity Building and Training Indicators for Climate Change Adaptation Mainstreaming				
Indicators	International partnership	New knowledge	Integration	Stakeholders	Dissemination
Workshops/seminars	O	O	O	O	O
Books/websites	O	O	O	O	O
Decision support systems	O			O	
Joint cross sectoral actions			O	O	O
Guidelines/handbooks		O	O	O	O
Joint studies	O	O	O	O	O
Graduate courses		O		O	O

Source: Prepared by the author.

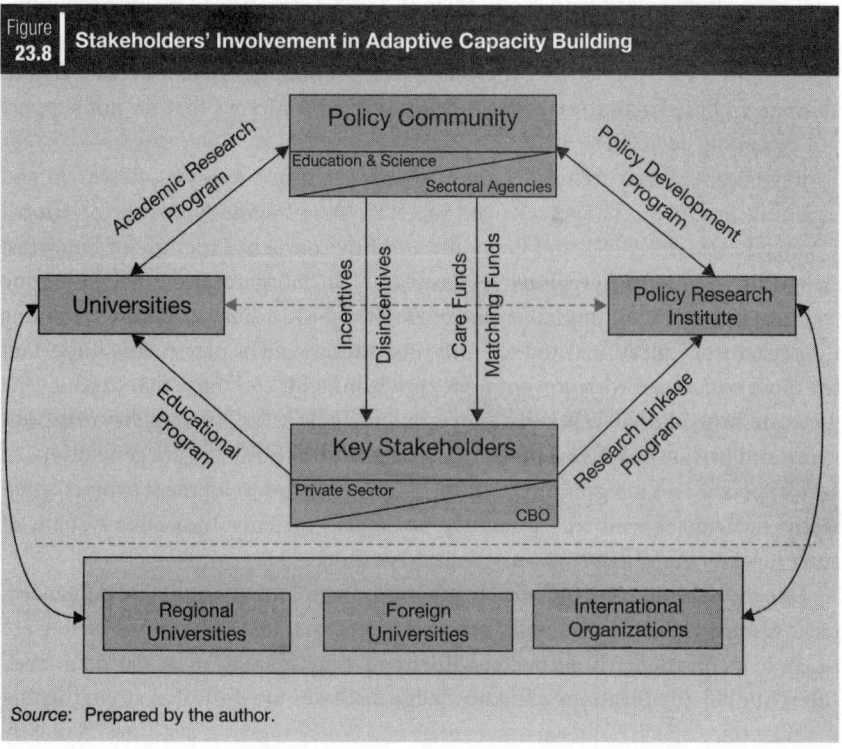

| Figure 23.8 | Stakeholders' Involvement in Adaptive Capacity Building |

Source: Prepared by the author.

Open governance of research and education networks is essential. These networks play an important role in giving consistency to the best practices communicated by national governments and/or to issues that demand intellectual dynamism to generate the wide array of insights required to catalyze effective strategies for climate adaptation.

Such research and educational networks have unique regional and cross sectoral engagement capacities, as they are able to identify multiple points of entry or leverages, and are capable of engaging in and replicating the results from learning strategies. Given the current status of the scientific capacity for climate change adaptation (Box 23.3), strengthening such networks and the institutions is likely to have higher and more consistent returns than focusing on individual organizations.

Box 23.3 | Current Status of Scientific Capacity for Climate Change Adaptation

- Lack of programming skills—most research and educational activities undertaken are on a short-term basis. Long-term planning is not thought of to continue develop solutions to solve the problems.
- Inadequate monitoring and evaluation—keen on activities but not enthusiastic in monitoring the impact of research and education.
- Inadequate communication skills—for example, downscaling the climate forecast at subregional level and communicating with decision makers.
- Lack of effective networking, experience sharing, and dissemination skills.
- Inadequate leadership, governance, and management capacities—undefined roles of team members and accountability.
- Inadequate capacity to raise adequate international resources, mobilize local resources, manage finances, and effective reporting.

Source: Anbumozhi (2009).

23.6.3 Private Sector

The role of the private sector in promoting innovative pilot projects to strengthen the access and delivery of climate related information through communication strategies needs to be further explored. The involvement of the private sector on adaptation measurements is essential, since major plans will rely heavily on activities that fall within the private sector actions. Such involvement will flow most naturally from research processes that lead to courses of action that reflect the core business interests and models on which private sector activity is based (Figure 23.9).

Direct business interests are the core reasons why the insurance and energy industries are heavily involved in work on climate change while it has proved difficult to involve other private sector actors. Identifying points of entry for other

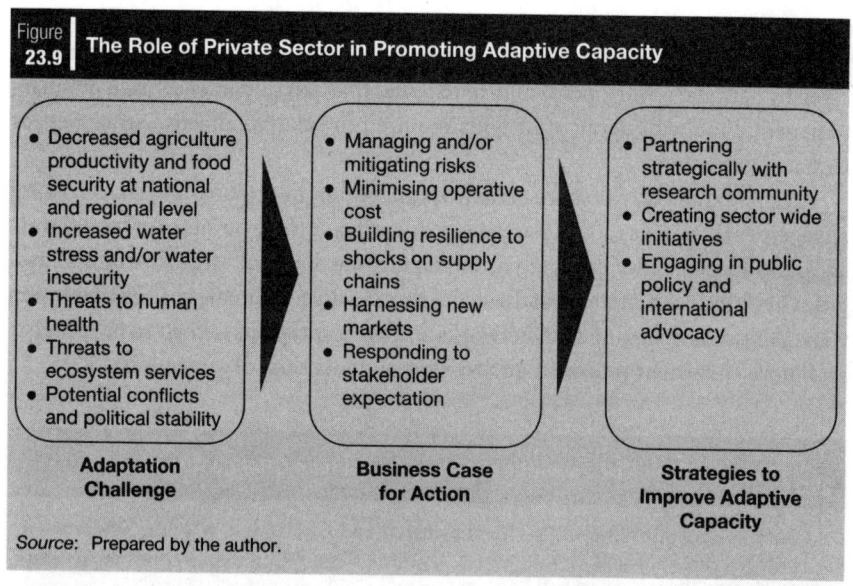

Figure 23.9 The Role of Private Sector in Promoting Adaptive Capacity

Source: Prepared by the author.

business such as infrastructure and health, needs special policy instruments to encourage their participation in climate mainstreaming processes. From this perspective perhaps the most important private sector groups to engage with are those involved in designing climate-proofed infrastructure and financing adaptation practices. Business incubator programs have specific experience in taking small innovative initiatives and driving them to scale using appropriate operational models. They also have specific skills in the innovation and incubation of businesses so that their products and services can be marketed at scale.

23.6.4 Civil Society and Community-based Organizations

The region is rich with examples of community-led natural resources management and development programs (see example in Box 23.4). Although not designed as climate change adaptation initiatives, they are indicative of adaptive mechanisms that could potentially be undertaken in drought, flood prone, coastal, and mountain or upland areas. Social institutions such as cooperatives can also play a significant role in strengthening links with markets for better returns in small-scale enterprises.

Bilateral and multilateral agencies support many national and international nongovernment organizations (NGOs) in the region. Their work has focused on development- and livelihood-related initiatives particularly linked to disaster risk reduction. These organizations play a particularly critical role in the innovation and incubation of new adaptation measures, strategies, and pilot initiatives to support climate adaptation that can then be replicated at scale through private sector

Box 23.4	Mediating Role of Institutions in the Context of Climate Impact—NGOs in the Philippines

Local institutions play a key role in recovery after disasters by shaping the direction, effectiveness, and allocation of external assistance. An example of their critical role can be found among the work of NGOs in the Philippines. Between 1995 and 2000, more than 75% of the disasters and 95% of disaster-related deaths in the Philippines were because of climate hazards: typhoons and tornadoes, flooding, and landslides being the most prominent hazards.

Many NGOs in the Philippines have integrated relief and rehabilitation strategies into their action programs. These strategies include socioeconomic projects to reduce local vulnerability, mediation of the flow of government and international assistance, community-based disaster management, small scale infrastructure development, and training for capacity building. In one interesting case, NGO staff focused on vulnerable communities to identify local leaders, conducted hazard and vulnerability analysis, initiated training related to disaster management, and established village level committees to foster effective disaster responses. Other NGOs have provided financial and technical assistance to help in community based disaster management activities. These examples show the critical role of local institutions in any area-based effort to undertake adaptation measures.

Source: Luna (2001).

business models or public sector interventions. They also play a critical role in the development of climate related social protection initiatives that would otherwise fall below the radar screen of national governments and do not generate the profits required to catalyze private sector investment.

23.6.5 International Organizations

Effective adaptation response requires international support. Three broad approaches for the Asia and Pacific region are:

- Adaptation under the UN Convention. Mechanisms and support for proactive adaptation by facilitating comprehensive national strategies and committing reliable funding for high priority implementation projects.
- Integrating adaptation capacity building programs with development aid. Factoring adaptation into development assistance through measures such as mandatory climate risk assessments for projects financed by multilateral and bilateral lenders.
- Climate insurance. Committing public and private funds to support climate relief or insurance-type approaches in vulnerable countries for losses resulting from both climate change and climate variability.

Each of these approaches, pursued independently, could contribute to national level capacity to reduce or cope with climate risks. Together, these three strategies could be seen as complementary elements of a comprehensive international effort. The first, supporting proactive planning and high priority implementation; the second, promoting integration with the broader development agenda; and the third, providing a safety net to ameliorate unavoidable impacts.

23.7 Opportunities for Innovation and Experimentation

There are many starting points for innovation and experimentation on the measures that can strengthen adaptive capacity for addressing impacts of climate change. The large amount of resources currently being used to improve the adaptive capacity is not well harmonized with efforts to increase country preparedness to act on climate information or adaptation strategies. This is especially true among the most vulnerable sectors. The current emphasis on climate prediction and risk forecasting offers development planners the opportunity to simultaneously embrace the recommendations of perfect information, vulnerability to hazards, and vulnerability to outcome approaches through communication. Even though they reflect very different views of political economy, the policies implied by these approaches are neither inconsistent nor mutually exclusive.

The dissemination of predictions at sectoral and local levels could be embedded in a larger process aimed at (i) facilitating the flow of available climate information, and identifying critical economic and social aspects of climate phenomena with better accuracy, (ii) identifying and addressing the bottlenecks in the potential use of climate information, and (iii) exploring the opportunities to address the root cause of implementing adaptation measures. This integration can lead synergies between three policy approaches of climate information, decision capacity, and financing as illustrated in Figure 23.10.

There is an opportunity to integrate these three approaches, bringing together all levels of analysis in a search for short- and long-term risk reduction. The objective should be to foresee climate related threats and reduce their negative effects, as well as to reduce the numerous other causes that make direct climate events become disasters. Climate predictions when combined with analysis of likely socio-economic costs are particularly well suited for attracting the attention of top level policymakers who tend to be both sensitive to and to have reason to be interested in making decisions on appropriate adaptive measures at a scale. Participatory workshops involving planning ministries provide an opportunity for different sectoral agencies to come together, learn the risks and benefits, and identify and prioritize actions. There is evidence that this participatory approach can lead to significantly better decisions. Ultimately this approach could be expanded in its scope,

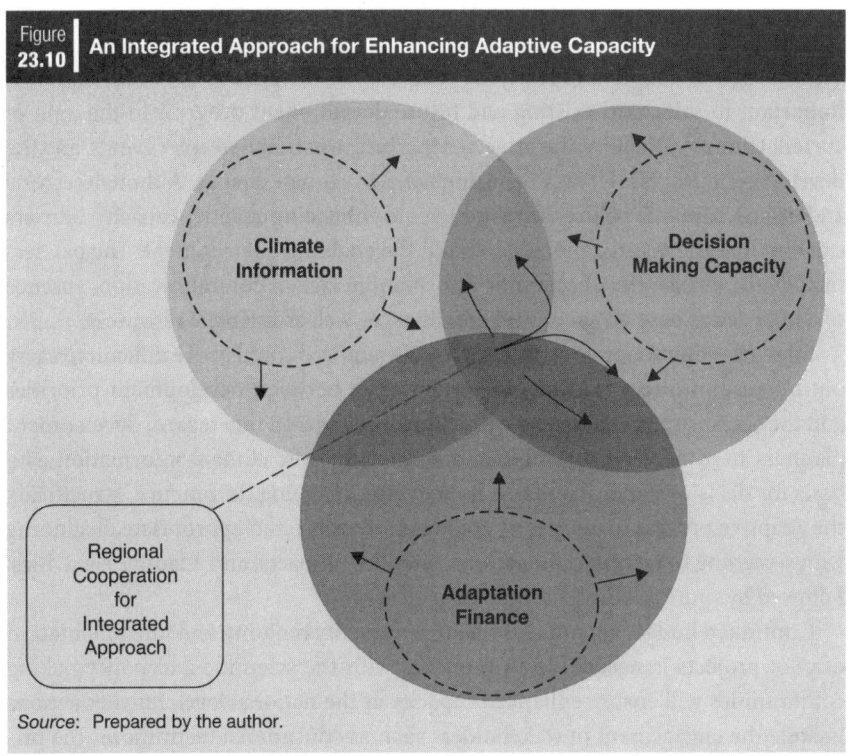

Figure 23.10   An Integrated Approach for Enhancing Adaptive Capacity

Source: Prepared by the author.

aiming to build countrywide resilience, as well as get the attention of international organizations for funding.

The synergies created by integrating the three policy streams could provide positive feedback. For example, if a potential decision maker of climate information identifies ways to put that information into developing a new adaptation strategy, they may become advocates for development of climatic knowledge and add strength to efforts aimed at securing resources for research and scientific capacity building. Public and private financiers who learn about useful climate information and economic soundness of adaptation measures may also want to increase their access to other forms of information for their financial products, therefore improving the overall flow of information and decision-making capacity.

To facilitate such a change, regional cooperation efforts need to be improved through the provision of expertise in climate information. Sharing country experiences will help to build analytical, monitoring, and decision-making capacities. Such efforts should include long-term economic strategies such as creating a regional fund for adaptive capacity building.

23.8 Conclusion

Enhancing the adaptive capacity of countries in the Asia and Pacific region is important to safeguard existing and future development progress in the light of current climate variability, the projected increase in extreme weather events, and the development progress already being impacted by climate change. Although sectoral and macroeconomic policies are conducive to enhancing adaptive capacity, barriers exist both at the organizational level and the enabling environment. The barriers include the availability of scientific information, lack of communication, absence of a knowledge base on successful measures, as well as financial resources.

Strengthening adaptive capacity at a scale required could prove difficult to carry out because of direct tradeoffs in certain cases between development priorities and the actions required to deal with climate change. In this regard, development planners need access to credible and context-specific climate information as a basis for decisions and one that is linked with financing. In practice, structuring the adaptive process to a series of graduated steps is often appropriate, beginning with screening to identify exposure, sensitivities, impacts, and adaptive capacities, followed by more detailed analyzes in critical areas.

Continued budget support for scientific capacity building and implementation of pilot projects is needed. Joint meetings with the scientific and policymaking communities will ensure enhanced capacity at the national level. Further actions include the engagement of stakeholders such as educational institutions, the private sector, and community-based organizations in supporting climate change adaptation projects and promoting comprehensive capacity building programs. Bilateral and multilateral development partners are well positioned with finance and knowledge to play a catalytic role in strengthening adaptive capacity of sectors. Regional cooperation has a role to play in facilitating effective sharing of climate information, supporting institutional coordination, and moderating the required resources for enhancing the adaptive capacity.

References

Adger, W. N. 2001. Scales of Governance and Environmental Justice for Adaptation and Mitigation of Climate Change. *Journal of International Development.* 13 (7). pp. 921–931.

Asian Development Bank (ADB). 2009. *The Economics of Climate Change in Southeast Asia: A Regional Review.* Manila: ADB.

———. 2010. *Building Climate Resilience in the Agriculture Sector in Asia and the Pacific.* Manila: ADB.

ADB Institute (ADBI). 2009. *Questionnaire on Climate Change Adaptation.* Survey conducted during the ADBI Workshop on Mainstreaming Climate Change Adaptation into Development Planning. April 14–17.

Anbumozhi, V. 2009. *Staged Approaches to Capacity Building in Support of Climate Change Adaptation, UN–IR3S Consultative Conference on Higher Education for Climate Change Adaptation.* United Nations University, Tokyo, Japan. June, 10–12.

Anbumozhi, V., E. Yamaji, J. Sato, and K. Ozawa. 2001. Interdisciplinary Research in Agricultural Engineering: An Alternative to Address Agro-environmental Issues. *Agricultural Engineering Journal.* 10 (1 and 2). pp. 91–103.

Anbumozhi, V., V. R. Reddy, L. Yao-chi, and E. Yamaji. 2003. The Role of Crop Simulation Models on Agricultural Research and Development: A Review. *Agricultural Engineering Journal* 12 (1 and 2). pp. 1–18.

Gagnon-Lebrun, F. and S. Agrawala. 2006. Progress on Adaptation to Climate Change in Developed Countries: An Analysis of Broad Trends. ENV/EPOC/GSP(2006)1/FINAL. Paris: OECD.

Luna, E. 2001. Disaster Mitigation and Preparedness: The Case of NGOs in the Philippines. *Disasters.* 25 (3). pp. 216–226.

Network of Asian River Basin Organizations (NARBO). http://www.narbo.jp/ (accessed October 2010).

Organization for Economic Co-operation and Development (OECD). 2006. Putting Climate Change Adaptation in the Development Mainstream. Policy Brief. http://www.oecd.org/dataoecd/57/55/36324726.pdf

PICCAP. 2005. *Climate Change: The Fiji Islands Response—Fiji's First National Communication under the Framework Convention on Climate Change.* Suva: Government of the Fiji Islands.

Reddy, V. R., V. Anbumozhi, and K. R. Reddy. 2005. *Achieving Food Security and Mitigating: Global Environmental Change—Is There a Role for Crop Models in Decision Making?* Paper Presented at International Agricultural Engineering Conference, Bangkok, Thailand. December, 6–9.

Ribot, J. C., A. Najam, and G. Watson. 1996. Climate Variation, Vulnerability and Sustainable Development in the Semi-arid Tropics. In J. C. Ribot, A. R. Magalhães, and S. S. Panagides, eds. *Climate Variability, Climate Change and Social Vulnerability in the Semi-Arid Tropics.* Cambridge, pp. 13–54. New York and Cambridge: Cambridge University Press.

United Nations Conference on Environment and Development (UNCED). 1992. *Agenda 21*–Chapter 37. http://habitat.igc.org/agenda21/

United Nations Framework Convention on Climate Change (UNFCCC). 2001. *The Marrakech Accords and the Marrakech Declaration Addendum. Part Two: Action taken by the Conference of the Parties. Volume I.* Decision 2 and 3/CP.7. http://unfccc.int/resource/docs/cop7/13a01.pdf

Chapter 24
Current Status of Adaptation Planning in the Region

Tomonori Sudo

24.1 Introduction

Climate change is closely related to human life and its development not only for the current generation but also for future generations. Addressing climate change is a high priority for all countries regardless of whether they are developed or developing. Having said that, developing countries, especially the least developed and small island countries, are not able to cope with current weather conditions. Under such circumstances, even if developing countries make a great effort to develop, climate change brings their effort to nothing. Therefore, developing countries need to address climate change adaptation along with development.

Multilateral and bilateral development agencies take climate change concerns into consideration in their development assistance policies. The Asian Development Bank (ADB) has conducted several works on climate change. ADB (2005) recognized the importance of taking "climate proofing" into consideration when they support infrastructure development. As a part of a series of studies, ADB (2009b) analyzed the economic impact of climate change in Southeast Asia and East Asia. The World Bank and United Nations Development Programme (UNDP) also focused on climate change in their flagship documents, the *World Development Report* (World Bank 2009) and *Human Development Report* (UNDP 2009).

The OECD Ministerial Meeting of Environment and Development held in 2006 decided to start work on developing guidance for integration of climate change adaptation into development cooperation. As a result of the work on climate change adaptation by the joint task force of the Development Assistance Committee and the Environment Policy Committee, the OECD policy guidance on integration of climate change adaptation into development cooperation was developed and endorsed at the High Level Meeting on Environment and Development in 2009 (OECD 2009a, 2009b).

In January 2008, the Government of Japan announced a new initiative on climate change named "Cool Earth Partnership" that assists mitigation and adaptation efforts of developing countries by providing financial assistance and technical cooperation. Even though the Japanese administration changed, the initiative has continued as the "Hatoyama Initiative" as an expanded scale of assistance. The Japan International Cooperation Agency (JICA), as Japan's official development assistance executing agency established its operations direction on climate change. JICA also focuses on climate change adaptation as well as mitigation through its cooperation projects (JICA 2007, 2008). According to Atteridge et al. (2009) among donor agencies, JICA, KfW, and Agence Française de Développement (AFD) are the largest contributors in climate change finance—they share almost three quarters of climate change finance by the public sector.

While climate change adaptation is focused on developing countries and development agencies, it is not necessarily recognized as an important development issue. This chapter focuses on the relationship between development and climate change adaptation and how climate change could be integrated into development policies and programs.

24.2 Climate Change and the Development Agenda

One of the main issues for developing countries, especially the least developing countries, is poverty alleviation. The main objective of the Millennium Development Goals (MDGs) is to reduce poverty to a significant level by 2015. To achieve the MDGs, developing countries need to address a variety of development issues for their sustainable development. These include stable and enough energy and food supplies, provision of basic human needs such as education and health, urban and rural development to improve quality of life and business circumstances, and natural resources management such as water and biological resources. Some countries may need to promote industrialization to increase employment. By addressing these development issues, developing countries may be able achieve economic development, hence alleviating poverty.

However, if developing countries continue to rely on fossil fuels for their energy, economic development will increase greenhouse gas (GHG) emissions that will lead to climate change. When climate change and/or climate related extreme weather events happen, they may impair accumulated developmental benefits that have so far been generated, and will affect the poorest people who are most vulnerable to climate change (Figure 24.1). In addition, even people who have moved out of poverty may lapse into poverty again due to climate change.

Climate change is deeply related to development issues. Developing countries, while continuing economic development to alleviate poverty, need to seek new development pathways toward a low carbon society by limiting GHG emissions and in climate resilient development way by securing their accumulated development benefit and an opportunity for the poor to move out of poverty.

Figure
24.1 Climate Change and Development

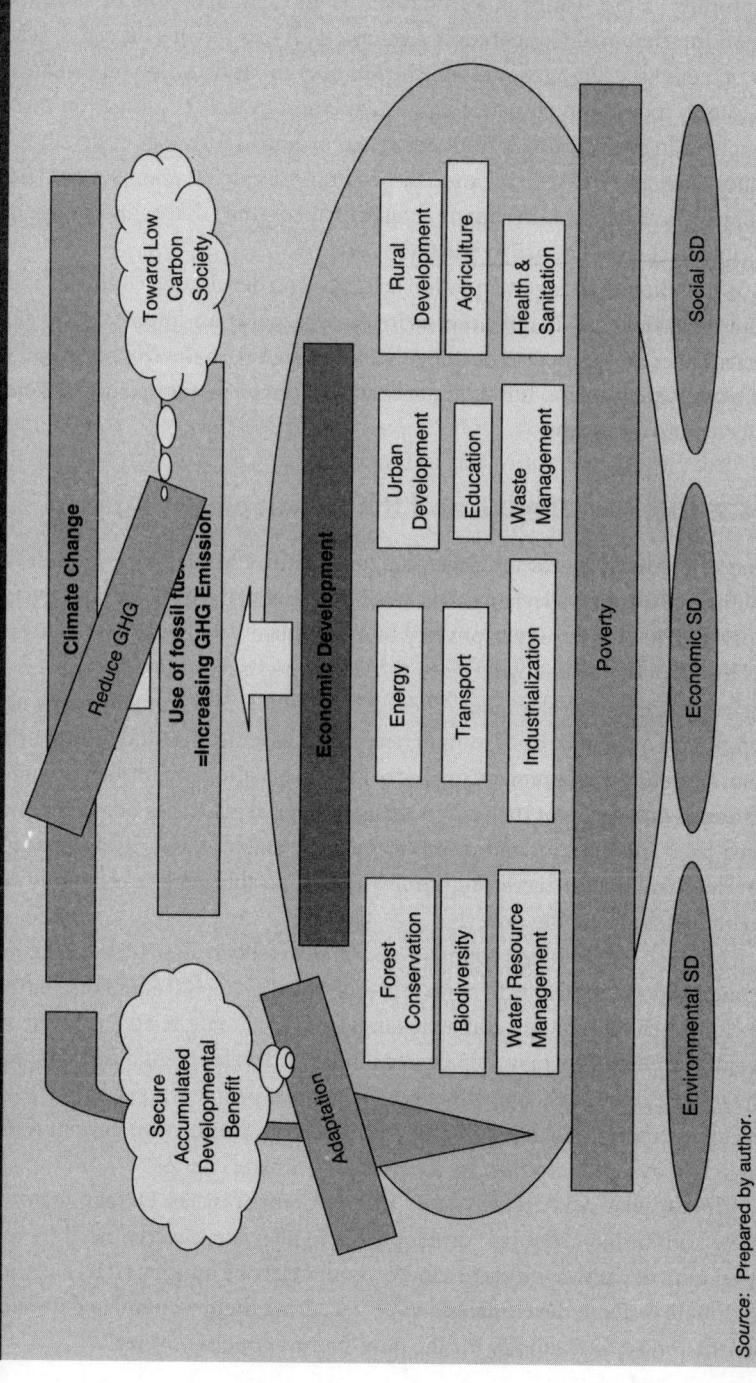

Source: Prepared by author.
Notes: GHG = greenhouse gases; SD = sustainable development.

24.3 Climate Change Adaptation as National Security

Developing countries face many risks to their sustainable development. The risks are not only at the national and local levels but also at the international level, including financial turmoil, and increases in fuel and food prices. Climate change is a serious threat to sustainable development, since, as discussed in Section 2, when climate change and/or climate change related extreme weather events happen, accumulated development benefits in the form of assets may be damaged and some people may lapse back into poverty. It means climate change will be a threat to human security. Thus, climate change should be dealt with as a national security issue, along with issues such as food security and energy security.

Climate change, as a national security issue, is complex due to uncertainties and external factors. Uncertainties of climate change vary since it is caused by complex climate systems on the earth and many factors related to climate change. Human activity is only a small factor. However, even if the human factor contributes only a small part, that factor relates to other factors that could further impact the climate system. In addition, impact of human activities on climate change depends on people's lifestyles. Unfortunately, nobody knows what sort of lifestyle future generations will choose. At least the current generation can make best efforts to mitigate human-induced climate change impact by shifting lifestyles to a more climate-friendly or low carbon environment, and to a readiness to manage climate change incidents. Readiness may be similar to national defense, that is, to protect lives and assets from unexpected incidents caused by climate change.

24.4 Integration of Climate Change Agenda into Development Actions

According to ADB's regional study on *Economics of Climate Change* (2009b), Asian developing countries are particularly vulnerable to the negative impacts of climate change. The study suggests it is necessary for them to develop and implement their climate change adaptation actions.

In general, developing countries do not necessarily have enough capacity to adapt to climate change, and needs of capacity development are emphasized. It may be true, but it does not mean that Asian developing countries have not addressed climate change at all so far.

The IGES (2006) pointed out the case of Indonesia,

> whether intended or not, Indonesia has taken adaptation measures in several sectors, including water resource management, agriculture, coastal defense, damage control for extreme weather events and health care. However, these measures and policies need to be reinforced further to cope with the future impacts of climate change.

Basher (2006) pointed out, "capacity needs for future weather risks are similar to those for today's risk."

Thus, development actions are not necessarily recognized as adaptation measures and, accordingly, governments have not prioritized appropriate climate change adaptation actions. But, at least, some developing countries in Asia have included adaptation coping actions within their development activities.

Figure 24.2 shows the framework of development policies and actions. Each developing country develops its own mid-term development policies and strategies based on their long-term vision, that is, the future aspects of the country. These policies and strategies consist of sector development plans reflecting characteristics of the sectors in each country. Based on the plans, each line ministry will formulate and implement development programs consisting of several projects and actions by themselves and/or with financial and technical assistance from donors. By implementing these programs and projects, developing countries can achieve development toward their vision.

Development planning ministries, ministries of finance, and line ministries may implement development programs and projects without climate change concerns and sometimes those actions can result in maladaptation (OECD 2009a). That is why climate change concerns need to be integrated in each level from development plan to development project.

Although the least developed countries in Asia have already developed their national adaptation programs of action (NAPAs), other countries have not prepared their own NAPAs. However, most countries have included some adaptation related development programs in their national development plans and/or strategies. Figure 24.3 shows some examples of risks due to climate change and adaptation measures.

Figure 24.2	**Framework of Development Policies and Actions**

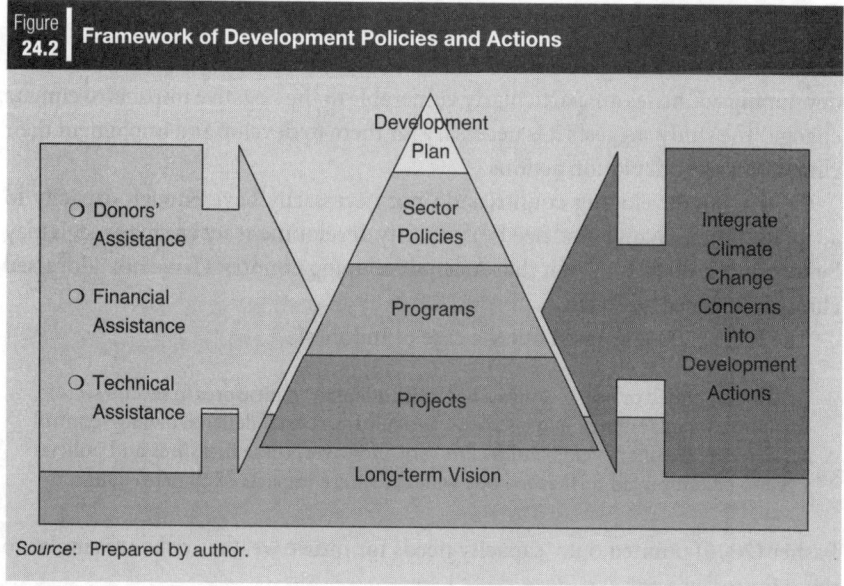

Source: Prepared by author.

Figure 24.3 National Development Plans of Asian Developing Countries

Sectors	Risks due to climate change	Possible Countermeasures
Water Resources	Decrease of water due to drought / Deterioration of water quality	Water supply projects, water conservation measures, waste water treatment
Disaster Prevention	Natural disasters such as flood damage and landslide disaster	Disaster prevention measures, coastal protection measures
Agriculture	Effects on agricultural products due to drought and other reasons	Enhancement of irrigation facility, promotion of water-saving agriculture, changing cropping pattern
Forests and Ecosystem	Decrease of forests due to drought and progressive desertification / Death of coral reefs due to rising sea water temperature	Prevention of desertification, afforestation, sustainable forest, and ecosystem management
Health and Sanitation	Expansion of distribution range of infectious disease transmitting species	Countermeasures against infectious diseases, control of epidemic diseases
Society and Economy	Loss of assets (including infrastructure) / Social instability	Improve social and economic resilience to extreme weather

Source: Prepared by author.

In general, there are two types of adaptation measures: "hard adaptation measures" and "soft adaptation measures." Hard adaptation measures mean construction of adaptive infrastructure such as irrigation systems, dykes, and disaster prevention infrastructure. Soft adaptation measures mean development of capacity through training, preparation of hazard maps, among others. Since Asia is an area with severe weather conditions such as typhoons, draughts, and floods, people are already prepared for extreme weather conditions to some extent.

In addition, when we look at the national development plans of Asian developing countries, we find many programs and projects that look like the measures shown in Figure 24.3. This shows that Asian developing countries have already implemented some adaptation measures.

24.5 Conclusion and the Way Forward

Many adaptation related policies, programs, and projects have already been included in the national development plans and strategies of Asian developing countries. These programs will assist countries to improve their resilience to climate change to some extent. However, measures may not be implemented due to reasons such as lack of budget, and low priority. This may make developing countries delay their readiness to climate change.

Policymakers should review their development plans and/or strategies through climate lens. When they do, policymakers can explicitly identify the role of these actions as climate change adaptation, and will give them an opportunity to reconsider the linkage between development and climate change.

References

Asian Development Bank (ADB). 2005. *Climate Proofing: A Risk-based Approach to Adaptation. Pacific Studies Series.* http://www.adb.org/Documents/Reports/Climate-Proofing/main-report.asp
———. 2009a. *Climate Change—ADB Programs: Strengthening Mitigation and Adaptation in Asia and the Pacific.* http://www.adb.org/Documents/Brochures/Climate-Change/default.asp
———. 2009b. *The Economics of Climate Change in Southeast Asia: A Regional Review.* http://www.adb.org/Documents/Books/Economics-Climate-Change-SEA/default.asp
ADB, Japan International Cooperation Agency, and World Bank. 2010. *Climate Risks and Adaptation in Asian Coastal Mega-Cities: A Synthesis Report.* http://www-wds.worldbank.org/external/default/WDSContentServer/WDSP/IB/2010/10/20/0003330 38_20101020234456/Rendered/PDF/571100WPOREPLA1egacities01019110web.pdf
Atteridge, A., C. K. Siebert, R., J. T. Klein, C. Butler, and P. Tella. 2009. Bilateral Finance Institutions and Climate Change: A Mapping of Climate Portfolios.*Stockholm Environment Institute Working Paper Series.* http://sei-international.org/publications?pid=1324

Basher, R. 2006. *Making Disaster Reduction an Adaptation Policy*. Presentation at Integrated Development and Climate Policies: How to Realize Benefits at National and International Levels, MNP/RIVM-IDDRI-IGES Workshop. Paris. September, 20–22.

Institute for Global Environmental Strategies (IGES). 2006. *Asian Perspectives on Climate Regime Beyond 20—Concerns, Interests and Priorities*. http://enviroscope.iges.or.jp/ modules/envirolib/view.php?docid=169

Japan International Cooperation Agency (JICA). 2007. *JICA's Assistance for Adaptation to Climate Change*. http://www.jica.go.jp/english/publications/reports/study/topi- cal/climate_2/

———. 2008. *Direction of JICA Operation Addressing Climate Change*. http://www.jica. go.jp/environment/pdf/info080501_en.pdf

Organisation of Economic Cooperation and Development (OECD). 2009a. *Integrating Cli- mate Change Adaptation into Development Co-operation: Policy Guidance*. http://www. oecd.org/env/cc/adaptation/guidance

———. 2009b. Declaration on Green Growth. C/MIN(2009)5/ADD1/FINAL. Adopted at the Council Meeting at Ministerial level, June 25, 2009. http://www.oecd.org/officialdocu- ments/displaydocumentpdf/?cote=C/MIN(2009)5/ADD1/FINAL&doclanguage=en

United Nations Development Programme (UNDP). 2009. *Human Development Report 2007/2008—Fighting Climate Change: Human Solidarity in a Divided World*. http://hdr. undp.org/en/reports/global/hdr2007-2008/

World Bank. 2009. *World Development Report 2010: Development and Climate Change*. http://econ.worldbank.org/WBSITE/EXTERNAL/EXTDEC/EXTRESEARCH/EXT- WDRS/EXTWDR2010/ 0,,contentMDK:21969137~menuPK:5287816~pagePK:64167 689~piPK:64167673~theSitePK:5287741,00.html

Chapter 25

Mainstreaming Climate Change Adaptation into Development Planning

Youssef Nassef

25.1 Introduction

According to the Intergovernmental Panel on Climate Change (IPCC 2007), adaptation is the "[a]djustment in natural or human systems in response to actual or expected climatic stimuli or their effects, which moderates harm or exploits beneficial opportunities."

This definition alludes to the fact that adaptation is contextual. It is determined by the particular "natural or human systems" within which it will take place, as well as by the "actual or expected climate stimuli or their effects" which it aims to mitigate. Not only is adaptation action planned and implemented according to its context, but action is also determined to be termed "adaptation" based on context. Action that constitutes adaptation in one setting, may not necessarily serve as such in another, and could even constitute maladaptation in yet another setting. One should label action as "adaptive" only after ascertaining that the context denotes it as such.

To elaborate further, adaptation is usually considered within different contexts and through different lenses. It differs according to sector, level, type of ecosystem and so forth, as elaborated below:

- Level: community, subnational and/or provincial, national, regional, international
- Sector: agriculture, water, health, tourism, labor
- Ecosystem: mountainous, coastal, marine, arid, mangrove, delta
- Region: Africa, Latin America, Europe
- Grouping: Annex I;[1] Non-annex I;[2] least developed countries (LDCs); small island developing states (SIDS); particularly vulnerable countries; countries

---

[1] Parties included in Annex I to the UNFCCC (includes industrialized countries that were members of the OECD in 1992 and countries with economies in transition).

[2] Parties not included in Annex I to the UNFCCC (mostly developing countries).

prone to natural disasters; countries in Africa affected by drought, desertification and floods

- Delivery mechanism: finance, insurance, technology, capacity building, awareness raising, education
- Discourse/entry point: development, disaster risk reduction, food security, ecosystem conservation, integrated water management, poverty reduction, integrated coastal zone management, humanitarian relief

25.2 Adaptation under the UNFCCC

The United Nations Framework Convention on Climate Change (UNFCCC) incorporates articles that deal with adaptation at the international level. Article 4.1 (b) is a mandatory commitment to plan and implement adaptation: "All Parties shall: ... formulate, implement, publish and regularly update ... measures to facilitate adequate adaptation to climate change". A commitment for cross-country cooperation is found in Article 4.1 (e): "All Parties shall: ...Cooperate in preparing for adaptation to the impacts of climate change ..." (UNFCCC 1992).

The Convention also states that the "developed country Parties ... shall also assist the developing country Parties ... in meeting costs of adaptation to those adverse effects" in Article 4.4, committing developed countries to provide resources for adaptation. Other commitments are mandated in Article 4.8, which states that the "Parties shall give full consideration to actions necessary to meet the specific needs and concerns of developing countries arising from the adverse effects of climate change..." committing Parties to support vulnerable countries through funding, technology transfer and insurance. It then delineates particular groups of countries, including:

- small island countries;
- countries with low-lying coastal areas;
- countries with arid and semi-arid areas, forested areas and areas liable to forest decay;
- countries with areas prone to natural disasters;
- countries with areas liable to drought and desertification;
- countries with areas of high urban atmospheric pollution;
- countries with areas with fragile ecosystems, including mountainous ecosystems;
- countries whose economies are highly dependent on income generated from the production, processing and export, and/or on consumption of fossil fuels and associated energy-intensive products; and
- landlocked and transit countries.

It also places special attention on the LDCs in Article 4.9 by mandating that the "Parties shall take full account of the specific needs and special situations of the least developed countries in their actions with regard to funding and transfer of technology."

25.3 Climate Risks in Land and Water Resources

Non-Annex I Parties have, in their National Communications, prioritized agriculture and water resources as key vulnerable sectors (Figure 25.1 and Table 25.1), and these are also the main sectors for which projects have been identified by LDCs in their national adaptation programs of actions (NAPAs) to address urgent and immediate adaptation needs (Figure 25.2).

25.4 Ongoing Adaptation Work under the Convention

There are diverse channels of work and support for adaptation under the UNFCCC process, established mainly in response to the needs of the developing country Parties. This is a continuous, learning-by-doing effort, which includes the following programs:

- The Buenos Aires Programme of Work (BAPW) on Adaptation and Response Measures (Decision 1/CP.10 of the UNFCCC Conference of the Parties):

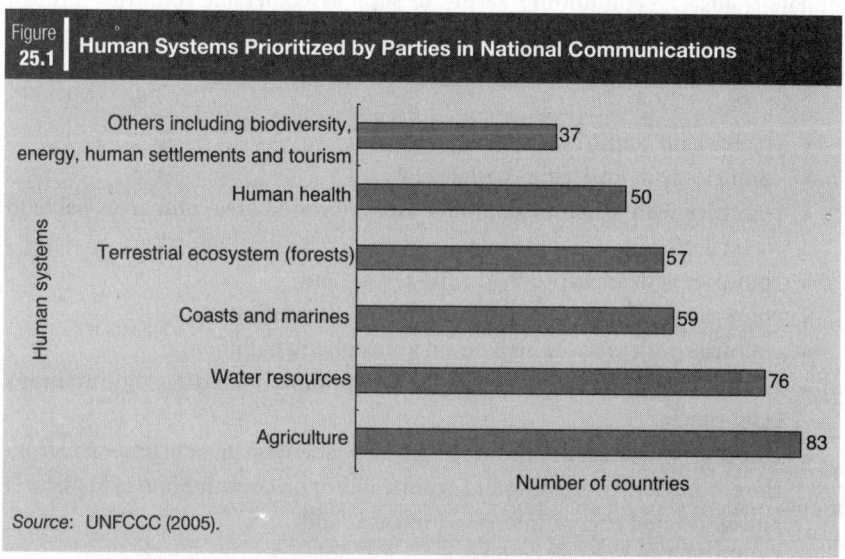

Figure 25.1 | Human Systems Prioritized by Parties in National Communications

Source: UNFCCC (2005).

Table 25.1	Regional Prioritization of Key Vulnerable Sectors	
Region	**Need to adapt**	**Key vulnerable sectors**
Africa	Very high	• Agriculture • Water resources
Asia	High	• Agriculture • Terrestrial ecosystems
Latin America and the Caribbean	High	• Agriculture • Water resources
Small island developing states	Very high	• Water resources • Coastal zone (sea-level rise)

Source: UNFCCC (2005).

The BAPW responds to the adverse effects of climate change in line with Article 4 of the Convention. It focuses on information exchange and integrated assessments, in order to assist Parties in identifying region-specific gaps, needs, and concerns (various workshops and synthesis reports are available on the UNFCCC website).

- National adaptation programs of actions (NAPAs): NAPAs are a process for the assessment and communication of urgent and immediate adaptation needs in LDCs, drawing on existing information, a bottom up methodology and community-level input. As of January 2010, the UNFCCC secretariat had received 43 NAPAs.

- Nairobi Work Programme on Impacts, Vulnerability and Adaptation to Climate Change (NWP): The Nairobi Work Programme is implemented by Parties, intergovernmental and nongovernmental organizations, the private sector, communities, and other stakeholders, to assist all Parties, but particularly the most vulnerable, to improve their understanding and assessment of impacts, vulnerability, and adaptation to climate change; and to make informed decisions on practical adaptation actions and measures to respond to climate change on a sound scientific, technical and socioeconomic basis, taking into account current and future climate change and variability. Figure 25.3 graphically depicts how the NWP functions.

At the 2010 Climate Change Conference in Cancun, Parties adopted the Cancun Adaptation Framework as part of the broader Cancun Agreements with the objective of enhancing action on adaptation, including through international cooperation and coherent consideration of matters relating to adaptation under the Convention.

Figure
25.2 | Key Sectors Identified in the NAPA Projects as of July 2011

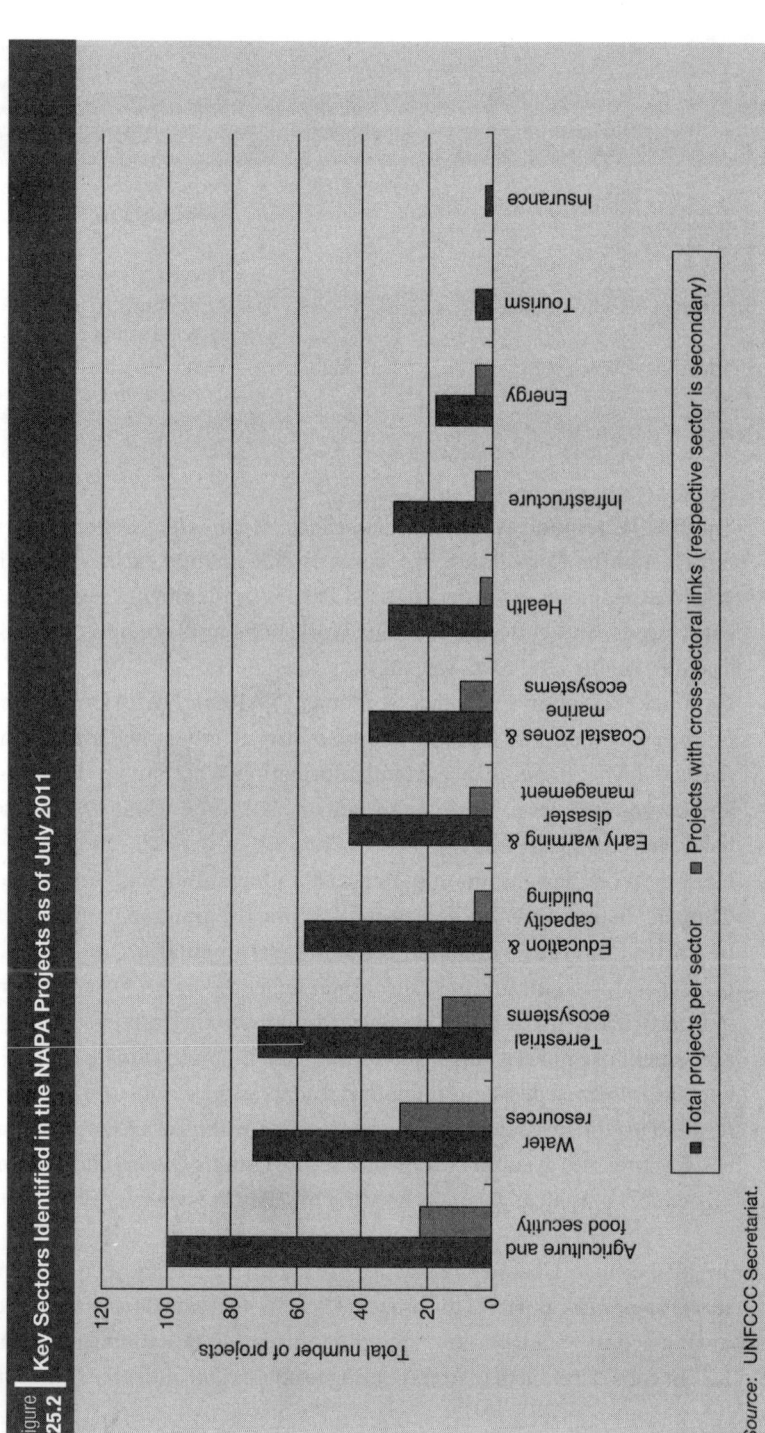

■ Total projects per sector   ■ Projects with cross-sectoral links (respective sector is secondary)

Source: UNFCCC Secretariat.

Figure
25.3 **Nairobi Work Program on Impacts, Vulnerability, and Adaptation to Climate Change**

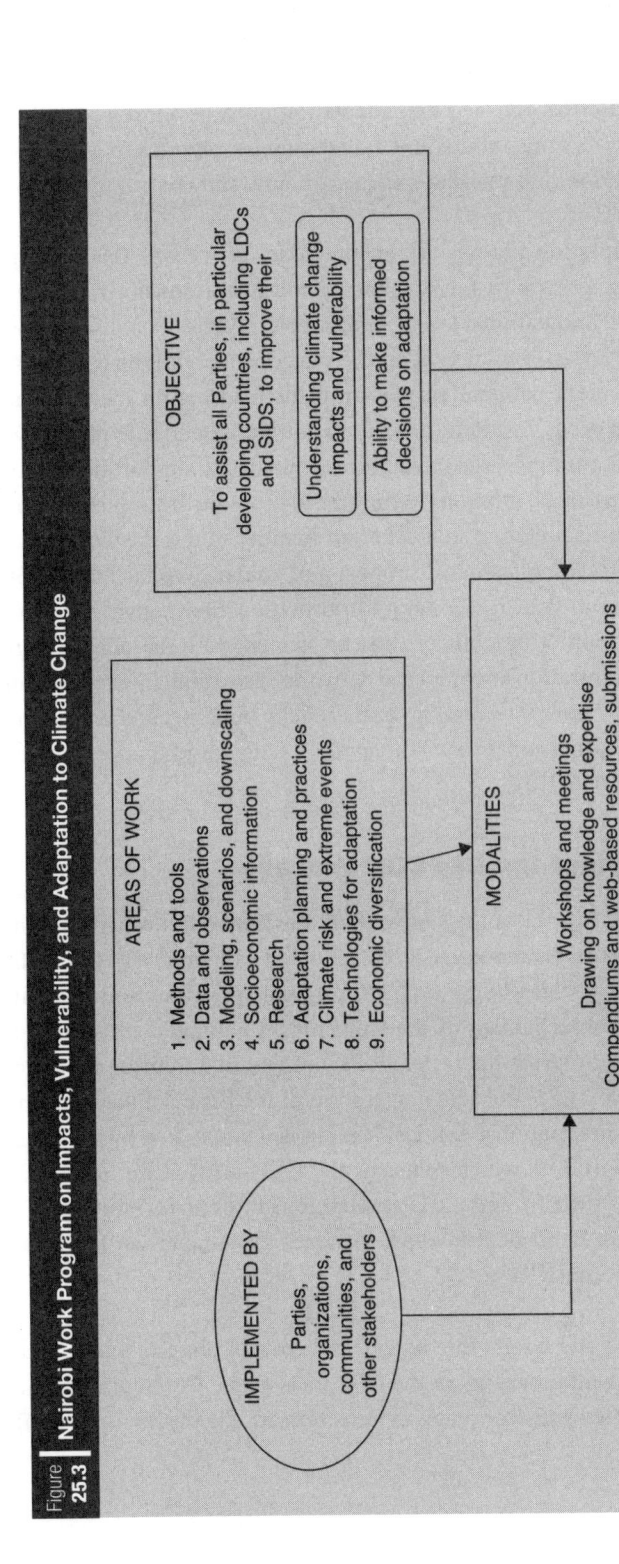

Source: Acosta (2009).

In the Agreements, Parties affirmed that adaptation must be addressed with the same level of priority as mitigation and that it should be undertaken with a view to integrating adaptation into relevant social, economic and environmental policies and actions.

The Cancun Adaptation Framework encompasses provisions related to the implementation of, and support for, adaptation and institutional arrangements at the global, regional and national levels. In particular, Parties established a process to enable LDC Parties—building upon their experience with the NAPAs—to formulate and implement national adaptation plans focusing on medium- and long-term adaptation needs. In addition, an invitation was extended to other developing country. Parties to employ the modalities formulated to support those plans. Parties also established an Adaptation Committee to promote the implementation of enhanced action on adaptation in a coherent manner under the Convention, including through providing technical support and guidance to facilitate implementation of adaptation; sharing of relevant information, knowledge, experience and good practices; promoting synergy and engagement with national, regional and international organizations, centers and networks; providing information and recommendations on finance, technology and capacity building; and considering information on monitoring and review of adaptation actions and support provided and received (UNFCCC 2010).

25.5 Funding under the UNFCCC Process

In accordance with Article 11 of the Convention, the financial mechanism functions under the guidance of, and is accountable to, the Conference of the Parties (COP), and the COP decides on its policies, program priorities and eligibility criteria. The financial mechanism of the Convention is mandated to have an equitable and balanced representation of all Parties within a transparent system of governance (Article 11.2). The actual operation of the financial mechanism is entrusted to existing international entities. It is currently operated by the Global Environment Facility (GEF), which manages the GEF Trust Fund, the Special Climate Change Fund (SCCF), and the Least Developed Countries Fund (LDCF); and by the Adaptation Fund Board, which manages the Adaptation Fund (AF) under the Kyoto Protocol (KP).

- The GEF Trust Fund covers the incremental costs of projects that generate global environmental benefits in the GEF focal areas. Developing country Parties and Parties with economies in transition are eligible for funding. The

Strategic Priority for Adaptation was established as part of the GEF Trust Fund and has supported demonstration and pilot projects that address adaptation needs. The funds for the incremental costs of the projects are derived from programs under the GEF climate change focal area, as well as other areas, while the other project costs are funded through co-financing by the host country or other bilateral and multilateral sources.

- The SCCF supports four work areas, with adaptation being a top priority of the fund. The fund supports developing countries that are Party to the UNFCCC by funding the project component that constitutes the additional cost of adaptation. Its modalities include a sliding scale developed as a proxy for additionality, to be used optionally. Smaller sized projects receive a higher percentage of GEF support. The SCCF adaptation program includes finance for adaptation measures in sectors including agriculture and food security, water resources, and infrastructure; coastal zone management; and disaster preparedness.
- The LDCF supports projects identified in the NAPAs of LDCs, with only LDCs eligible to apply. It funds the project component that constitutes the additional cost of adaptation, and its modalities also include an optional sliding scale developed as a proxy for additionality.
- The AF is operated by the Adaptation Fund Board, and developing country Parties to the Kyoto Protocol that are particularly vulnerable to the adverse effects of climate change are eligible to apply for funding. The funding is generated from a 2% levy on CDM projects, as well as from other sources. Support is channeled through programs and projects; with applicants to the AF choosing to develop and implement activities either through direct access with an in-country executing entity (that meets agreed due diligence standards for financial management), or to go through an implementing entity (recognized by the Adaptation Fund Board).

Institutional arrangements under the Convention, including the arrangements for funding, are complex. In order to illustrate how they relate to each other, they are depicted in Figure 25.4.

In Cancun, developed country parties also committed to a goal of mobilizing jointly US$100 billion per year by 2020 to address the needs of developing countries. Parties also decided that a significant share of new multilateral funding for adaptation should flow through the newly established Green Climate Fund, which would be an additional operating entity of the financial mechanism of the Convention under Article 11 (UNFCCC 2010).

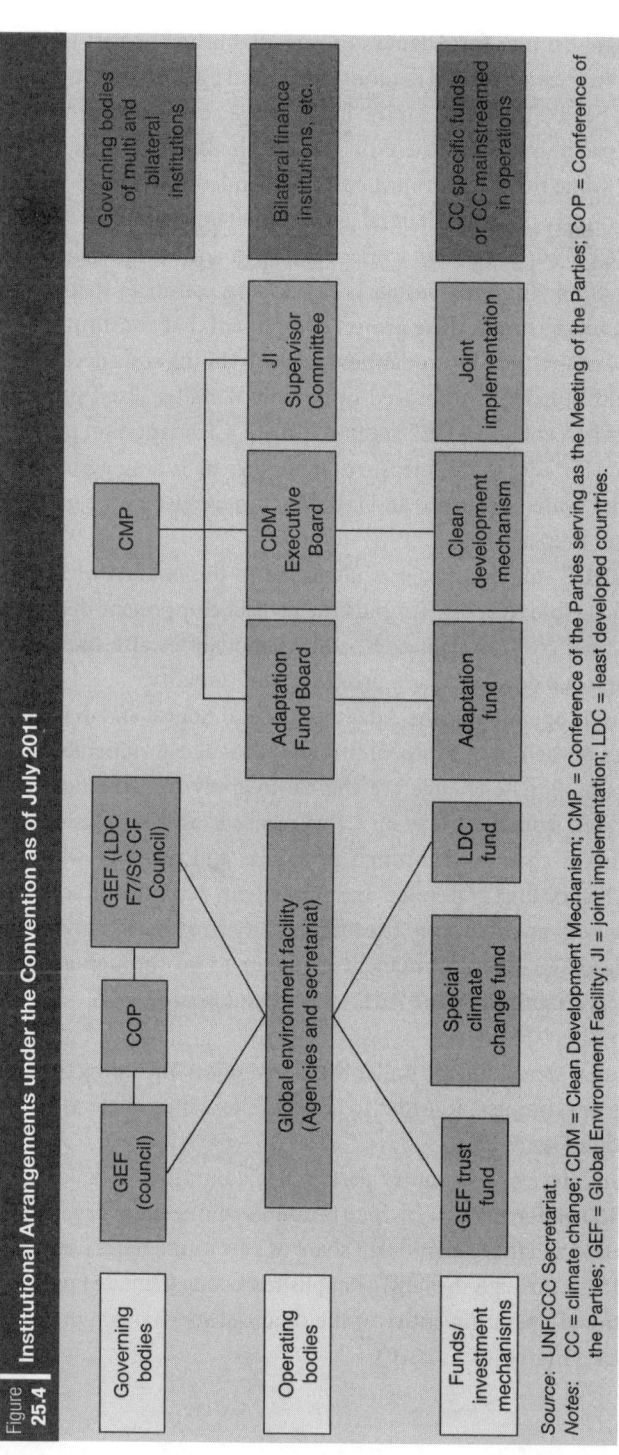

Figure 25.4 Institutional Arrangements under the Convention as of July 2011

Source: UNFCCC Secretariat.

Notes: CC = climate change; CDM = Clean Development Mechanism; CMP = Conference of the Parties serving as the Meeting of the Parties; COP = Conference of the Parties; GEF = Global Environment Facility; JI = joint implementation; LDC = least developed countries.

References

Acosta, R. 2009. *Adaptation Needs for Climate Information and Services in the Context of the UNFCCC World Climate Conference Three (WCC-3): A Pathway for Improved Climate Services for Adaptation.* Bonn, Germany, March 31.

Intergovernmental Panel on Climate Change (IPCC). 2007. Climate Change 2007: Impacts, Adaptation and Vulnerability. Contribution of Working Group II to the Fourth Assessment. In M. L. Parry, O. F. Canziani, J. P. Palutikof, P. J. van der Linden, and C. E. Hanson, eds. *Report of the Intergovernmental Panel on Climate Change.* Cambridge, UK: Cambridge University Press.

United Nations Framework Convention on Climate Change (UNFCCC). 1992. *The United Nations Framework Convention on Climate Change.* Geneva: United Nations.

———. 2010. *Report of the Conference of the Parties on its Sixteenth Session. Cancun 29 November to 10 December 2010. Part Two: Action Taken by the Conference of the Parties at its Sixteenth Session.* Geneva: United Nations.

Conclusions, Policy Implications, and the Way Forward

Robert Dobias, Venkatachalam Anbumozhi, Vangimalla R. Reddy, and Meinhard Breiling

With the focus of mainstreaming climate change adaptation into development planning, the workshops covered sectors that are the most vulnerable to climate change. The implications allow drawing of policy recommendations and conclusions of broad relevance. The analytical framework used in this book is a combination of sector-wide reviews (using secondary sources on issues, policies, and institutions) and case studies with implementation experiences. Ranging across several countries, the case studies helped to gain a deeper understanding of the barriers, as well as the contributors, to better adaptive capacity. The general conclusions are discussed below.

1 Impacts

Climate change is a reality. The Asia and Pacific region is projected to be the most affected by the impacts of climate change. This region is home to 903 million people living below US$1.25 a day, implying that there are vast needs that have to be met (ADB 2010). The poor are at a higher risk to future climate change, given their heavy dependence on agriculture, strong reliance on ecosystem services, high concentration of population and economic activity in coastal areas, and poor health services.

Recent Asian Development Bank (ADB) studies (ADB 2007, 2009a, 2009b, 2009c, 2009d) on climate change, particularly *Building Climate Resilience in the Agriculture Sector* and *Climate Change and Migration in Asia and the Pacific*, found that more people than originally thought in the Asia and Pacific region are vulnerable to climate change. The affected population in the agriculture sector accounts for more than 60% of the economically active population and their dependents, amounting to 2.2 billion affected people. The studies indicated that agriculture is the most vulnerable sector; for instance, in the Greater Mekong region alone, climate change is projected to increase the price of rice by up to 37%. Water scarcity in the Asia and Pacific region is expected to affect a decline in rice yield in the range of 14%–20% over the next 40 years. The number of malnourished children in Southeast Asia is projected to increase by 16%, to 11 million. A warmer and drier climate, and more frequent and intense extreme weather events, is projected to reduce the gross domestic product (GDP) of all countries in Asia. This is particularly the case in South and Southeast Asia. Adding to this scenario, an insufficient capacity to adapt to climate change impacts, inadequate infrastructure, meager household income and savings, and limited support from public services, could lead to disastrous consequences.

2 Adaptation

Adapting to climate change is a serious development problem. Given that efforts to adapt to the changing climate are connected to the many aspects of development, implementation of adaptation activities is connected to a wide range of other activities, including natural resources management, agriculture technology, disaster preparedness, infrastructure improvement, health systems, and poverty alleviation. Furthermore, effects of climate change vary over time and place, creating unique, dynamic adaptation needs in each country. Each country's unique institutional and socioeconomic circumstances affect its adaptation ability, at least when dealing with the physical impacts. In this context, how to design adaptation measures, who does it, and where to prioritize investments, all become challenging questions.

Fortunately, many of the policies that are good for economic development in general, also offer effective strategies for adapting to climate change. Such no-regret strategies include the following:

(i)   investment in adaptive agricultural research,
(ii) improved water management to deal with extreme rainfall events and glacial melt,
(iii) governance of common natural resources,
(iv) transportation and communication infrastructure as well as regional and international trade facilitation,
(v)   private-sector participation in insurance and credit markets, and
(vi) facilitation of migration to allow the poor to take full advantage of changes in climate and economic landscapes.

There is an urgent need for these climate change adaptation strategies to be integrated into sectoral and regional development programs. Policies on adaptation that include long-term weather forecasting, dissemination of technology, and creating drought- and flood-resistant crop varieties will require national and international planning and investment. Almost 70% of the water in the Asia and Pacific region is used for irrigation. Therefore, improving water management by understanding water flow and water quality, improving rainwater harvesting, water storage, and the diversification of irrigation techniques is critical. Greener practices, better erosion control and soil conservation measures, agroforestry and forestry techniques, as well as better town planning are some other steps that can be initiated to blunt the impacts of climate change. Since the affected communities are often constrained by access to credit, facilitating better access to credit is a related area that needs attention. Catastrophic or weather-risk insurance and index insurance, insurance linked to a particular index, such as rainfall, humidity, or crop yields, rather than actual loss can be used as new climate risk management tools. These improvements and building climate-resilient rural roads in the region will cost about US$3.0 billion to US$3.8 billion annually from 2010 to 2050, as predicted by ADB studies.

To date, a number of countries, including Cambodia, Bangladesh, Lao People's Democratic Republic, and Nepal, among others, have made pilot initiatives to promote mainstreaming, and efforts are underway to build national action programs for adaptations (NAPAs) through more comprehensive planning documents. There has also been speculation as to whether environmental impact assessments and poverty reduction strategy papers could provide effective vehicles for mainstreaming adaptation into sectoral planning. However, most of these efforts remain in the early stages. Likewise, the national climate change plans recently released have not yet been operationalized, and it is not clear how they will interact with other national planning efforts. The multilateral funds that have been pledged for climate change adaptation in developing countries currently amount to about US$400 million (UN 2010)—a sharp contrast to the US$4 billion to US$109 billion needed annually, as estimated by experts and aid agencies (Smith et al. 2011).

3 Adaptive Capacity

Policymakers concerned about the impact of climate change cannot wait for the academic and international communities to resolve the uncertainties that currently exist. There is little to be gained by waiting before taking concrete steps to deal with these issues—particularly in countries where extreme events are already imposing severe burdens on poor communities and economic growth. The crucial role of climate information in developing adaptation solutions must be emphasized. Available climate information must be considered to ascertain where systematic observation needs are most pressing.

Collaboration between national and international providers of climate information, the research community, users in all sectors, the decision makers, and generating awareness among different user communities of the usefulness of such information, is crucial. Climate change assessment tools that are more geographically precise as well as more useful for sectoral policy making and reviewing program and scenario assessment are urgently needed. Economic diversification to reduce dependence on climate-sensitive resources could also be an important adaptation strategy. Improved food security through crop diversification, developing local food banks for people and livestock, improving local disaster preparedness, and food preservation capabilities needs to be encouraged. One area that has been neglected is gender diversification, which needs to be tackled to bring wider perspectives into decision making, since climate change and natural disasters have gender-differentiated impacts. Women can contribute significantly to this process, particularly in domestic (home) disaster prevention activities of everyday life, and in recovery activities.

This book discusses methods for measuring climate impacts, concepts of adaptation, and cases of mainstreaming. As discussed in Parts II, III, and IV, these measures have been applied in many countries over several years. One lesson for policymakers is that adaptation policies must be tailored to local conditions, because the local impacts of climate change vary a great deal across space. Policymakers must be careful when transferring interventions from one country to another to make sure they are appropriate in each place. Technologies, management practices, and crop varieties that are proven successful in one country need to be carefully evaluated before being introduced in another country. The level of infrastructure development and human capacity are factors that may support or prevent good practices from being extrapolated from one country to another.

In some circumstances, developing countries of the Asia and Pacific region may benefit from policies that have been designed and implemented in developed countries. Best management practices and new technologies may be transferrable from developed to developing countries. Water saving technologies, stress-resistant crop varieties, early warning systems, and innovative financing systems created in developed countries could be modified for the conditions existing in developing countries. They then could be introduced by having full support systems (such as institutions and finance) in place.

In large countries, such as the People's Republic of China (PRC), India, and Mongolia, the climate, landscape, and institutional and local capacity vary a great deal across the country. Such countries need to take care when designing different policies for different regions within the country. Even small countries must be careful not to adopt uniform policies in different regions within the country. One way to address physical and structural differences in a country is to develop different policies depending on a set of climatic conditions and existing infrastructure and institutions in the various regions. Further, the economic and institutional ability to implement adaptation measures may also vary. It is possible that communities facing similar climate situations may be affected differently, depending on other physical and economic or institutional conditions they face. Both physical and economic conditions can affect the type of adaption relevant for each location and the ability of the community residing in each location to adapt. Therefore, policymakers should consider climate information and planning tools that are tailored to the assistance they receive.

4 The Way Forward

While the impetus for reacting to climate change and more effective adaptive action is building and becoming recognized, albeit to varying degrees at all levels and by all players, there is a serious gap in public awareness for constructive actions with respect to addressing very complex climate change adaptation issues. Increasing speculation makes the much-needed mainstreaming difficult to agree upon and implement, further exacerbating climate risks. There is an urgent need to start working toward developing a commonly shared vision on the way forward, involving all principal stakeholders and reconciling diverse perspectives.

Managing expectations from the public and decision makers regarding the adaptation process is important. Successful adaptation does not just happen. One key recommendation is to carefully plan and execute long-term national programs for supporting public participation in climate change adaptation aimed at educating and building capacity of all stakeholders. The first step could be to develop detailed guidelines as well as provide training on public participation for both environmental authorities and sectoral agencies, adjusted to specific needs of the sector. Attention should be given to building capacity at local level to help communities understand the climate risks and links to sector activities, and thus effectively gain participation in public forums. Overall, the programs should be designed and targeted according to the diversity of stakeholders.

Effective mainstreaming requires informed consensus on climate change risks, objectives, and policies that are based on a good understanding of the shared roles and responsibilities of all players, including sectoral agencies, ministries of environment, ministries of planning, and the affected community. This fundamental notion of shared responsibility is currently challenged in the Asia and Pacific region, by the general perception among the public, project proponents, and development authorities alike, that climate change is the sole responsibility of environmental agencies to effectively implement necessary adaptation measures. As economies in Asia and the Pacific continue to accelerate growth rates, the responses to climate change will come under increased scrutiny and pressure. The cases discussed in this book, however, observed that unless an increasing demand for mainstreaming is matched by adequate capacity building, it would be naive to expect substantial progress and unfair to solely blame the sectoral agencies.

There are significant capacity constraints of sectoral agencies to meet their existing mandates, as well as the need to introduce new adaption programs and tools and improve the effectiveness of existing ones. We recommend that ministries of environment, education, and sectoral agencies, using recent examples of good practices, develop medium-term capacity strengthening action plans to meet the current and projected needs, including financing requirements. These plans should first explore the possible capacity gains through (*i*) rationalizing decision-making processes; *(ii)* upgrading climate information; *(iii)* decentralizing responsibilities to regional offices, along with staff, resources, and equipment, and outsourcing certain noncore functions; and *(iv)* training to upgrade skills. It would conclude with a staffing plan including the need for additional positions to meet the core requirements, upon exhausting all options for improvements in processes and efficiency. The plans could be used for negotiations with planning and financing agencies, subject to making a strong and verifiable case.

There is also a fundamental need for sectoral agencies to facilitate better climate proofing of individual projects, more sustainable development of sectors as a whole, and greater cross-sectoral coordination, particularly at the planning stage. Case studies and sector reviews show that environmental monitoring and enforcement of specific sources of risks after implementation can do very little to improve the situation on the ground. Environmental factors are not considered at the time of location decisions, spatial planning, project design, and in technology and infrastructure selection.

There is a general consensus that all institutions can play a key role in strengthening the knowledge base and technical capacity that are important in adapting to climate change. Unless steps are taken to initiate and strengthen cooperation among academic and research institutions, regional and international organizations, and nongovernmental organizations to provide opportunities for strengthening the knowledge base, dealing with climate change impacts may be unmanageable. Some relevant information is already available to various institutions, and it is important to focus future efforts on

(i)   disseminating information more evenly across the country;
(ii) providing high and comparable quality sector-specific training across states and organizations; and
(iii) developing targeted, well-designed, and well-delivered programs for community learning.

The lack of effective mechanisms for interagency coordination is often a barrier for mainstreaming. It is, thus, critical for sectoral, environmental, and financing authorities to evaluate, share, and promote national best-practice examples of policies and institutional mechanisms, as well as relevant international experiences that enable meaningful participation of environmental agencies in the planning and design of infrastructure development projects. New priorities and programs will require even greater cross-sectoral cooperation and integration within particular spatial zones. Appropriate local governments are the best positioned to have the right incentives to ensure the coordination needed. Therefore, it would be important to provide them with sufficient authority and capacity to forge such coordination. Devolving more powers to and building capacity of local governments is necessary for developing and implementing climate change adaptation programs aimed at measurable improvements of climate risk reduction in the areas of their jurisdiction, with the participation of all concerned sectors, as well as affected communities.

The emerging adaptation agenda is large. The necessary institutional changes and large-scale improvements on the ground will require national commitment and consensus on a program of actions spanning over the short to long term. Many measures would involve further examination, design, as well as consultations with the public, other government agencies, and the affected community.

In the short to medium term, irrespective of whether a project or program approach is pursued, a set of key policy choices that can build the adaptive capacity of countries shall be proposed. The core of such practices should be strategized as:

- Moving up in the policy ladder, integrating sectoral policies, and linking them with economic policies and national planning.
- Moving down into specific investment plans now, to avoid economic and environmental costs later. Countries also need to move forward to improve energy and agricultural efficiency, and lift productivity. To stand still will mean to fall behind as others move on to a low carbon, lower cost, more competitive, and secure sources of energy and food.
- Moving together. Global problems start at home; therefore, global solutions require local action. Domestic leadership now will shape regional and international thinking.

It would be important to move quickly toward reaching broad agreement with all major stakeholders on the priority actions, starting with the identified list (Table 1), and develop a medium- to long-term program of implementing the agreed actions, supported by necessary resources, monitorable targets, and clear accountability mechanisms.

Those policy options by no means represent the full spectrum of criteria that could be considered in shaping the mainstreaming. Most policymakers will also learn to develop a range of other specific criteria relevant to their country's particular climate impacts and development concerns. However, these policy choices can support the development of an enabling institutional and policy environment that builds capacity over time and fosters mainstream adaptive actions by a range of stakeholders—academia, private sector, local governments, civil society—and helps successful imitative replication.

An enormous agenda is not new for Asia and the Pacific, which has on numerous occasions risen to meet such challenges. Encouragingly, many steps and initiatives in setting the right direction have been taken recently, by various players.

With its focus on adaptation concepts and cases, this book lays the ground for more structured and systematic analysis of how the Asia and Pacific region can

Table 1 | **Key Proposed Actions for Improving the Adaptive Capacity and the Role of Different Stakeholders**

Key issue	Strategic policy choices and actions	Responsible institution	Timeline
Promote Public Awareness	Develop national and subregional programs on climate change impacts, their causes, and best adaptation practices.	Sectoral agencies	Short to medium term (1–3 years)
	Develop sectoral guidelines and training on public participation in adaption programs.	Ministries of environment	Short term (1 year)
	Devise gender-specific strategies to deliver climate risk information.	Sectoral agencies	Short term (1–3 years)
	Share local knowledge with environmental and sectoral agencies to disseminate examples of when public participation improves adaptation responses.	Media, civil society	Short term (1–2 years)
Improve Scientific Capacity	Develop and update public online databases on climate risk indicators.	Ministries of environment	Short to medium term (1–5 years)
	Upgrade and expand targeted research and educational programs and/or sectoral research and training scientists, institutions, etc.	Sectoral ministries	Short term, then continues
	Publicize the regional knowledge centers and create satellite offices to disseminate relevant information to affected communities.	Academia, civil society	Short term (1–3 years)
	Maximize the effectiveness of current acts and programs by developing clear procedural guidelines regarding climate change adaptation add-ons.	Sectoral ministries, local governments	Continuous
Set Feasible Standards and Benchmarks for Structural Measures	Review best international practice procedures for infrastructure standards setting and develop national guidelines, strengthen and/or expand the application of zoning concepts in setting national standards.	Ministries of environment, academia	Medium term (1–5 years)
	Strengthen the instruments of social and economic impact assessment of new infrastructures by developing a clear methodology drawing on best international practices and adjusted to national and local contexts.	Sectoral ministries	Short to medium term (1–7 years)
	Provide necessary climate and economic information, collaborate on the analysis, and facilitate consultation with industry.	Planning commissions, sectoral agencies	Short term (1–3 years)
	Provide information on social and community impacts of the proposed standards.	Civil society, academia	Short term (1–3 years)

(Continued)

(Continued)

Key issue	Strategic policy choices and actions	Responsible institution	Timeline
	Develop a focused and well-packaged program for most vulnerable locations that integrate targeted structural measures with nonstructural measures, including a funding mechanism for scaling up.	Planning commissions, sectoral agencies	Medium term (1–7 years)
Develop New Programs to Strengthen Nonstructural Measures	Develop a set of regulatory incentives to support voluntary initiatives, using existing good practices.	Ministries of environment	Continuous
	Provide training and capacity building to policymakers and private-sector operators for better no-regret adaptation management focusing on international best practices that are locally appropriate.	Sectoral ministries, academia	Medium term (1–7 years)
	Periodically update sectoral guidelines for monitoring and adding new sectors of growing impact.	Local governments, NGOs	Continuous
	Strengthen existing formal mechanisms, such as environmental impact assessment (EIA) statements and poverty reduction strategy papers (PRSP), and involve environmental authorities in designing structural and nonstructural measures.	Ministries of environment	Short term (1–3 years)
Improve Cross-sectoral Coordination	Coordinate the development of strategic adaptation framework for using global environmental financing instruments.	Ministries of environment	Medium term (2–4 years)
	Empower local governments to oversee regional climate change adaptation programs and foster cross-sectoral coordination.	Sectoral agencies, civil society	Short term (1–3 years)
	Develop sectoral guidelines to overcome specific identified gaps and facilitate uptake of best practices.	Sectoral agencies	Short term (1–3 years)

Category	Action	Responsible party	Timeframe
Augment Financial Resources	Explore innovative financing instruments including insurance programs, catastrophe bonds, and other risk transfer products to support future developments via a global climate change agenda.	International donors	Short term (1–3 years)
	Develop a consistent budgetary framework for integrating climate risks and set it as input into a consistent and realistic delivery mechanisms related to most vulnerable sectors, communities, or households in a transparent way.	Ministries of finance	Short term (1–3 years)
	Link trade and business promotion incentives to adaptation financing, make heavy representation within regional and/or international adaptation-funding institutions, and help shape allocation decisions.	Planning commissions	Short term (1–3 years)
	Develop and implement medium-term capacity strengthening action plans, as well as training and staffing plans to meet growing mandates.	Line agencies	Medium term (1–5 years)
Strengthen the Capacity for International Cooperation	Introduce an enhanced methodology for climate prediction at regional level, strengthen early warning systems for international river basins, and economic impact assessment of collective cross-border actions.	Academics from advanced economies	Short term (1–3 years)
	Share and promote regional best-practice examples of mainstreaming adaptation practices in sectoral planning.	Regional associations	Short to medium term (1–7 years)
	Provide technical and human resources for effective management of cross-border climate change impacts and make clear the roles and responsibilities of all parties involved for collective actions.	Ministries of environment and foreign affairs	Short to medium term (1–5 years)
	Develop a network of regional centers within appropriate existing institutions to provide high-quality training and knowledge to ensure high standard of professionalism across countries.	Regional associations	Continuous

Source: Prepared by authors.

build the capacity for mainstreaming adaptation concerns in the short and medium term. Subsequent work will need to enrich the portfolio of policy choices, adaptation measures, methods, and cases presented here, by giving special attention to economic cost analysis for each action and developing monitorable targets. This will be particularly relevant to those who devise adaption strategies, local stakeholders as well as policymakers, and the international community. We invite them and the academia to combine forces in a collective endeavor to address these issues that affect the social fabric, the economic base, and the ecosystem that will ultimately shape the future of 2.2 billion people living in the Asia and Pacific region.

References

Asian Development Bank (ADB). 2006. *Urbanization and Sustainability in Asia: Good Practice Approaches in Urban Region Development.* Manila: ADB.
———. 2007. *Investing in Clean Energy and Low Carbon Alternatives in Asia.* Manila: ADB.
———. 2009a. *Building Climate Resilience in the Agriculture Sector of Asia and the Pacific.* Manila: ADB.
———. 2009b. *Climate Change and Migration in Asia and the Pacific.* Manila: ADB.
———. 2009c. *The Economics of Climate Change in Southeast Asia: A Regional Review.* Manila: ADB.
———. 2009d. *Improving Energy Security and Reducing Carbon Intensity in Asia and the Pacific.* Manila: ADB.
———. 2010. *Addressing Climate Change in Asia and the Pacific Impacts on Food, Fuel, and People.* Manila: ADB.
Smith, J. B., T. Dickinson, J. D. B. Donahue, I. Burton, E. Haites, R. J. T. Klein, and A. Patwardhan. 2011. Development and Climate Change Adaptation Funding: Coordination and Integration. *Climate Policy.* 11: 987–1000. http://web.me.com/rjtklein/Site/Home_files/11sddbhkp_cpol.pdf

Glossary

Adaptation refers to adjustments or management strategies to deal with climate risks and their effects. It relates to practices, processes, and structures that moderate harm or realize opportunities associated with climate change. It is a very broad concept and can be used in a variety of ways. It can be anticipatory and reactive, autonomous and planned, and can be implemented by both public and private actors. Private actors include individuals, households, communities, commercial companies, and NGOs. Public actors include government bodies at all levels.

Adaptive capacity is defined as the ability of people and systems to adjust to climate change, for example, individual or collective coping strategies for the reduction and mitigation of risks or by changes in practices, processes or structures of systems. Adaptive capacity cannot be easily measured and is not well understood. It is related to general levels of sustainable development such as political stability, economic well-being, human and social capital, and climate specific aspects.

Carbon sequestration is the process of removing carbon from the atmosphere and depositing it in a reservoir.

Climate can be understood as average weather. It represents the state of the climate system over a given time period and is usually described by the means and variation of variables such as temperature, precipitation, and wind, most commonly associated with weather.

Climate variability refers to variations in the mean state and other statistics of the climate on all temporal and spatial scales beyond that of individual weather events. Variability may be due to natural internal processes within the climate system or to variations in natural or anthropogenic external forcing.

Climate change refers to any change in climate over time, whether due to natural variability or as a result of human activity that alters the composition of the global atmosphere and which is in addition to natural variability observed over comparable time periods.

Disaster is a serious disruption of the functioning of a community or a society causing widespread human, material, economic, or environmental losses that exceed the ability of the affected community or society to cope using its own resources.

Disaster prevention includes all activities undertaken to avoid the adverse impact of hazards and related environmental, technological, and biological disasters.

Disaster risk reduction represents the systematic development and application of policies, strategies, and practices to minimize vulnerabilities and disaster risks throughout a society, and to avoid or to limit adverse impact of hazards, within the broad context of sustainable development.

Mainstreaming refers to the incorporation of climate-change–adaptation initiatives (measures, options, strategies) into other existing policies, programs, management systems or decision-making structures that are not necessarily about climate or climate change.

Maladaptation refers to development resulting in actions that do not succeed in reducing vulnerability of systems and social groups.

Models are a representation of a real system, and usually describe the structure or function of that particular system.

Mitigation entails all human interventions that reduce the sources or enhance the sinks of greenhouse gases. Carbon dioxide is the major greenhouse gas.

Preparedness includes all activities and measures taken in advance to ensure effective response to the impact of disasters, including the issuance of timely and effective early warnings and the temporary removal of people and property from a threatened location.

Sustainable development is defined as development that meets the needs of the present without compromising the capacity of future generations to meet their own needs.

Tools refer to approaches and instruments that can be employed to analyze and plan for and/or implement mainstreaming of climate change adaptation into development planning.

Vulnerability is a more dynamic concept that encompasses exposure to risks, hazards, shock and stress, difficulties in coping with contingencies, and access to assets. In the context of climate change, vulnerability is used in this report to mean that the risk of climate change will cause a decline in the well-being of poor people and poor countries. This refers to the degree to which a system is susceptible or unable to cope with the adverse effects of climate change, including climate variability and extremes. This vulnerability is a function of the character, magnitude, and rate of climate variation to which a system is exposed, as well as its adaptive capacity.

About the Editors and Contributors

Editors

Venkatachalam Anbumozhi, Capacity Building Specialist, Asian Development Bank Institute, Japan.

Meinhard Breiling, Senior Researcher, Technology, Tourism, Landscape, Interfaculty Cooperation Centre, Vienna University of Technology, Austria.

Selvarajah Pathmarajah, Senior Lecturer, Faculty of Agriculture, Department of Agricultural Engineering, University of Peradeniya, Sri Lanka.

Vangimalla R. Reddy, Research Leader, USDA-ARS, Crop Systems and Global Change Laboratory, United States.

Contributors

Midori Aoyagi, Chief of the Environmental Planning Section, Social and Environmental Systems Division, National Institute for Environmental Studies, Japan.

Agastin Baulraj, Associate Professor of Economics, St John's College, Manonmaniam Sundranar University, India.

Zhijun Chen, Water Resources Development and Conservation Officer, FAO Regional Office for Asia and the Pacific, Thailand.

Robert Dobias, Senior Advisor, Asian Development Bank, Philippines.

David H. Fleisher, Agricultural Engineer, USDA-ARS, Crop Systems and Global Change Laboratory, United States.

Serena Fortuna, Program Officer, Climate Change Adaptation and Disaster Risk Reduction Regional Office for Asia and the Pacific, United Nations Environment Programme, Thailand.

Kazuhiko Fukami, Leader, Hydrologic Engineering Research Team, International Centre for Water Hazard and Risk Management under the auspices of UNESCO, Japan.

Sevi Govindaraj, Tamil Nadu Agricultural University, India.

Shigeko Haruyama, Professor, Mie University, Graduate School of Bio Resources and University of Tokyo, Graduate School of Frontier Science, Japan.

Srikantha Herath, Senior Academic Program Officer, Institute for Sustainability and Peace, United Nations University, Japan.

Chu Thai Hoanh, Senior Officer, International Water Management Institute, Southeast Asia Regional Office, Lao People's Democratic Republic.

Tae Yong Jung, Principal Climate Change Specialist, Asian Development Bank, Philippines.

Takahiro Kawakami, Researcher, Hydrologic Engineering Research Team, International Centre for Water Hazard and Risk Management under the auspices of UNESCO, Japan.

Ikuyo Kikusawa, Visiting Researcher, Program Management Office, Institute for Global Environmental Strategies, Japan .

Hideki Kimura, Manager, Mitsui & Co. Ltd., Japan.

Masanori Kobayashi, Coordinator, Program Management Office, Institute for Global Environmental Strategies, Japan.

Guillaume Lacombe, Researcher-Hydrologist, International Water Management Institute, Southeast Asia Regional Office, Lao People's Democratic Republic.

Jostacio M. Lapitan, Technical Officer, Urbanization and Emergency Preparedness, WHO Centre for Health Development, Japan.

Jun Magome, Researcher, Japan Water Agency, Japan.

Worapot Manupipatpong, Former Director, Capacity Building and Training, Asian Development Bank Institute.

Yuri Murayama, Researcher, IR3S, University of Tokyo, Japan.

Seishi Nabesaka, Researcher, Water-related Hazard Research Group, International Centre for Water Hazard and Risk Management under the auspices of UNESCO, Japan.

Youssef Nassef, Regional Manager, Adaptation Technology and Science Program, UN Framework Convention for Climate Change Secretariat, Germany.

Bui Duong Nghieu, Senior Researcher, Institute of Financial Science, Ministry of Finance, Viet Nam.

Toshio Okazumi, Director, River Management Office, River Improvement and Management Division, River Bureau, Ministry of Land, Infrastructure, Transport, and Tourism, Japan.

Taikan Oki, Professor, Institute of Industrial Science, University of Tokyo, Japan.

Eiji Otsuki, Director, International Water Management Coordination, River Planning Division, River Bureau, Ministry of Land, Infrastructure, Transport and Tourism, Japan.

Go Ozawa, Researcher, Hydrologic Engineering Research Team, International Center for Water Hazard and Risk Management under the auspices of UNESCO, Japan.

Kuppannan Palanisami, Director, TATA Policy Research Program, International Water Management Institute, Hyderabad Office, India.

Coimbatore Ramarao Ranganathan, Professor, Tamil Nadu Agricultural University, India.

K. Raja Reddy, Professor, Department of Plant and Soil Sciences, Mississippi State University, United States.

Samiappan Senthilnathan, Assistant Professor, Tamil Nadu Agricultural University, India.

Tomonori Sudo, Advisor, Office for Climate Change, Japan International Cooperation Agency, Japan.

Tomonobu Sugiura, Researcher, Japan Water Agency, Japan.

Thada Sukhapunnaphan, Director, Hydrology and Water Management Center for Upper Northern Region, Royal Irrigation Department, Thailand.

Dennis J. Timlin, Soil Scientist, USDA-ARS, Crop Systems and Global Change Laboratory, United States.

Thierry Valéro, Institute of Research for Rural Development, Vientiane, Lao People's Democratic Republic.

Tsugihiro Watanabe, Professor, Research Institute for Humanity and Nature, Japan.

Yang Yang, Research Associate, Wye Research and Education Center, University of Maryland, United States.

Harumi Yashiro, Manager, Risk Modeling Group, Tokyo Marine and Nichido Risk Consulting Co. Ltd., Japan.

Fan Zhai, Managing Director, Asset Allocation and Strategic Research Department, China Investment Cooperation, People's Republic of China.

Juzhong Zhuang, Assistant Chief Economist, Asian Development Bank, Philippines.

Index